LOW TEMPERATURE STRESS IN CROP PLANTS

THE ROLE OF THE MEMBRANE

ACADEMIC PRESS RAPID MANUSCRIPT REPRODUCTION

Proceedings of an International Seminar on Low Temperature
Stress in Crop Plants Held at the East-West Center, Honolulu,
Hawaii, March 26-30, 1979

Sponsored by

The United States National Science Foundation
The Australian Department of Science
The College of Agricultural and Environmental Sciences,
University of California at Davis

LOW TEMPERATURE STRESS IN CROP PLANTS
THE ROLE OF THE MEMBRANE

Edited by

JAMES M. LYONS

Department of Vegetable Crops
University of California at Davis
Davis, California

DOUGLAS GRAHAM

JOHN K. RAISON

Plant Physiology Unit
CSIRO Division of Food Research
and School of Biological Sciences
Macquarie University
North Ryde, N.S.W., Australia

ACADEMIC PRESS
A Subsidiary of Harcourt Brace Jovanovich, Publishers
New York   London   Sydney   Toronto   San Francisco   1979

ACADEMIC PRESS, INC.
111 Fifth Avenue, New York, New York 10003

United Kingdom Edition published by
ACADEMIC PRESS, INC. (LONDON) LTD.
24/28 Oval Road, London NW1 7DX

LIBRARY OF CONGRESS CATALOG CARD NUMBER:

80–10820

ISBN 0–12–460560–5

PRINTED IN THE UNITED STATES OF AMERICA

80 81 82 83   9 8 7 6 5 4 3 2 1

CONTENTS

THE PRIMARY TEMPERATURE SENSOR

A.   Characterization of Fluidity Parameters

Contents

SPECIAL TOPICS RELATED TO THE USE OF ARRHENIUS PLOTS

PARTICIPANTS

David J. Bagnall Division of Plant Industry, CSIRO, Canberra City, ACT 2600, Australia

David G. Bishop, Plant Physiology Unit, CSIRO, Division of Food Research and School of Biological Sciences, Macquarie University, North Ryde, N.S.W. 2113, Australia

R. William Breidenbach, Plant Growth Laboratory, University of California, Davis, California 95616

Michael J. Burke, Department of Horticulture, Colorado State University, Ft. Collins, Colorado 80521

Meryl N. Christiansen, Plant Stress Laboratory, Plant Physiology Institute, USDA–RS, Beltsville, Maryland 20705

Adam W. Dalziel, Plant Growth Laboratory, University of California, Davis, California 95616

Philip J. Dix, Department of Genetics, Ridley Building, Claremont Place, University of Newcastle Upon Tyne, Newcastle Upon Tyne, NEI 7RU, Great Britain

David C. Fork, Department of Plant Biology, Carnegie Institute of Washington, 290 Panama Street, Stanford, California 94305

Melvin P. Garber, Tree Propagation Scientist, Weyerhaeuser Company, Southern Forestry Research Center, P. O. Box 1060, Hot Springs, Arkansas 71901

Douglas Graham, Plant Physiology Unit, CSIRO Division of Food Research and School of Biological Sciences, P. O. Box 52, North Ryde, N.S.W. 2113, Australia

Larry V. Gusta, Crop Development Center, University of Saskatchewan, Saskatoon, Canada S7N OWO

Reinfriede Ilker, General Foods Corporation, Technical Center, Tarrytown, New York

Paul H. Jennings, Department of Plant and Soil Sciences, Bowditch Hall, University of Massachusetts, Amherst, Massachusetts 01002

Frank Katterman, Department of Plant Sciences, University of Arizona, Tucson, Arizona 85721

Alec D. Keith, Department of Biochemistry & Biophysics, The Pennsylvania State University, University Park, Pennsylvania 16802

Paul H. Li, Laboratory of Plant Hardiness, Department of Horticultural Science, 286 Alderman Hall, University of Minnesota, 1970 Folwell Avenue, St. Paul, Minnesota 55108

James M. Lyons, Department of Vegetable Crops, University of California, Davis, California 95616

James R. McWilliam, Department of Agronomy & Soil Science, University of New England, Armidale, N.S.W. Australia

E. J. McMurchie, Plant Physiology Unit, CSIRO Division of Food Research and School of Biological Sciences, Macquarie University, North Ryde, N.S.W. 2113 Australia

Paul Mazliak, Laboratorie de Physiologie Cellulaire, Universite Pierre et Marie Curie, 4, Place Jussieu, Tour 53, 75005 Paris, France

Norio Murata, Department of Biology, University of Tokyo, College of General Education, Komaba, Meguro-Ku, Tokyo 153, Japan

Takao Murata, Laboratory of Postharvest Physiology and Preservation of Fruits and Vegetables, Faculty of Agriculture, Shizuoka University, Oya, Shizuoka 422, Japan

T. Nikki, Institute of Low Temperature Science, Hokkaido University, Sapporo, Japan 060

Brian D. Patterson, Plant Physiology Unit, CSIRO Division of Food Research and School of Biological Sciences, P. O. Box 52, North Ryde, N.S.W. 2113, Australia

Robert Paull, Department of Botany, University of Hawaii at Manoa, 3190 Maile Way, Honolulu, Hawaii 96822

Tim Peoples, Department of Plant Sciences, University of Arizona, Tuscon, Arizona 85721

Carl Pike, Department of Plant Biology, Carnegie Institute of Washington, 290 Panama Street, Stanford, California 94305

John K. Raison, Plant Physiology Unit, CSIRO Division of Food Research and School of Biological Sciences, Macquarie University, North Ryde, N.S.W. 2113, Australia

Eric W. Simon, Botany Department, The Queen's University of Belfast, Belfast BT7 INN, North Ireland

Robert M. Smillie, Plant Physiology Unit, CSIRO Division of Food Reserach and School of Biological Sciences, P. O. Box 52, North Ryde, N.S.W. 2113, Australia

Peter L. Steponkus, Department of Agronomy, Cornell University, Ithaca, New York 14853

Judith B. St. John, Plant Physiologist, USDA, RS, AEQU, Weed Science Laboratory, Beltsville, Maryland 20705

F. Tagawa, Institute of Low Temperature Science, Hokkaido University, Sapporo, Japan 060

Yasuo Tatsumi, Laboratory of Postharvest Physiology and Preservation of Fruits and Vegetables, Faculty of Agriculture, Shizuoka University, Oya, Shizuoka 422, Japan

Guy A. Thompson, Jr. Department of Botany, The University of Texas, Austin, Texas 78712

C. Eduardo Vallejos, Department of Vegetable Crops, University of California, Davis, California 95616

Neil L. Wade, N.S.W. Department of Agriculture, CSIRO Division of Food Research, P. O. Box 52, North Ryde, N.S.W. 2113, Australia

Alan J. Waring, Johnson Research Foundation G4 and Physical Chemistry, University of Pennsylvania, Philadelphia, Pennsylvania 19174

Mary E. Willcox, CSIRO Division of Mathematics and Statistics, located at CSIRO Division of Food Research, P. O. Box 52, North Ryde, N.S.W. 2113, Australia

Claude Willemot, Station de Recherches, Agriculture Canada, 2560, Boul. Hochelaga, Ste-Foy, Quebeck, Canada GIV 2J6

John M. Wilson, School of Plant Biology, Memorial Building, University College of North Wales, Bangor LL572UW, Great Britain

Joe A. Wolfe, Department of Applied Mathematics, Research School of Physical Science, Australian National University, Canberra City, ACT 2600, Australia

Shizuo Yoshida, Institute of Low Temperature Science, Hokkaido University, Sapporo, Japan 060

Ferenc Zsoldos, Department of Plant Physiology, Attila Jozsef University, H-6701 Seged, P. O. Box 428, Hungary

PREFACE

This volume is based on the proceedings of an international seminar on "Low Temperature Stress in Crop Plants: The Role of the Membrane," which was held at the East–West Center, Honolulu, Hawaii, March 26–30, 1979. It contains a series of articles which focused on exploration of the fundamental mechanisms involved in the temperature response of crop plants.

The seminar was sponsored by the United States National Science Foundation and the Australian Department of Science under the auspices of the United States–Australian Cooperative Science Program, as well as the College of Agricultural and Environmental Sciences, University of California at Davis. The primary focus of the seminar was to examine the hypotheses related to the primary temperature sensor in crop plants and to evaluate evidence in support of current concepts in the mechanisms of low temperature injury. As food reserves approximate the amount of annual production, the impact of fluctuating weather and climate become increasingly important. Selection of plant materials and breeding of improved varieties has historically been done against a background of climatic fluctuations and environmental variables. Intolerable low temperature is one of the climatic variables that imposes stress on crop production. Avoiding low temperature damage by growing shorter-season varieties on the edges of more suitable production areas is one approach to the solution of variable low temperatures. However, the necessity to incorporate increased genetic potential for resistance to these low temperatures remains as a high priority for improving food production.

It is hoped that the discussions that are included in this volume will make a significant contribution to the researchers involved in understanding and attenuating crop losses by low temperature and that specific research questions have been identified which should provide the best approaches to planning research strategies and finding solutions that will aid in expanding the limits of crop production.

We would like to express our appreciation to Mr. Jim McMahon and his staff at the East–West Center for their superb management of the conference facilities during the week's program. Because of the fine facilities and ease in presentation, a very full week was easily handled.

We would also like to acknowledge Ms. Betty Perry, Department of Vegetable Crops, University of California at Davis, for her impeccable secretarial work and preparation of the camera-ready manuscript.

Symposium participants and contributors (left to right). First row: David C. Fork, Eric Simon, Yasuo Tatsumi, T. Nikki, Robert M. Smillie, Paul H. Li, Tim Peoples, Eduardo Vallejos, Philip Dix, Alec Keith, F. Zsoldos, Guy A. Thompson, Shizuo Yoshida, F. Tagawa. Second row: Paul Mazliak, Norio Murata, Takao Murata, David G. Bishop, Carl Pike, A. Dalziel, Reinfriede Ilker, T. McMurchie, John Raison, Douglas Graham, Jim Lyons, M. Garber, Judith St. John, Bill Breidenbach, Claude Willemot. Third row: Alan Waring, Joe Wolfe, Robert Paul, F. Katterman, L. Gusta, J. McWilliam, Michael Burke, Meryl Christiansen, Brian Patterson, David Bagnall, John Wilson, Peter Steponkus, Paul Jennings.

THE PLANT MEMBRANE IN RESPONSE
TO LOW TEMPERATURE: AN OVERVIEW

J. M. Lyons

Department of Vegetable Crops
University of California
Davis, California

J. K. Raison

CSIRO Plant Physiology Unit
School of Biological Sciences
Macquarie University
North Ryde, N.S.W. Australia

Peter L. Steponkus

Department of Agronomy
Cornell University
Ithaca, New York

I. INTRODUCTION

The world's need for food increases every year. Currently, food for some 4 billion inhabitants is required. In the year 2000 food for at least 6 billion will be needed. As food reserves approximate the amount of annual production, the impact of fluctuating weather and climate become increasingly important. Efficiency and production of food have improved steadily to meet man's needs. Selection of plant materials and breeding of improved varieties has historically been done against a background of climatic fluctuations and environmental variables. Although there is the general consensus that environmental stresses limit crop distribution and production, it is usually accompanied by an attitude of accepting rather than addressing the devastating effect of climatic extremes. Within the plant kingdom there is an extreme range of genetic diversity in the capacity to withstand low temperatures. While many species of tropical and subtropical regions are injured by low temperatures above $0^{\circ}C$, numerous temperate zone species can withstand $-196^{\circ}C$.   Exploitation of the genetic diversity for low temperature tolerance is however hampered by a universal obstacle. A complete understanding of what constitutes low temperature resistance or even how low temperatures result in injury is lacking. In order to ultimately improve plant resistance to low temperature one must 1) characterize the stress imposed on the plant, 2) determine the repercussion of the stress on the cellular environment and architecture, and 3) determine what constitutes injury at the cellular and molecular level.

When such information is available, the physical and biochemical aspects of adaptation, either constituitive or facultative, will be better understood. With such an approach, procedures for the assessment and improvement of low temperature resistance will emerge for use in evaluation of germplasm, in breeding programs, and in formulation of appropriate cultural and management practices.

This "Overview" summarizes some aspects of what is known about the mechanism(s) of how low temperature controls both chilling and freezing processes leading to injury and identifies questions and areas where inadequate information or understanding exists. Hopefully, these questions will aid in focusing research strategies to develop a better understanding and contribute to finding solutions to low temperature limitations of world food production.

A.   Chilling Injury

Many important crop species of tropical or subtropical origin (rice, corn, tomato, cotton, soybean, etc.) are sensitive to low temperatures in the range of 20° down to about $0^{\circ}C$. Below some critical temperature, they are sharply restricted in germination, growth, reproduction, and postharvest longevity. This physiological damage to plant tissues not involving freezing is commonly referred to as "chilling injury." Such chilling sensitive species are dramatic contrasts to species of temperate origin which do not manifest this sensitivity to low temperature, and in some cases have temperature optima in this range.

The physiological dysfunctions that occur upon exposure of chilling-sensitive species to low temperatures lead to a variety of visible symptoms.   The extent of this visible injury is a function of the temperature extreme, duration of exposure to cold conditions, plant species, and morphological and physiological condition of the plant material at time of exposure. The observations include effects on plant materials at all stages of development.   To offer a few examples: Low temperatures have been shown to be harmful in the germination of sensitive species such as cotton (10), muskmelons and peppers (27) tomatoes (99, 113), wild tomato species (79), and many other species (112).   If exposure to cold occurred after the plant has become established, the most striking symptom in a number of herbaceous plants such as the cucumber and bean is water loss and wilting (67, 127, 128).

Leaf stomates in some instances are unable to close at chilling temperatures and the ability of leaves to transport water is impaired. In sorghum, night temperatures of 13°C or less during meiosis induces male sterility (8). The problem of low temperature lesions has been studied intensively by those interested in the postharvest storage and handling of these chilling sensitive commodities, since low temperature is the most effective means of extending storage life. A number of physiological problems develop when chilling-sensitive species experience either a pre- or postharvest exposure to low temperatures. For example, the failure of tomatoes to ripen normally at low temperature (90), the surface pitting and accelerated decay of cucumber fruits (21), discoloration of latex vessels and sub-epidermal discoloration in bananas (38), and many similar aberations are all common occurrences following a chilling treatment. In addition many of the temperate fruits display undesirable symptoms when held at low temperatures. Apple scald (superficial scald), a brown discoloration giving the skin a cooked appearance occurs in some varieties upon long-term storage at 0-4°C (123). Low temperature breakdown of apples, also caused by storage at such temperatures, occurs in the cortex and in certain varieties involves most of the flesh (125).

Another feature of the chilling phenomena is its reversibility following short-term exposure to a chilling treatment. The physiological dysfunction resulting from the molecular changes ·induced at low temperature can be reversed (or repaired) if the tissue is returned to non-chilling temperatures before actual injury occurs (12, 50). This reversibility can be accomplished experimentally by alternating low/high temperatures over some short time cycle (116) but it also reflects field conditions where, in the highland regions of the tropics for example, each cold night is followed by a warm day (see Vallejos, this volume). Thus, while physiological dysfunction is initiated below some critical temperature, it does not lead to visible manifestation of injury or effects on parameters such as growth rate and development because the dysfunction has been reversed before it could become persistent.

Conditioning or hardening chilling sensitive tissues by exposure to temperatures just slightly above the chilling range for some period of time before placing them at an injurious temperature can prepare them to withstand low temperatures for a somewhat longer period before showing visible symptoms (3). This conditioning also induces changes in composition of membrane lipids (116, 118).

This brief review presents some of the phenomenology associated with the chilling response and includes a number of physiological events which have been studied in an endeavor to understand the events leading to the observations. For the sake of clarity, the following definitions will be used:

chilling - the exposure of plant material to some low temperature (above freezing).

chilling injury - a damage to plant tissues, cells, or organs which results from the imposition of low temperature (i.e., chilling, or a chilling treatment) for a period of time sufficient to cause permanent or irreversible damage.

dysfunction - a disordered or impaired function in response to temperature which, if allowed to occur for *sufficient time,* will cause an injury.

primary response to temperature - the primary event in the cell that causes some dysfunction which, in time, leads to injury.

The purpose of this overview is to discuss mechanisms and evidence which focus on the *primary response to temperature* in chilling sensitive plant species. As the reader will discover, there is a bias in this overview toward reducing the possibilities available to a physical effect either on a protein, on the cellular membranes, or on the interaction of the two. The causes of *chilling injury* are the many sequential events which result from a physiological or biochemical *dysfunction* occurring as a result of *chilling* over some time period.

B. Freezing Injury

While a distinction between chilling sensitive and resistant species can be made, chilling resistant species can be further distinguished on the basis of their resistance to freezing temperatures. Freezing injury may be incurred over a broad spectrum of sub-zero temperatures depending on the species in question and the stage of acclimation. In a non-acclimated condition, most plants will be killed by freezing to -1° to -3°C. In the acclimated condition, some species will only withstand freezing to temperatures slightly below the freezing point while others will withstand freezing to -196°C. Within this spectrum, examples of intermediate degrees of freezing resistance are readily documented. Although there is extreme diversity in the range of freezing temperatures which can be tolerated, there is however one common feature of freezing resistance.

Membrane damage is a universal manifestation of freezing in biological systems and is commonly inferred to be the primary cause of injury. The flaccid, water-soaked appearance of various plant tissues and organs following thawing strongly suggests that exposure to lethal low temperatures results in membrane disruption. The rapidity with which injury is manifested further suggests that injury is not primarily due to metabolic dysfunction (as in chilling injury). Additionally, nearly all of the methods used to evaluate tissue viability following a freeze-thaw cycle (plasmolysis and deplasmolysis, vital staining, and solute leakage) are based on the retention of intact cellular membranes (102) even the metabolic reduction of triphenyl tetrazolium chloride (104). Despite the overwhelming circumstantial evidence to warrant the conclusion that membrane damage is the primary cause of injury, Levitt and Dear (46) have emphasized that ". . .categorical proof of direct injury to the plasma membrane during extracellular freezing is still lacking."

An understanding of what constitutes freezing injury at the cellular and molecular levels is important for the formulation of mechanisms of freezing injury. Most hypotheses regarding the mechanism of freezing injury have evolved from the analysis of the physico-chemical events occurring during freezing and only projections of potential membrane damage have been provided [see Mazur (57, 58, 59, 69); Meryman (65); Meryman *et al.* (66) for reviews]. Only recently has attention been focused on what constitutes freezing damage to cellular membranes [for reviews see Steponkus *et al.* (107), Steponkus (103, 104, 105), Steponkus and Weist (106)]. Of utmost concern is the demonstration of a membrane lesion resulting from a freeze-thaw cycle which quantitatively accounts for the observed cellular injury.

As cellular membranes are inferred to be the primary site of freezing injury, cold acclimation must involve cellular alterations which allow the membranes to survive subzero temperatures. Such alterations may be in the cellular environment or in the membranes *per se.* Alterations in the cellular environment may alter the stresses that occur during freezing or may result in direct protection of the membrane. Both possibilities were identified by early investigators [see Chandler (9)] and are still viable today. For instance, Heber and Santarius (31) consider that one component of cold acclimation involves the formation of protective compounds which can result in protection through non-specific colligative dilution of toxic compounds. Alternatively, cryoprotective proteins may confer protection in a non-colligative manner (29). Olien (72, 73) has suggested that the degree of hardiness depends on arabo-xylan polymers which modify the freezing stresses. Alterations in the membrane may influence the stresses resulting from freezing or may render the membrane more tolerant of the freezing stresses. As early as 1936, Scarth and Levitt (93) proposed that cold acclimation could result in an alteration in membrane water permeability to permit the rapid removal of water to extracellular sites of ice nucleation. This provided an explanation for the observation that intracellular ice formation occurred at slower freezing rates in nonacclimated tissues than in acclimated tissues. However, subsequent work by Stout *et al.* (110) indicates that there may not be a direct cause and effect relationship between the two observations. Evidence that cold acclimation results in changes in the tolerance of the plasma membrane has been provided by Scarth *et al.* (94) and Siminovitch and Levitt (98). Clearly, many alterations are involved in the process of cold acclimation, and elucidation of the specific physiological and biochemical aspects requires an understanding of the mechanism of freezing injury.

As early as 1912, Maximov (56) concluded that freezing damage was the result of freeze-induced removal of water from the surface of the plasma membrane. In 1940, Scarth *et al.* (94) addressed changes in plasma membrane structure in relation to cold acclimation and freezing injury. On the basis of several previous publications (47, 48, 94), they indicated that both intracellular and extracellular ice formation result in damage specifically to the plasma membrane, albeit for different reasons. Moreover, while extracellular ice formation results in cellular dehydration and plasmolysis followed by deplasmolysis upon thawing, plasmolysis *per se* was not the injurious event. Rather, rupture of the plasma membrane occurred upon deplasmolysis. The significance of these observations were, however, overshadowed by numerous incursions into the cytoplasm to explain the mechanisms of freezing injury and cold acclimation. In 1970, Levitt and Dear (46)) indicated that attention was again being directed to the plasma membrane, however, interest in membranes was stimulated by studies of alterations in mitochondrial (41, 42) and chloroplast membranes (28, 30), rather than studies of the plasma membrane directly. Numerous papers by Heber and coworkers [see Heber and Santarius, (31)] provided a wealth of information on the effects of freezing on the function of chloroplast thylakoids. Subsequently, Garber and Steponkus (23, 24) and coworkers (105, 107) extended this work providing information on the repercussions of freezing and cold acclimation on thylakoid structure and function at the molecular level. To date information at a comparable level for the plasma membrane awaits elucidation.

II   PROPAGATION OF THE TEMPERATURE RESPONSE

As described above, a number of physical, physiological and biochemical phenomena have been closely correlated with low temperature stress in plant species. Each of these phenomena will be examined with respect to interpretation of the data as it might relate to an understanding and assignment of events at the molecular level.

A.   Chilling Injury

1. Cytological Responses. An array of pathological responses have been observed at the electron microscope level when sensitive plant tissues have been exposed to a chilling treatment. For example, apparent loss of cell turgor, vacuolization, reduction in the apparent volume of both the cytoplasm and the vacuole or protein bodies, apparent deposition of new material in the cell walls, general disorganization of organelles. and a general loss of cytoplasmic structure have been reported (36). Giaquinta and Geiger (25) showed that an increased vesiculation of the sieve tube cytoplasm, disruption of the sieve tube reticulum, and the appearance of extraneous masses inside vacuoles occurred in bean petioles after only 30 min at $0°C$. A recent study on the time-course of the effects of low temperature on cellular ultrastructure in the chilling sensitive tomato (see Ilker, et al, this volume) demonstrated that membrane alterations preceded other cellular changes, and different lengths of exposure were required to damage different types of organelles. Surprisingly, the plasmalemma appeared relatively more resistant to chilling than the other membranes in these studies. As little as 2 hours of chilling at 5°C induced observable changes in ultrastructure in some but not all cells.

Patterson, et al (this volume) also observed vesiculation at the light microscope level in living trichomes of several chilling sensitive species. These observations correlated with differences observed in protoplasmic streaming. They showed that the rate of streaming at 5°C was greatest from ecotypes of the wild tomato L. hirsutum collected from the highest altitudes, suggesting that the rates correlated with genetic adaptation to the low temperatures experienced at higher altitudes (76). These wild tomato species native to Equador and northern Peru have adapted to a wide range of temperature extremes (88). These observations present a fairly clear picture of the changes in cellular organization in response to low temperature but do not afford the opportunity to differentiate the primary temperature response and how it is propagated into the obvious disorganization that occurs.

2. *The Plasma Membrane.* Another physiological event observed as the result of a chilling experience that has focused attention on membrane integrity has been the loss of electrolytes from tissue injured by low temperature. An enhanced loss of potassium from chilled slices of sweet potato roots (50), carbohydrate and amino acids from the roots of cotton seedlings at 5°C (11) of other electrolytes from chilled leaf tissues (12, 129) and seedlings (114) has been observed. This loss of electrolyte from low temperature damaged tissues has been used in attempts to quantitatively measure the extent of chilling injury. For example, in the assessment of chilling of cucumber leaves (128) and as a measure of differences between *Passiflora* species tolerant to a range of climactic environments (78). Each of these observations can be correlated to the physical effects of chilling temperatures on changes in membrane permeability and/or membrane integrity.

The plasma membrane structure and function in response to low temperature is a key issue as it controls the uptake or loss of various electrolytes as described above. For example, the efflux of rubidium has been observed for roots of chilling-sensitive species below 10°C (131) and differences in the composition of both the cell walls and membranes of root cells of thermophylic plants differed from non-thermophylic plants in their ion-exchange properties, especially in the apical region of the root tip (132). The plasma membrane has also been shown to play a key role in freezing injury of temperate species (see Steponkus and Wiest, this volume, for review). Membrane permeability to water and non-electrolytes can also be modified by Ca^{++} (101). This effect is suggested to be the result of a binding of Ca^{++} to the polar head groups of the membrane phospholipids which widen the distance between the phospholipid molecules and allows more water molecules to be located between the hydrocarbon ends. It would be interesting to examine these Ca^{++} effects on membrane phospholipids in sensitive species with esr and other techniques described in this volume.

     3.   Mitochondrial Membranes. The response of mitochondrial membranes in chilling sensitive plant tissues during a low temperature treatment has been recently reviewed by Raison (81). Plants that are tolerant to low temperatures exhibit a decrease in respiratory activity as a result of the general loss in kinetic energy and this reduction is similar to the decrease in activity of other metabolic reactions. Thus for most plants, the respiratory activity at low temperatures remains in balance with glycolysis and other closely associated reactions and there is little change in the respiratory quotient. In contrast, the respiratory activity of the chilling sensitive plant species decreases disproportionately when the temperature is lowered below about $10^{o}C$ (120, 121). This disproportionality has also been inferred from non-linear Arrhenius plots of respiratory rate as a function of temperature for both intact plant parts and isolated mitochondria (83). It has been suggested that a change in molecular ordering of the membrane lipids is the primary event in the low temperature response of these chilling sensitive species and that this change in physical state alters the conformation and hence activity of the respiratory enzymes of the mitochondria which leads to a disproportionate decrease in the rate of mitochondrial respiration relative to the rate of glycolysis [cf., Raison (80), Lyons and Raison (54), Raison and Chapman (83)]. The metabolic imbalances, and effects on growth and development which occur subsequent to the alteration in membrane structure and the eventual irreversible loss of cellular integrity of the tissues maintained at chilling temperatures, are viewed by these workers as secondary events dependent on the primary structural change in the membrane.

     In related experiments Dix and Street (19) showed that the Arrhenius plot for oxygen uptake by mitochondria extracted from a chilling tolerant line of *Capsicum annuum* exhibited a continuous linear function with temperatures between 0° and $30^{o}C$, unlike that for unselected lines which showed a discontinuity around $10^{o}C$ typical of chilling-sensitive materials. While it remains questionable as to the ability to regenerate such cell lines, it appears as though the potential is sufficient to warrant further exploration with these techniques. (For additional discussion see Dix, this volume).

     4.   Chloroplast Membranes. One of the early responses observed in chilling-sensitive plants growing under field conditions when exposed to low temperature is a decrease in photosynthetic activity (14, 111). A number of studies have been undertaken to examine the impact of low temperature on chloroplast function and particularly membrane related responses. For example, an Arrhenius plot of the photoreduction of $NADP^{+}$ from water by chloroplasts isolated from tomato and bean shows a marked discontinuity at about $12^{o}C$ whereas a similar plot for chilling resistant lettuce and pea shows a constant slope over the temperature range from near 0 to 25°C (97). A deviation from this correlation was observed by Nolan and Smillie (70). They observed that the uncoupler stimulated photoreduction of 2,6-dichlorophenolendophenol by chloroplasts shows an increase in Ea below about 12°C with both chilling sensitive and chilling tolerant plants. These data question the necessity of postulating the existence of membrane changes below 10°C - 12°C to explain inhibition of growth at chilling temperatures (Smillie, this volume). Studies on the molecular architecture of thylakoid membranes of a number of plants in response to temperature alterations has been recently reviewed (82). These data indicate there is a correlation between a physical change in the membrane system in response to low temperature and changes in the physiological responses of intact tissues.

B.   Freezing Injury

1.   *The Freezing Process.* The physico-chemical events which occur during the freezing of a cell suspension and the redistribution of water with respect to both its physical state and location have been detailed by Mazur (57, 58, 59, 60). Briefly when cooled below 0°C, the cells and the surrounding medium initially remain unfrozen due to depression of the freezing point by the solutes present and, to some extent, due to supercooling. Ice nucleation will initially occur in the extracellular solution between -2° and -15°C, depending on the solute concentration and extent of supercooling. The intracellular solution remains unfrozen and supercooled, presumably because the plasma membrane prevents the growth of ice crystals into the cell. This results in a disequilibrium in the chemical potential of water in the intracellular solution relative to the chemical potential of the partially frozen extracellular solution. The lower chemical potential of water molecules in ice is a direct function of temperature.

Equilibrium may be achieved either by cellular dehydration and continued extracellular ice formation or by intracellular ice formation. The *manner* of equilibration is influenced by the rate at which the cell is cooled and the minimum temperature achieved relative to the capacity for water flux out of the cell. The *amount* of water which must be removed to achieve equilibrium by cellular dehydration will depend on the initial osmolality of the intracellular solution and the minimum temperature to which the cell suspension is cooled. *Whether* this amount of water will be removed depends primarily on the permeability of the plasma membrane and the surface area available for efflux relative to the cell volume in relation to the cooling rate and temperature. These factors interact to determine the magnitude of disequilibrium that will result at any state in the disequilibrium continuum. This disequilibrium is manifested as super-cooling of the intracellular solution, and the probability of intracellular ice formation becomes greater as the extent of supercooling increases. In other words, if water flux is adequate relative to the cooling rate, excessive supercooling of the intracellular solution will be precluded and the cell will dehydrate and ice formation will only occur extracellularly. Alternatively, if water flux is not adequate, heat transfer mechanisms will dominate over mass transfer mechanisms, and excessive supercooling of the intracellular solution will result in intracellular ice formation.

The manner of equilibration is of considerable concern because intracellular ice formation is generally considered to be lethal, although there are some reported exceptions [see Mazur (59)]. Extracellular ice formation and attendant cell dehydration may or may not be lethal, depending on other factors. The manner in which extracellular ice formation causes cellular injury, particularly to the plasma membrane, is of foremost concern to the field of cryobiology.

  2. *Repercussions of Freezing on the Cellular Environment.* The repercussions of the freezing process on the cellular environment of the cell and its components are numerous. These changes include the obvious decrease in temperature, the presence of ice crystals, and dehydration of the cell. There is a general concensus that, under conditions of extracellular ice formation, decreases in temperature or the presence of ice crystals *per se* are not the primary causes of freezing injury (31) and that the process of cellular dehydration is the most disruptive and injurious repercussion of the freezing process (58). There are, however, several consequences of dehydration which include: a reduction in cell volume and surface area, an increase in concentration of solutes, precipitation of some buffering salts resulting in pH changes, and removal of water of hydration of macromolecules. Mazur (57, 58) refers to these as 'solution effects' and notes that all occur as a monotonic function of temperature.

It is within this array that there is a great divergence in hypotheses on the mechanism of freezing damage. All of the repercussions, either singularly or in various combinations, have served as the basis for mechanisms of freezing damage [see Mazur (57, 58, 59. 60), Heber and Santarius (31), Meryman (65), Meryman *et al.* (66) for reviews]. Some of these hypotheses have failed to provide an adequate explanation: for instance, the sulfhydryl hypothesis of Levitt (45) has been gradually diminished [see Mazur (57, 58) and Heber and Santarius (31) for criticisms]. Others remain partially attractive (52, 64), but fail to explain all of the observations [see Mazur (60)]. The very attempt to explain all of the manifestations of freezing injury may be the major shortcoming of each.

Since freezing results in a multitude of stresses, it is reasonable to assume that the overall mechanism of freezing injury is a composite of the many hypotheses put forth and they should not be considered as mutually exclusive. It might be more appropriate to view the freezing process as a sequential series of potentially lethal stress barriers. Thus, any one single freezing stress is a potentially injurious factor--depending on the immediate conditions and the successful surmounting of previously encountered stress barriers.

*In summary,*the phenomenology involved in chilling injury and freezing stress has been briefly described. It is now appropriate to examine evidence which might elucidate the primary temperature sensor. In addition, we will attempt to define areas where future experimentation should focus, and to define certain terminology which should be used in describing phenomena associated with low temperatures and membrane lipid order.

III. PRIMARY TEMPERATURE SENSOR

A. *Membrane Fluidity and Chilling Injury*

Over the past ten years or so there have been a number of reports that have related physiological events in low temperature stress in plants to membrane "phenomonology." A part of this stimulus was derived from the observation that Arrhenius plots of oxygen uptake by mitochondria isolated from chilling sensitive plant tissues demonstrated a non-linear temperature dependency or "break" at the temperature critical for chilling injury in the intact plant (54). Furthermore, application of esr techniques provided a physical method which showed that the intact membranes and the lipid components of the membranes of these mitochondria underwent a change in molecular order at the same temperature as that of the "break" in the Arrhenius plot for oxygen uptake (85). These results suggested that the temperature-induced change in temperature-sensitive membranes is an intrinsic property of the membrane lipid and a direct correlation between the physical state of membrane component and enzyme activity was inferred. These observations established physiological dependency on membrane structure and led to articulation of a hypothesis to explain the mechanism of the primary temperature sensor leading to chilling injury in sensitive plants---this model was based on the assumption that "cellular membranes in sensitive plants undergo a physical-phase transition from a normal flexible liquid-crystalline to a solid gel structure at the temperature critical for chilling injury" (53, 85). The change in molecular ordering of membrane lipids at about $10\text{-}12^{\circ}C$ in sensitive species was viewed as the primary response to low temperatures and ultimate dysfunction and injury was the net result of a series of subsequent sequential events.

This model, based on the molecular structure of membrane systems, was a refinement of the historical belief that "solidification of the protoplasmic lipids" could account for the observed death or injury of sensitive plants at low temperatures.(110). Reference to a phase transition from "a liquid-crystalline structure to a solid gel" in the model was derived from the studies of Luzzati and Husson (55) employing model lipid systems. Application of their terminology to complex biological membranes from higher plants was, in retrospect, an oversimplification that has led to controversy, but it provided a certain utility at the time.

1.   Characterization of Fluidity Parameters. A number of power-ful techniques - - esr, nmr, fluorescence, X-ray scattering, dsc --are available to discern physical phenomena in biological membranes (2). The task before us is to attempt to correlate these different techniques and different results into a common picture of molecular events in membrane systems of plants sensitive to low temperatures and to develop a common language that will reduce a certain amount of confusion that exists in the literature.

For example, the esr data from the first studies on membranes derived from chilling sensitive plants were used to infer "melts" in the hydrocarbon zones of phospholipids (85). The abrupt, temperature-induced alterations in the molecular ordering of membrane lipids discerned by esr were referred to as membrane "phase changes." Application of these techniques to the chilling sensitive mung bean showed that two changes in molecular ordering of the membrane lipids of mitochondria and chloroplasts could be detected, one at about $28^{\circ}C$ and one at about $15^{\circ}C$ (83). The terminology developed from studies of binary mixtures of phospholipids and membranes of *E. coli* was adopt-ed, designating the change in physical state of the membrane lipids observed at the lower temperature as T_s and at the higher temperature T_f (95). Hence, the changes occurring in the membranes of mung bean tissue were viewed as a change from a fluid to a mixture of fluid + gel below $28^{\circ}C$ (T_f) and from a fluid + gel to a predominantly gel phase below $15^{\circ}C$ (T_s) and T_s and T_f, were considered the temperature for the initiation and termination of a thermal phase transition in the membrane lipids.

Further studies on mitochondria extracted from wheat also indicated two phase changes, one at about $0^{\circ}C$ and at about $30^{\circ}C$ (84). Again, these changes were interpreted to show that the membrane lipids undergo a melting transition, and that within the temperature range of growth for wheat, the membrane lipids were in a mixed liquid-crystalline, gel form. Much of our knowledge about the physical state of biological membranes is derived from extrapolation of results obtained with mixtures of synthetic or natural phospholipids dispersed in water and from the observation that phospholipid bilayers undergo a reversible, thermotropic gel-liquid-crystalline phase transition which is derived from a cooperative melting of the hydrocarbon chains in the interior of the bilayer. These physical transformations in phospholipid bilayers have been correlated to growth of *Acholeplasma laidlawii* where   it   was reported that both the absolute rates and the temperature coeffients of cell growth were similar in cells whose membrane lipids existed entirely in the liquid-crystalline state, but their absolute growth rates declined rapidly and temperature coefficients increased when most of the membrane lipids became solidified (61). A subsequent study using differential thermal analysis did not support the observations with electron spin resonance spectroscopy that the minimum and maximum growth temperatures of the organism might be directly determined by the solid/fluid phase transition boundaries of the membrane lipids (62).

Their new evidence suggested that the electron spin resonance techniques did not detect the gel to liquid-crystalline phase transition of the bulk membrane lipids as could be discerned by differential thermal analysis. Several studies have shown that at least half the lipids in the membrane phospholipids of *E. coli* and *A. laidlawii* must be in the disordered state to allow for normal growth but *E. coli* is able to grow with almost all of its lipid in the disordered state which suggests that the breadth of the lipid phase transition is due to the diversity of fatty acyl chains normally found (13). A recent review by Cronin (13) has shown that a number of studies using freeze-fracture electron microscopy have been interpreted in terms of the segregation of membrane protein in patches of fluid or disordered lipid. The smooth areas discerned are seen as patches of ordered lipid and the aggregates as membrane proteins that have been excluded from the ordered lipid patches into patches of fluid lipids. Physical isolation and analysis of the different regions has been achieved and shown that these smooth areas have a somewhat higher proportion of saturated fatty acids than do the particle rich areas [see Cronin (13) for reviews].

While it has been tempting to interpret the descriptive aspects of the physical events determined using membrane probes, caution must be exercised in extending the results obtained with bilayers composed of binary mixtures of phospholipids or with bacterial membranes to the complete lipids in plant membranes.

As will be reported in subsequent chapters in this volume (see Dalziel and Breidenbach, Waring and Glatz, Bishop, *et al*), it is clear that the events observed with esr probes do not reflect a phase-transition from "a liquid-crystalline to gel" transformation or "melts" of the bulk lipids; but rather some change in molecular ordering of discrete domains within the membrane.

Further elucidation of the molecular architecture of the membrane and/or membrane components in response to temperature is one of the key areas that should attract research emphasis in the field of low temperature stress in plants.

2.   *Control of Molecular Ordering of Membrane Lipids.* If the molecular ordering of membrane lipids in response to temperature is critical in determining the chilling response, then any of the factors altering properties of the membrane and the transition temperature must be understood. Because of the problems associated with localizing discrete pools or areas of specific lipid composition within membranes, considerations based on analysis of bulk lipids extracted from plant tissues or extrapolations from theoretical consideration of pure compounds or model systems (see Bishop *et al.* and Willemot, this volume) as an explanation of the control of membrane fluidity by temperature must be treated with caution.

a.   *Acyl chain composition.* Much of our knowledge on the role of fatty acid composition in determining thermotrophic phase transitions in membranes comes from studies with microorganisms. For example, McElhaney (61) showed that growth of *A. laidlawii* in the presence of specific fatty acid supplements shifted the membrane phase transition to a temperature above the absolute minimum growth temperature and this shift was clearly controlled by the fatty acid composition of the cell membrane. Furthermore, permeation rates of non-electrolytes passively diffusing into synthetic liposomes, *A. laidlawii* cells, or liposomes derived from *A. laidlawii* membrane lipids, were strongly dependent on the chemical structure and chain length of their fatty acids (16, 63, 89). Incorporation of branched-chain, more unsaturated, or shorter-chain fatty acids, which increase membrane fluidity, all increased non-electrolyte permeability to a similar extent in both cells and liposomes. Active transport of glucose by *A. laidlawii* (86) and the β-galactoside and β-glucoside transport systems of *E. coli* (51, 71) have been clearly shown to depend on the acyl chain composition of the membrane lipids. Baldassare *et al* (4) have shown that the particular fatty acid species present in the membrane lipids of *E. coli* determine the function of several membrane associated enzymes.

Alteration of the fatty acid composition of the membranes of a number of organisms in order to maintain membrane fluidity at different growth temperatures has relevance to the problem of cold sensitivity. For example, Paton *et al* (75) have recently reported that a decrease in growth temperature of *Bacillus amyloliquefaciens* was accompanied by an increase in the ratio of branched-to straight-chain fatty acids and marked increase in the level of unsaturation of the branched-chain fatty acids and they suggested a direct correlation between membrane fluidity and the susceptibility to cold shock. Another example of the influence of fatty acid composition on membrane fluidity is found in the series of reports on membrane properties of *Tetrahymena pyriformis* during temperature acclimation (40, 43). These studies indicated that the activity of membrane associated fatty acid desaturases were controlled by membrane fluidity in response to low temperature rather than temperature *per se*. In addition to the relative amount of saturation/unsaturation and the branching of the acyl chains, the chain length of the fatty acyl moieties of the phospholipids is also a major determinant in the fluidity and phase transition of lipids [see review by Cronin, (13)].

Membrane phospholipid acyl chain composition can reflect in part the composition of the fatty acids available to the organism. Animals fed corn oil (68), microorganisms cultured in the presence of fatty acids (34), yeast grown in medium containing fatty acids (1), and potato tuber tissue slices incubated in medium containing fatty acids and fatty acid derivatives (117) all show changes in the fatty acyl chain composition of the organisms phospholipids to approximate that of the supplement. Fatty acid composition of the membrane lipids can be altered by means other than direct supplementation. Pyridazinone derivatives (108, 124) have been reported to interfere with fatty acid biosynthesis and alter the acyl chain composition of plant membrane lipids. Similarly, treatment of tomato seedlings with ethanolamine-tween-oleate induced altered phospholipid composition and somewhat diminished the symptoms of chilling (118).

Can the fatty acyl chain composition have any influence over membrane fluidity in higher plants? It is clear that molecular ordering of membrane lipids in some systems can be altered by variations in fatty acyl composition. It remains to be shown if a similar situation exists in higher plants, particularly those sensitive to chilling temperatures.

   b.   Polar head group composition. The role of phospholipid polar head groups in regulating the fluidity of membranes of higher plants has not been ascertained. It has been shown that there is more than a 30°C difference in the temperature of the transition from the gel to liquid-crystalline state for aqueous dispersions of pure phosphatidylcholine and phosphatidylethanolamine having the same saturated acyl chain composition (13). That the polar head group of phospholipids can exert an influence on the activity of associated enzymes is clear (69, 115) but the relative importance in relation to other factors is yet to be determined in relation to the chilling response.

   c.   Sterol composition. In model membrane systems, adding cholesterol to the phospholipids alters their molecular packing causing them to be more condensed above and more fluid below their transition temperature [see review by Demel and De Kruyff, (18)].   Because of those unique properties it has been proposed that cholesterol and other sterols such as those common in plants may function as regulators of membrane fluidity in eucaryotes and in procaryotes that have sterols in their membranes (61). Cholesterol content has been shown to influence the phase-transition temperature for membrane-associated enzymes (17, 63, 91). These reports indicate that factors other than phospholipids and their acyl chain composition can be involved in membrane phase transitions and by inference could be involved in chilling sensitivity.

   It is of interest to note, however, the almost precise correlation that exists in the temperature at which discontinuities in Arrhenius plots of oxygen uptake, in the temperature coefficient of spin label motions with intact membranes, and with vesicles of phospholipids extracted from those membranes free of sterols (and proteins for that matter). Furthermore, since the phospholipid vesicles are also free of proteins, there is strong indication that the lipids are the components sensing the temperature change.

   d.   Membrane protein. The intrinsic proteins of membranes do not appear to influence the temperature at which the membrane lipids undergo a change in molecular ordering which would suggest only a small proportion of the membrane lipids are involved in hydrophobic interactions with protein (80). This view is supported by the fact that the temperature of the phase change in mitochondrial membranes is the same regardless of whether the protein is heat denatured or in a native state and coincides with the temperature of the phase change in extracted phospholipids.   Furthermore, a comparison of the heat of transition for the lipids of *M. laidlawii* membranes with that for the extracted lipids indicates that 90% of the membrane lipids are in a bilayer configuration and only 10% are involved in hydrophobic interactions, presumably with protein (87). Sharp breaks in the temperature profiles of activities of many enzymes coincides quite closely with the phase transition detected in the isolated lipids, implying that the membrane phase transition is unaffected by the presence of the protein (71). Studies with ATPase activity however have shown that the annulus (a layer of about 30 lipid molecules immediately surrounding the enzyme proteins) exerts the predominant effect and suggests that the fluidity gradient and lipid chains in the immediate vicinity of the protein may be substantially less than the gradient observed in the bulk membrane (119).

   e. *Chemical perturbers.* In addition to chemical treatments directed toward modifying membrane phospholipids and acyl chain composition *per se,* some chemicals have been shown to directly perturb the membrane and maintain fluidity without chemical modification of membrane components. For example, butylatedhydroxytoluene (BHT), because of its high lipid solubility and extremely low aqueous solubility results in a strong interaction with the hydrocarbon zones of membranes and perturbs membrane-associated events (22, 100). Similarly, adamantane, another lipophylic molecule shifts the membrane phase change in yeast cells and as a consequence affects biological function (22). In addition to these compounds which have been shown to directly perturb the membrane, certain antioxidants have been used to reduce the symptoms of chilling injury. For example, the use of diphenylamine, either as a coating on the apple skin or in wrappers around the fruit prevented superficial scald in apples after prolonged storage at $0\text{-}4^{\circ}C$ (35). Jones *et al* (38) tested a number of chemical agents for their ability to prevent chilling injury of bananas and showed that postharvest treatment with dimethylpolycyloxane, safflower oil, and mineral oil were effective in preventing under-peel discoloration of bananas from a chilling treatment. Similarly, Wang and Baker (116) could prevent some of the symptoms of chilling injury using sodium benzoate and ethoxyquin, compounds which scavange free-radicles. Whether these compounds act by preventing lipid peroxidation to maintain membrane fluidity, or act additionally as membrane perturbers is not clear at this time.

  In summary, it is both tempting and difficult to extrapolate from the results with model binary mixtures of phospholipids and microorganisms, having relatively simple phospholipids in their membranes, to the complex membrane of higher plants. However, information about the factors controlling membrane fluidity is important in understanding the correlation between chilling sensitivity and membrane fluidity. Recent reports that show a correlation between chilling sensitivity in avocado fruit and lipid composition as a function of ripening (44), and the relationship between altered mitochondrial activity and changes in the fatty acid composition of mitochondrial lipids in mango fruit stored at low temperature (39) are in contrast to the results obtained with lipids extracted from *Passiflora* species varying in their resistance to chilling where there was no correlation with the degree of unsaturation of the lipids (77).

A better understanding of the factors controlling membrane fluidity is essential in order to assign observed physiological phenomena to the correct molecular events.

B. Membrane-Cytoplasmic Interactions

While changes in the physical nature of the membrane can influence directly those functions that are an integral part of that structure, eg., a membrane-bound enzyme, consideration must be given to the interaction between the membrane bilayer and the cytoplasm and cytoskeletal elements of the cell. For example cessation or impairment of protoplasmic streaming in chilling-sensitive species is one of the more immediate effects that has been reported as the result of exposure to chilling temperatures. As early as 1864, Sachs (92) reported that protoplasmic streaming ceased at about 10 to 12°C in root hairs of cucumber and tomato plants, while those of chilling tolerant species continued streaming down to or near 0°C. Lewis (49) found that streaming ceased or was just perceptible after one or two minutes at 10°C in petiole trichome of chilling sensitive plants and ceased promptly at 5°C or 0°C. These observations were confirmed by Wheaton (122). In contrast to these results where 10°C has been shown to be the lower limit of streaming for the plants studied, Patterson and Graham (76) have reported that streaming did not actually cease at some critical temperature but that the temperature coefficient for rate of streaming markedly differed between sensitive and tolerant species. Their results were in agreement with similar work on the chilling sensitive tobacco (15).

In eukaryotic cells, cytoplasmic microtubules containing tubulin, cytoplasmic filaments containing actin and associated myosin, are considered to be the principal components of the cytoskeleton. Cytoplasmic microtubules are involved in maintaining cell shape, in the intracellular stratification of organelles, in transport and secretive processes, and cell surface topography. The gelation of actins, myosins, and actin-binding proteins is temperature and pressure dependent, and the gel is destroyed by low temperatures above 0°C (74). The movement and mechanical work functions of many kinds of cells appear to depend on the presence of a gel-like cytoplasm, because low temperatures or extreme pressure destroy the gel and interfere with cyclosis and cytokinesis (74). The interference of low temperature with cyclosis in plants has been investigated extensively [for recent review see Patterson and Graham, (76)]. The distribution of tubulin-containing structures has been studied in cultured animal cells and in algae, but only to a very limited extent in higher plants (7). Microtubuler bridges to the plasma membrane have been described in ultrastructural studies of

plants [see Heppler and Palivitzh (33) for review] . Henry, *et al* (32)
employed water soluble spin labels to demonstrate that alterations in
membrane composition in microorganisms influence cytoplasmic
viscosity. Keith (unpublished data) has found that the viscosity of water
in the cytoplasm of chilling sensitive tissues as discerned by water-
soluble spin labels changes markedly below about 10° C and there is a loss
of diffusion barriers in the cytoplasm of the sensitive tissues not
observed in chilling resistant tissues. The exact mechanism controlling
cytoplasmic viscosity and the low temperature response is not clear.
However, it seems significant that the observation of the effects of low
temperature on protoplasmic streaming and chilling sensitive species
taken together with the known influence of temperature on reversible
dissassociation of the cytoskeletal polymers point to the interacting and
integrated response of the membrane cytoskeletal systems.

C.   *Direct Effect of Low Temperature on Proteins*

The discussion to this point has focused on the molecular ordering of
the membrane lipids as the controlling element in the low temperature
response. It is now appropriate to address the question of the role of
proteins in the temperature response.   Apparent discontinuities in
Arrhenius plots as the result of a conformational change or disassociation
of proteins have been reported for a number of systems (5, 26). While
some enzyme proteins may undergo these conformational changes, it is
very important to realize that there is neither any theoretical reason to
expect nor experimental evidence to indicate, that many enzymes
undergo conformational changes over the narrow range of temperature
observed for the chilling response. Further, with but few exceptions as
noted below, the critical temperature for physiological dysfunction has
not been correlated with any protein conformational change as has been
done with membrane fluidity changes. If protein conformational changes
are considered as a primary determinant of the critical temperature for
chilling, then the broad range of consequences to cellular function would
seem to require that the protein be present in major quantities in all of
the sensitive species, or, have a means of rapidly amplifying and
propagating its conformational change to affect many cellular aspects.
In addition some evidence should be presented to show that metabolic
dysfunction leading to a physiological event occurs at the step catalyzed
by the enzyme.
There is little evidence of any master protein or master reaction
that can explain the commonality of the physiological dysfunctions in
chilling sensitive species when temperatures are lowered below about 10-
12°C. In his review of genetic regulation of temperature responses in
microorganisms, Ingraham (37) indicated that while the chemical basis
for loss of function of a protein at high temperatures is relatively well
understood, (i.e., those chemical bonds which maintain the proper
secondary and tertiary structure of proteins become weakened at

elevated temperatures, resulting in denaturation and loss of function of the protein), the loss of function at low temperature is not readily apparent, (although weakening of hydrophobic bonds is involved). His research indicated that cold-sensitive mutants of *E. coli* owed their cold sensitivity to a slight alteration in the structure of enzymes involved in ribosome synthesis at low temperature. The enzymes in these mutants were more sensitive to feed-back inhibition at all temperatures.

With higher plants that are chilling sensitive, Yamaki and Uritani (129, 130) considered that the primary response was in the ability of membrane proteins to bind phospholipids as a result of chilling treatments. Furthermore, Shirahashi *et al* (96) have shown that pyruvate, orthophosphate dikinase from maize is cold labile and that this cold lability correlates with a sharp change in activation energy of the dikinase -catalyzed reaction observed near 12°C. Studies have shown that some enzymes associated with membrane systems in chilling sensitive plants undergo an anomalous change in rate at about the same temperature as that shown for the physical change in membrane characteristics; in contrast, non-membrane associated enzymes maintain a relatively linear relationship with decreasing temperature (6, 85). Downton and Hawker (20) have shown that the soluble enzymes responsible for starch synthesis in chilling sensitive maize, avocado and sweet potato exhibit a discontinuity in Arrhenius plots at about 12°C which correlate with physiological dysfunction in those tissues. In contrast similar enzymes from the chilling resistant potato exhibit a linear plot in response to temperature over the range from 23°C down to 0°C. Their results indicated that there was an association between starch synthetase and the lipid, lysolethicin, and they suggested that the discontinuities observed in the plots from chilling sensitive plants reflected a transition in lipid at the critical temperature similar to that observed with the membrane-bound enzymes reported previously for those species. These studies of Downton and Hawker (20) focus attention on the lipid rather than expecting the protein enzymes from the different species to respond differently to temperature.

Again, assignment of the primary temperature response to a direct effect on a protein would have to provide a mechanism to explain how that response could be propagated throughout a number of crop species at a relatively narrow temperature range focusing around $10-12^\circ$C.

In summary, this volume is intended to present what is or might be known about the primary molecular events involved in the response of plants to low temperature and, as such, will hopefully enhance our ability to plan appropriate research to fill knowledge gaps where they might exist.

IV. REFERENCES

1. Ainsworth, P. J., Tustanoff, E. R. *Biochem. Biophys. Res. Comm.* *47*, 1299-1305 (1972).
2. Anderson, H. C. *Ann. Rev. Biochem. 47*, 359-383 (1978).
3. Apeland, J. *Int. Inst. Refrig., Bull. 46, Annexe 1*, 325-333 (1966).
4. Baldassare, J. J., Brenckle, G. M., Hoffman, M., Silbert, D. F. *J. Biol. Chem. 252*, 8797-8803 (1977).
5. Brandts, J. F. "Thermobiology" pp. 25-72. (A. H. Rose, ed.). Academic Press, NY (1967).
6. Breidenbach, R. W., Wade, N. L., and Lyons, J. M. *Plant Physiol. 54*, 324-327 (1974).
7. Brinkley, B. R., Fuller, G. M., and Highfield, D. P. *In*"Microtubules and Microtubule Inhibitors." pp. 297-312. (M. Borgers and M. DeBander, eds.). Amsterdam: North-Holland Publishing Company (1975).
8. Brooking, I. R. *Aust. Jour. Plant Physiol. 3*, 589-596 (1976).
9. Chandler, W. H. *Mo. Agr. Expt. Sta. Bull. 8*, 141-309 (1913).
10. Christiansen, M. H. *Plant Physiol . 43* 743 (1968).
11. Christiansen, M. H., Carns, H. R., and Slyter, D. J. *Plant Physiol. 46*, 53-56 (1970).
12. Creencia, R. P., and Bramlage, W. J. *Plant Physiol. 47*, 389-392 (1971).
13. Cronan, J. E., Jr. *Ann. Rev. Biochem. 47*, 163-189 (1978).
14. Crookston, R. K., O'Toole, J., Lee, R., Ozbun, J. L., and Wallace, D. H. *Crop Sci. 14*, 457-464 (1974).
15. Das, T. M., Hildebrand, A. C., and Riker, A. J. *Amer. J. Bot. 53*, 253-259 (1966).
16. De Gier, J., Mandersloot, J. G., Hupkes, J. V., McElhaney, R. N., and Van Beek, W. P. *Biochem. Biophys. Acta 233*, 610-618 (1971).
17. DeKruyff, B., Van Dijck, P. W. M., Goldbach, R. W., Demel, R. A., and Van Deenen, L. L. M. *Biochim. Biophys. Acta 330*, 269-282 (1973).
18. Demel, R. A., and DeKruyff, B. *Biochim. Biophys. Acta 457*, 109-132 (1976).
19. Dix, P. J. and Street, H. E. *Ann. Bot. 40*, 903-910 (1976).
20. Downton, W. J. S., Hawker, J. S. *Phytochemistry 14*, 1259-1263 (1975).
21. Eaks, I. L., Morris, L. L. *Proc. Amer. Soc. Hort. Sci. 69*, 388-399 (1957).
22. Eletr, S., Williams, M. A., Watkins, T., and Keith, A. D. *Biochim. Biophys. Acta 339*, 190-201 (1974).
23. Garber, M. P., and Steponkus, P. L. *Plant PHysiol. 57*, 673-680 (1976a).
24. Garber, M. P., and Steponkus, P. L. *Plant Physiol. 57*, 681-686 (1976b).
25. Giaquinta, R. T., and Geiger, D. R. *Plant Physiol. 51*, 372-377 (1973).

26. Hardy, R. W. F., Holsten, R. D., Jackson, E. K., and Burns, R. C. *Plant Physiol. 43,* 1185-1207 (1968).
27. Harrington, J. E. and Kihara, G. M. *Proc. Amer. Soc. Hort. Sci. 75,* 485-489 (1960).
28. Heber, U. *Plant Physiol. 42,* 1343-1350 (1967).
29. Heber, U. *In* "The Frozen Cell."   pp. 175-188.   (G. E. W. Wolstenholme and M. O'Connor, eds.). J. A. Churchill (1970).
30. Heber, U. W., and Santarius, K. A. *Plant Physiol. 39,* 712-719 (1964).
31. Heber, U., and Santarius, K. A. *In* "Temperature and Life."   pp. 232-263. (H. Precht, J. Christopherson, H. Hensel and W. Larcher, eds.). Springer-Verlag, New York (1973).
32. Henry, S. A., Keith, A. D., and Snipes, W. *Biophys. J. 16,* 641-653 (1976).
33. Hepler, P. K., and Palevitz, B. A. *Ann. Rev. Plant Phys. 25,* 309-362 (1974).
34. Huang, L., Jaquet, D. D., and Huag, A. *Can. J. Biochem. 52,* 483-490 (1974).
35. Huelin, F. E., Coggiola, I. M. *J. Sci. Food Agr. 21,* 44-48 (1970).
36. Ilker, R., Waring, A. J., Lyons, J. M., and Breidenbach, R. W. *Protoplasma 90,* 229-252 (1976).
37. Ingraham, J. L. *In* "Temperature and Life." pp. 69-77. (H. Precht, J. Christophersen, H. Hensel, and W. Larcher, eds.). Springer-Verlag, New York (1973).
38. Jones, R. L., Freebairn, H. T., and McDonnell, J. F. *J. Amer. Soc. Hort. Sci. 103,* 219-221 (1978).
39. Kane, O., Marcellin, P., Mazliak, P. *Plant Physiol. 61,* 634-638 (1978).
40. Kasai, R., Kitajima, Y., Martin, C. E., Nozawa, Y., Skriver, L., and Thompson, G. A., Jr. *Biochem. 15,* 5228-5233 (1976).
41. Kenefick, D. G. *Agronomy Abstracts* p. 91 (1964).
42. Kenefick, D. G., and Swanson, C. R. *Crop Sci. 3,* 202-205 (1963).
43. Kitajima, Y. and Thompson, G. A., Jr. *J. Cell. Biol. 72,* 744-755 (1977).
44. Kosiyachinda, S. and Young, R. E. *Plant Physiol. 60,* 470-474 (1977).
45. Levitt, J.   Responses of Plants to Environmental Stresses. Academic Press, N.Y. 697 pp. (1972).
46. Levitt, J., and Dear, J. *In* "The Frozen Cell." pp. 149-174. (G. E. W. Wolstenholme and M. O'Connor, eds.). J. A. Churchill (1970).
47. Levitt, J., and Scarth, G. W. *Can. J. Res. C. 14,* 267-284 (1936a).
48. Levitt, J., and Scarth, G. W. *Can. J. Res. C. 14,* 285-305 (1936b).
49. Lewis, D. A. *Science 124,* 75-76 (1956).
50. Lieberman, M., Craft, C. C., Audia, W. V., Wilcox, M. S. *Plant Physiol. 33,* 307-311 (1958).
51. Linden, C. D., Wright, K. L., McConnell, H. M., and Fox, C. F. *Proc. Nat. Acad. Sci. 70,* 2271-2275 (1973).
52. Lovelock, J. E. *Biochim. Biophys. Acta 10,* 414-426 (1953).

53. Lyons, J. M. *Ann. Rev. Plant Physiol. 24,* 445 466 (1973).
54. Lyons, J. M., and Raison, J. K. *Plant Physiol. 45,* 386-389 (1970).
55. Luzzati, V., and Husson, F. *J. Cell Biol. 12,* 207-219 (1962).
56. Maximov, N. A. *Ber. Deut. Bot. Ges. 30,* 52-65, 293-305, 504-516 (1912).
57. Mazur, P. *Ann. Rev. of Plant Physiol. 20,* 419-448 (1969).
58. Mazur, P. *Science 168,* 939-949 (1970).
59. Mazur, P. *Cryobiology 14,* 251-272 (1977a).
60. Mazur, P. *In* "The Freezing of Mammalian Embryos." pp. 19-48. CIBA Foundation Symp. London (1977b).
61. McElhaney, R. N. *Jour. Supramolecular Structure 2,* 617-628 (1974).
62. McElhaney, R. N., Souza, K. A. *Biochim. Biophys. Acta 443,* 348-359 (1976).
63. McElhaney, R. N., DeGier, J., and Van der Neat Kok, E. C. M. *Biochim. Biophys. Acta 298,* 500 512 (1973).
64. Meryman, H. T. *Nature 218,* 333-336 (1968).
65. Meryman, H. T. *Ann. Rev. Biophys. 3,* 341-363 (1974).
66. Meryman, H. T., Williams, R. J., and Douglas, M. St. J. *Cryobiology 14,* 287-302 (1977).
67. Minchin, A., and Simon, E. W. *Jour. Expt. Bot. 24,* 1231-1235 (1973).
68. Monero, R. D., Blou, B , Farias, R. N., and Trucco, R. E. *Biochim. Biophys. Acta. 282,* 157-165 (1973).
69. Nagle, J. F. *Jour. Membrane Biol. 27,* 223-250 (1976).
70. Nolan, W. E., and Smillie, R. M *Plant Physiol. 59,* 1141-1145 (1977).
71. Overath, P., Schairer, H. U., and Stoffel, W. *Proc. Natl. Acad. Sci., USA, 67,* 606-612 (1970).
72. Olien, C. R. *Cryobiology 1,* 47-53 (1965).
73. Olien, C. R. *In* "Barley Handbook." pp. 121-127. U.S. Dept. Agric., Wash., D.C. (1968).
74. Pollard, T. P. *J. Cell Biol. 68,* 579-601 (1976).
75. Paton, J. C., McMurchie, E. J., May, B. K., and Elliott, W. H. *Jour. Bact. 135,* 754-759 (1978).
76. Patterson, B. D., and Graham, D. *J. Expt. Bot. 28,* 736-743 (1977).
77. Patterson, B. D., Kenrick, J. R.,and Raison, J. K. *Phytochem. 17,* 1089-1092 (1978).
78. Patterson, B. D., Murata, T., and Graham, D. *Austral. J. Plant Physiol. 3,* 435-442 (1976).
79. Patterson, B. D., Paull, R., and Smillie, R. M. *Aust. Jour. Plant Physiol. 5,* 609-617 (1978).
80. Raison, J. K. *J. Bioenergetics 4,* 357-381 (1972).
81. Raison, J. K. *In* "Encyl Plant Physiology." (P. K. Stumpf, ed.). (In press). (1979).
82. Raison, J. K., and Berry, J. A. *Carnegie Inst. Wash. Yearbook (In press) (1979).*

83. Raison, J. K., and Chapman, E. A. *Austral. J. Plant Physiol. 3,* 291-299 (1976).
84. Raison, J. K., Chapman, E. A., and White, P. Y. *Plant Physiol. 59,* 623-627 (1977).
85. Raison, J. K., Lyons, J. M., Mehlhorn, R. J., and Keith, A. D. *J. Biol. Chem. 246,* 4036-4040 (1971).
86. Read, D., McElhaney, N. *J. Bact. 123,* 47-55 (1975).
87. Reinert, J. C., Steim, J. M. *Science 168,* 1580-1582 (1970).
88. Rick, C. M. *In* "Genes, Enzymes and Populations." pp. 255-269. (A. M. Srb, ed.). Plenum Press, NY (1973).
89. Romijn, J. C., Van Golde, L. M. G., McElhaney, R. N., and Van Deenen, L. L. M. *Biochem. Biophys. Acta 280,* 22-32 (1972).
90. Rosa, J. T. *Proc. Amer. Soc. Hort. Sci. 23,* 233-242 (1926).
91. Rottem, S., Cirillo, V. P., DeKruyff, B., Shinitzky, M., and Razin, S. *Biochim. Biophys. Acta 323,* 509-519 (1973).
92. Sachs, J. von. Handbuch der Experimental Physiologie der Pflanzen. Engelman Press, Leipzig. 514 pp. (1865).
93. Scarth, G. W., and Levitt, J. *Plant Physiol. 12,* 51-78 (1937).
94. Scarth, G. W., Levitt, J., and Siminovitch, D. *Cold Spg. Harbor Symp. 8,* 102-109 (1940).
95. Shimshick, E. J., McConnell, H. M. *Biochemistry 12,* 2351-2360 (1973).
96. Shirahashi, K., Hayakawa, S., and Sugyama, T. *Plant Physiol. 62,* 826-830 (1978).
97. Shneyour, A., Raison, J. K., and Smillie, R. M. *Biochim. Biophys. Acta 292,* 152-161 (1975).
98. Siminovitch, D., and Levitt, J. *Can. J. Res. Sect. C 19,* 9-20 (1941).
99. Smith, P. G., and Millett, A. E. *Proc. Amer. Soc. Hort. Sci. 84,* 480-484 (1964).
100. Snipes, W., Person, S., Keith, A., and Cupp, J. *Science 188,* 64-66 (1974).
101. Stadelmann, E. J., and Lee, O. Y. *In* "Comparative Biochemistry and Physiology of Transport." pp. 434-441. (L. Bolis, K. Block, S. E. Luria, and F. Lynen, eds.). North-Holland Publ. Co., The Netherlands (1974).
102. Steponkus, P. L. *Morris Arboretum Bulletin 20,* 26-32 (1969).
103. Steponkus, P. L. *Cryobiology 8,* 570-573 (1971).
104. Steponkus, P. L. *Adv. in Agron. 30,* 51-98 (1978).
105. Steponkus, P. L. *In* "Stress Physiology of Crop Plants." (H. Mussell and R. C. Staples, eds.). (In press). Wiley-Interstate, New York (1979).
106. Steponkus, P. L., and Wiest, S. C. *In* "Plant Cold Hardiness and Freezing Stress - Mechanisms and Crop Implications." pp. 75-91. (P. H. Li and A. Sakai, eds.). Academic Press, New York (1978).
107. Steponkus, P. L., Garber, M. P., Myers, S. P., and Lineberger, R. D. *Cryobiology 14,* 303-321 (1977).
108. St. John, B. Judith, and Christiansen, Meryl N. *Plant Physiol. 57,* 257-259 (1976).

24   J. M. Lyons *et al.*

109. Stout, D. G., Steponkus, P. L., and Cotts, R. M. *Plant Physiol.* *60*, 374-378 (1977).
110. Tait, J. *Am. J. Physiol. 59*, 467 (1922).
111. Taylor, A. O., Rowley, J. A. *Plant Physiol. 47*, 713-718 (1971).
112. Thompson, P. A. *Nature 225*, 827 (1970).
113. Thompson, P. A. *Sci. Hort. 2*, 35-54 (1974).
114. Vancura, V. *Plant and Soil 27*, 319-328 (1967).
115. Walker, J. A. and Wheeler, K. P. *Biochim. Biophys. Acta 394*, 135-144 (1975).
116. Wang, C. Y., and Baker, J. E. *Plant & Cell Physiol. 20*, 243-251 (1979).
117. Waring, A. J., and Laties, G. G. *Plant Physiol. 69*, 11-16 (1977).
118. Waring, A. J., Breidenbach, R. W., and Lyons, J. M. *Biochim. Biophys. Acta 443*, 157-168 (1976).
119. Warren, G. B., Houslay, M. D., Metcalfe, J. C., Birdsall, N. J. M. *Nature 255*, 684-685 (1975).
120. Watada, A. E., Morris, L. L. *Proc. Amer. Soc. Hort. Sci. 89*, 368-374 (1966).
121. Watada, A. E., Morris, L. L. *Proc. Amer. Soc. Hort. Sci. 89*, 375-380 (1966).
122. Wheaton, T. A. Physiological Comparison of Plants Sensitive and Insensitive to Chilling Temperatures. Ph.D. thesis. University of California, Davis. 93 p. (1963).
123. Wilkinson, B. G. The Biochemistry of Fruits and Their Products. *Academic Press, N.Y. 1*, 537-554 (1970).
124. Willemot, C. *Plant Physiol. 60*, 1-4 (1977).
125. Wills, R. B. H., Scott, K. J. *Phytochem. 10*, 1783-1785 (1971).
126. Wilson, J. M. *New Phytol. 76*, 257-270 (1976).
127. Wright, M. *Planta 120*, 63-69 (1974).
128. Wright, M., and Simon, E. *J. Expt. Bot. 24*, 400-418 (1973).
129. Yamaki, S., and Uritani, I. *Agr. Biol. Chem. 36*, 47-55 (1972).
130. Yamaki, S., and Uritani, I. *Plant Cell Physiol. 15*, 385-393 (1974).
131. Zsoldos, F. *Plant and Soil 37*, 469-478 (1972).
132. Zsoldos, F. and Karvaly, B. Physiol. Plant 43, *326-330 (1978)*.

ADAPTATION TO CHILLING: SURVIVAL, GERMINATION,
RESPIRATION AND PROTOPLASMIC DYNAMICS

Brian D. Patterson and Douglas Graham

Plant Physiology Unit
CSIRO Division of Food Research
School of Biological Sciences
Macquarie University
North Ryde, N.S.W. 2113, Australia

Robert Paull

Department of Botany
University of Hawaii at Manoa
3190 Maile Way
Honolulu, Hawaii

I.   INTRODUCTION

In this paper we describe experiments with two sorts of plants with contrasting chilling-resistance. The first are varieties of the tropical species *Lycopersicon hirsutum,* a wild tomato, which is indigenous to the Andes of Ecuador and Peru. Varieties at the higher altitudes of its native habitat experience temperatures below $10^{\circ}C$ throughout the year (minimum night temperatures), while varieties growing near sea level rarely or never experience temperatures below $18^{\circ}C$ (12). Varieties from the lower altitudes are at least as chilling-sensitive as the domestic tomato *Lycopersicon esculentum* (9). The second group of plants come either from "temperate" climates or from the tropical lowlands. Most of the "temperate" group were domesticated in the winter rainfall areas of the Mediterranean. Such plants are better adapted to survive prolonged chilling than high altitude tropical plants. However, they have the disadvantage that they are not closely related to the sensitive tropical plants with which it is useful to compare them. It is therefore often difficult to decide which physiological differences between, say, the onion and the water melon are specifically adaptations to chilling. For instance, the greater frost resistance of the onion might

25

be reflected in a different membrane lipid composition. Hence the
usefulness of comparing the physiology of closely-related species such as
the tomatoes which vary in chilling sensitivity, although none is
completely chilling resistant. In these we have examined lipid
properties, respiration, germination and the diurnal fluctuation of
chilling resistance. We have then compared the immediate effect of cold
on the structure of the cytoplasm of the very chilling resistant onion
(Allium cepa) and the very chilling sensitive water melon *(Citrullus la-
natus)*.

II.   RESULTS AND DISCUSSION

A.   *Lipid Properties of Lycopersicon Varieties*

In previous studies (7, 8) *Passiflora* species were shown to differ in
chilling resistance. However, the composition of their polar lipids did not
vary appreciably. Differences in lipid unsaturation which would account
for differences in chilling resistance were not found. Differences in the
physical properties of the lipids was, however, suggested by electron spin
resonance studies (ESR). There were apparent changes in the
temperature coefficient of spin label motion at two temperatures, i.e.
there was an upper break temperature and a lower break temperature.
The results suggested a relationship between the temperature of the
lower break and the chilling sensitivity of the particular *Passiflora* species or hybrid. We therefore predicted that the membrane lipids of *L.
hirsutum* and those of the domestic tomato *L. esculentum* would show
corresponding break temperatures which would also relate to their
chilling sensitivity. Figure 1 shows a plot of the log of the spin label
motion parameter τ_o against the reciiprocal of absolute temperature
(Arrhenius plot) for vesicles of polar lipids from the domestic tomato *L.
esculentum*. It is often difficult to decide when to reject models for such
data which postulate straight lines separated by breaks, in favour of a
smooth curve. Therefore in this case, before fitting straight lines,
foreshortened plots which exaggerate curves but preserve linearity were
made by the method of Wilcox and Patterson (13). Figure 2 shows the
result: a smooth curve was at least as convincing a model for the data as
one based on three straight lines with breaks. Very similar results were
obtained for altitudinal varieties of *L. hirsutum*. A break in the
temperature coefficient of spin label motion in lipids from the tomato
has been reported at $13^{o}C$ (11). The changes in slope were similar to
those reported here for *L. esculentum* and *L. hirsutum*. Our results suggest that objective tests which permit the option of smooth curves should
be applied to ESR data. This is particularly important in view of the
prominence of breaks in ESR for the lipid theory of chilling sensitivity.

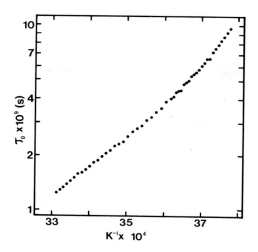

FIGURE 1. Spin label (12NSMe) motion in vesicles of polar lipids prepared from leaves of tomato (L. esculentum cv. Rutgers). Lipid preparation and ESR techniques were as described previously (7).

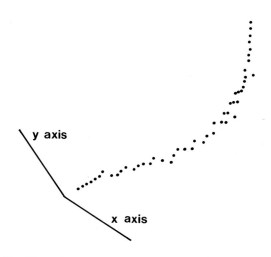

FIGURE 2. Visual demonstration that a curve model fits the ESR data shown in Figure 1. The plot was foreshortened by five times in order to magnify any changes in slope, as described elsewhere (13). The increased angle between the original axes is shown. No breaks are indicated.

B.   Temperature Effects on Respiration

The temperature dependence of respiration has been reported to show abrupt changes at specific temperatures ("breaks"), especially around 10°C in the case of chilling-sensitive plants such as cucumber (3). When macroscopic tissues are used (leaves, roots, fruits), it is quite difficult to make accurate measurements of the effect of temperature on plant respiration, especially at low temperatures, because small changes in oxygen concentration are difficult to measure in gas streams. However, pollen has a high rate of respiration which can be measured in an oxygen electrode. Tomato pollen grains have the dimensions of microorganisms, their nutrient requirements are simple, and they can be harvested as needed. Even close to 0°C, the reproducibility of respiration measurement is as high as at 20°C. Figure 3 shows the effect of temperature on the rate of respiration of pollen from two varieties of *L. hirsutum;* a relatively chilling resistant one from high altitude (3200m) and a chilling-sensitive one from low altitude (30m). Neither shows a significant change in the temperature coefficient of respiration between 0 and 20°C. This result might be expected for the high altitude form, for in its native habitat it shares the temperature environment of wild and domesticated potatoes, which are generally considered to be chilling-resistant (2). However, the form from low altitude is more chilling-sensitive, as would be expected for a plant from a lowland tropical environment. While these results refer quite specifically to pollen and do not necessarily reflect the respiration of other tissues, they show that abrupt changes in the temperature coefficient of respiration are not invariably associated with chilling sensitivity of the parent plant.

C.   Genetic Association of Chilling Adaptations

Chilling resistance could be a simple property, resulting say from the physical properties of membrane lipid, or it could be the manifestation of a whole series of unconnected adaptations to temperature as has been described for microorganisms (1). If it were a single property, it would be expected to be inherited in a unitary manner. Therefore the extent to which different adaptations to low temperature in *Lycopersicon* were associated was investigated. These were low temperature pollen germination and low temperature seed germination. Figure 4 shows the extent to which pollen from two altitudinal varieties of *L. hirsutum* germinated when held at different temperatures for 18 h. Appreciable differences in the minimum temperature of germination are shown. The temperature at which 50% germination was achieved was then plotted against a measure of low temperature seed germination for a number of altitudinal varieties of *L. hirsutum*. Table I shows that there is no apparent relation between these characters.

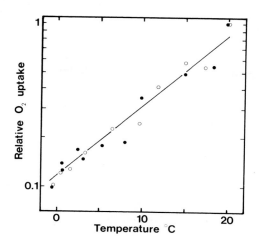

FIGURE 3. Pollen respiration. Pollen (1 mg.ml^{-1}) was suspended in 10% sucrose, 0.01% H_3BO_3 and the rate of oxygen uptake plotted as a proportion of the rate at $20^{\circ}C$ (log scale). Each point represents the initial rate of uptake determined from a time course. Oxygen concentration in the electrode cell was maintained at 250 ± 20 μmoles.ml^{-1}. (●), pollen from low altitude L. hirsutum; (O), pollen from high altitude L. hirsutum. Both plots show a constant Q_{10} of between 2 and 3.

TABLE I.   Ability of Seed to Germinate Quickly at Low Temperature and Ability of Pollen to Germinate at Low Temperature

Altitude of origin (metres)	Temperature at which 50% seed germination occurs after 10 days ($^{\circ}C$)	Temperature at which 50% pollen germination occurs after 15 h ($^{\circ}C$)
400	15.4	8.8
1000	15.2	7.0
1500	13.3	6.8
2100	11.0	7.5
2650	13.0	7.7
3200	10.8	7.2

Seed germination (radical extrusion) at different temperatures, and the temperature at which 50% germination occurred after 10 days was found. The temperature at which 50% germination of pollen occurred was found as described in Fig. 4.

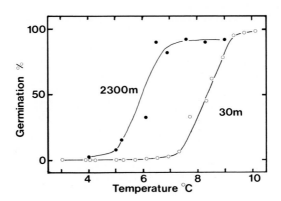

FIGURE 4. Minimum temperatures of germination of pollen for different varieties of L. hirsutum. Pollen was germinated on 10% sucrose, 0.01% boric acid, solidified with 1% agar. After 15 h, the pollen was fixed with formaldehyde vapour and photomicrographs made, from which percentage germination was assessed. For these and other altitudinal varieties of L. hirsutum, the temperature at which 50% germination occurred was used as a standard for comparison.

D.   Effect of Chilling on the Structure of Cytoplasm

Our results suggest that adaptation to chilling temperatures may require adaptation at many sites within a cell. We have therefore looked for effects of chilling which do not necessarily relate to the loss of turgor which eventually occurs and which is irreversible (see Section E). One such effect is the change in the structure of cytoplasm which occurs when chilling sensitive plants are chilled. We found that within seconds of cooling living trichomes of water melon, tomato or other chilling sensitive plants a distinctive change could be seen using phase contrast microscopy. Figure 5 shows the appearance of a water melon cell which has been cooled to 4°C. The normal thin strands of cytoplasm which span the vacuole have disappeared and have been replaced by spherical vesicles. Also shown is the same cell after it was warmed to 15°C. Between 12 and 15°C the vesicles disappeared and the thin strands of cytoplasm were reformed. Quite unrelated plants show the same effect as long as they are chilling sensitive, for instance corolla hairs of *Datura* (Reid, Patterson and Ferguson, unpublished film, 1978). In contrast, plants which can tolerate prolonged chilling, such as the onion, maintain normal cell structures down to 0°C (Fig. 5). The rapidity of the response of protoplasmic structure to cold suggests that we are observing something close to a primary effect on the cell. However, we have observed that the replacement of the fine strands of cytoplasm by vesicles is also a general response to stress. In both chilling sensitive and resistant cells, toxic agents, mechanical damage or bacterial infection produce a similar loss of linear structure and the simultaneous appearance of spherical vesicles. It has been reported that a similar effect occurs when fixatives are used in electron microscopy (4). The microtubule elements that partly make up the cytoskeleton of eukaryotic cells are known to be disrupted by chilling temperatures *in vitro* and *in vivo* (5) and this dissociation of the microtubular elements is reversible on warming. Therefore, our results are consistent with a model in which a cytoskeleton which supports linear strands of cytoplasm is dissociated at chilling temperatures into a pool of tubulin subunits. On rewarming, polymerisation is favoured, and the cytoskeleton is reformed, so that the cytoplasmic strands are stabilized. If this model is correct, it should be possible to demonstrate differences in the temperature dependence of microtubule formation for chilling-resistant and chilling-sensitive plants.

E.   Diurnal Variation in Chilling Sensitivity

When physiologists classify species according to their chilling sensitivity, they often use criteria which vary depending on whether they are referring to whole plants or their detached organs, such as tubers or fruit. In order to compare different kinds of wild and domestic tomato, the criterion of chilling resistance we have mainly used is the ability to survive exposure to 0°C (9). In order to standardize the physiological state of the plants, they were grown with a day length of 12 h and sampled for chilling half way through the light period. However, it was found that when plants were chilled in the dark, but beginning at different times in the day/night cycle, their capacity to survive chilling varied strikingly. Figure 6 shows the diurnal variation in the loss of turgor induced by chilling the seedlings at 0°C for 6 days. The environment of the seedling was kept saturated with moisture so that the loss of turgor was not a result of water loss. If plants were chilled from 1 h after the onset of the light period (7.00 h) there was a maximum sensitivity to chilling measured as loss of turgor. The plants were then warmed to 20°C in a humidified atmosphere, transferred to continuous light at 22°C and after 1 week the proportion of survivors assessed. Figure 6 shows that, using the criterion of survival, the plants are least sensitive to chilling when they are placed at 0°C 4 h after the onset of the dark period (22.00 h). The time at 0°C required to kill half the seedlings was 6 days for the least sensitive time of the day, but only 3 days for the most sensitive part of the day. We assume that the differences in chilling symptoms for different times of the day reflect differences in metabolism. The results are consistent with the hypothesis that chilling causes a metabolic imbalance (2, 10); in addition, a way to test the hypothesis is suggested. If a particular metabolite is primarily responsible for the chilling damage, it is reasonable to assume it must be produced to a greater extent when chilling is started at 7.00 h than when it is started at 22.00 h under our conditions. We are therefore investigating which metabolites do vary in the expected manner during chilling.

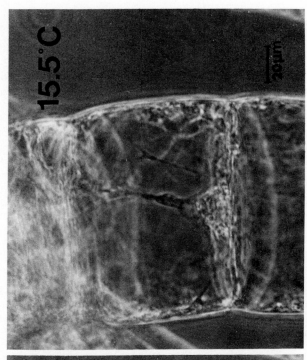

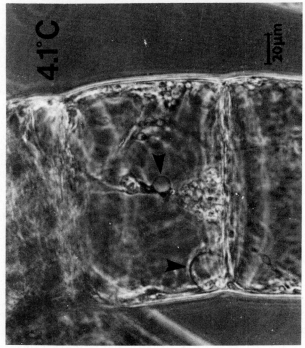

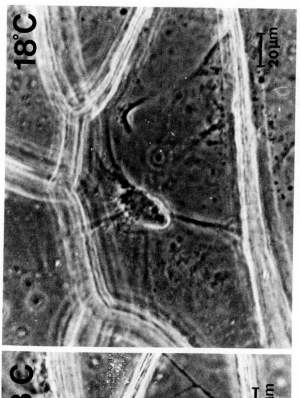
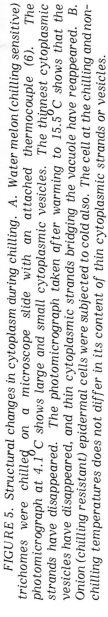
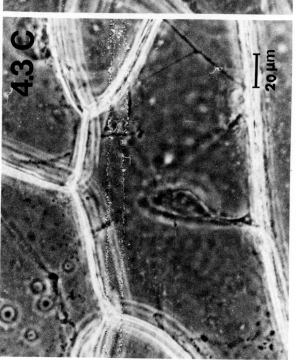

FIGURE 5. Structural changes in cytoplasm during chilling. A. Water melon (chilling sensitive) trichomes were chilled on a microscope slide with an attached thermocouple (6). The photomicrograph at 4.1°C shows large and small cytoplasmic vesicles. The thinnest cytoplasmic strands have disappeared. The photomicrograph taken after warming to 15.5°C shows that the vesicles have disappeared, and thin cytoplasmic strands bridging the vacuole have reappeared. B. Onion (chilling resistant) epidermal cells were subjected to cold also. The cell at the chilling and non-chilling temperatures does not differ in its content of thin cytoplasmic strands or vesicles.

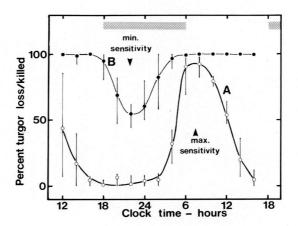

FIGURE 6.   Diurnal variation in chilling sensitivity of tomato seedlings. Seeds (L. esculentum cv. Grosse Lisse) were sown at intervals of 2 h through a growing cycle of 12 h day/night and $22^{\circ}C/18^{\circ}C$. After exactly 9 days, each group of seedlings was chilled to $0^{\circ}C$ at 100% R.H. After 6 days exactly, each group was examined for turgor loss (curve A). After warming to $20^{\circ}C$, plants were returned to conditions of constant light at $22^{\circ}C$, and assessed for % killed after 7 days (curve B). The shaded bars show the dark period.

II.   CONCLUSIONS

Using the tomato and its wild relative *L. hirsutum,* no evidence was found with ESR techniques for an abrupt change in lipid properties near $13^{\circ}C$.   Neither was the temperature dependence of pollen respiration found to change near this temperature, so that the Q_{10} for oxygen uptake was constant between 0 and $20^{\circ}C$.   Two different adaptations to chilling temperatures (low temperature germination of pollen and seed) were not necessarily associated in the same plant. Ability to survive chilling in tomato seedlings was not constant, but varied on a diurnal rhythm.   In a variety of chilling sensitive plants, chilling has a very rapid effect on cytoplasmic structure; this may be the result of a direct effect on the microtubules of the cytoskeleton.

ACKNOWLEDGMENTS

        We thank Mrs. L. Payne for technical assistance, Mrs. J. Kenrick for help with lipid preparation, and Mr. N. F. Tobin for ESR determinations. One of us (R.P.) was supported by a grant from the Rural Credit Development Fund of the Reserve Bank of Australia. We thank Professor C. M. Rick and Mr. E. Vallejos for the gift of altitudinal varieties of *L. hirsutum*.

IV. REFERENCES

1. Ingraham, J. L. *In* "Temperature   and   Life"   (H.   Precht,   J. Christophersen, H. Hensel and W. Lärcher, eds.), pp. 60-86. Springer Verlag, Berlin (date).

2. Lyons, J. M. *Ann. Rev. Plant Physiol. 24,* 445–466 (1973).

3. Lyons, J. M. and Raison, J. K. *Plant Physiol. 45,* 386–389 (1970).

4. Mersey, B. and McCulley, M. E. *J. Microsc. 114,* 49–76 (1978).

5. Olmsted, J. B. and Borisy, G. G. *Ann. Rev. Biochem. 42,* 507–539 (1973).

6. Patterson, B. D. and Graham, D. *J. Exptl. Bot. 28,* 736–743 (1977).

7. Patterson, B. D. Kenrick, J. and Raison, J. K. *Phytochem. 17,* 1089 (1978).

8. Patterson, B. D., Murata, T., and Graham, D. *Aust. J. Pl. Physiol. 3,* 435–442 (1977).

9. Patterson, B. D., Paull, R. and Smillie, R. M. *Aust. J. Pl. Physiol. 5,* 609–617 (1978).

10. Raison, J. K. *Symp. Soc. Exp. Biol. 27,* 485–512 (1973).

11. Raison, J. K. *In* "Mechanisms of Regulation of Plant Growth" (R. L. Bieleski, A. R. Ferguson and M. M. Cresswell, eds.), pp. 487–497. Bull. 12, Royal Soc. New Zealand, Wellington. (1974).

12. Schwerdtfeger, W. "World Survey of Climatology", Vol. 12 (Elsevier, New York) (1976).

13. Wilcox, M. E. and Patterson, B. D. This volume (1979).

SEED GERMINATION AT LOW TEMPERATURES

E. W. Simon

Department of Botany
The Queen's University of Belfast
Northern Ireland

The seeds of some species are able to germinate, albeit slowly, at temperatures near freezing, or even on ice. Others fail to germinate when held continuously at low temperatures, although they will germinate normally if transferred subsequently to warmer conditions. A third type of response is seen in those seeds said to exhibit chilling injury; if they are placed in cold water at the start of imbibition the seedlings formed during subsequent growth in the warm show symptoms of injury (3, 9).

It is the purpose of this paper to survey briefly what is known about this form of chilling injury and the failure of germination in the cold, and to assess the role in each response of a change in membrane fluidity.

I. CHILLING INJURY DUE TO IMBIBITION IN COLD WATER

When the seeds or embryos of certain species are exposed to low temperatures for a period at the start of imbibition they suffer chilling injury. Thus cotton (which makes no growth if held continuously at $5^{o}C$) will nevertheless germinate rather slowly if first exposed to water at $5^{o}C$ and then transferred to $31^{o}C$ - although under these conditions there is a tendency for the primary root to abort, lateral roots growing out instead (3, 5). As little as 30 minutes at $5^{o}C$ is enough to induce root abnormalities and reduce the speed of germination, but longer periods of chilling have a more pronounced effect. Likewise if soybean embryos are imbibed for 30 minutes at $12^{o}C$ or less, the percentage of embryos able to germinate and the subsequent growth of the axis are both reduced (1).

The consequences of a few days of chilling at 5 or $10^{o}C$ are still very evident in the plants long afterwards. Cotton plants, for instance, still grew more slowly than controls 4 weeks after a period of chilling at $10^{o}C$, dry weight and height being reduced in proportion to the number of days of chilling (4). Similarly soybean plants made less dry weight in 2 weeks and were less tall if they had been imbibed for 12 hours at $5^{o}C$ rather than

37

25°C (21). Even more remarkable are the data of Highkin and Lang (11) who germinated pea plants at a range of temperatures and then grew them on in a greenhouse. Seeds germinated at 3°C produced plants that were substantially smaller than those from seeds germinated at 27 or 31°C, with slower growth and fewer flowers, pods and seeds.

Exposure to chilling temperatures does not inevitably result in the development of injury symptoms even in sensitive plants. Thus cotton seeds and lima bean axes are rendered immune to cold water if they are first allowed to imbibe for a few hours at 25 or 31°C (6, 24); soybean embryos do not suffer chilling injury if they first take up enough warm water to reach a water content of 35% (1). Another way to prevent chilling injury is to expose seeds to moist air until the water content in cotton and soybean seed rises to at least 13% (7, 12), that of corn to 13-16% (2), and the axes of lima bean seed to 20% (23).

As chilling injury can be avoided by an initial period of imbibition with warm water or moist air, it follows that only when dry seeds imbibe cold water do they sustain injury. Evidence has been assembled elsewhere (28, 29) that in dry seed tissues, the water content is likely to be so low that the membrane phospholipids are forced into the hexagonal configuration and only come to adopt the familiar lamellar arrangement when the water content rises above about 20% (or possibly more relevant, when water potential rises above -80 bars (31). During the few seconds or minutes when the phospholipid molecules are reorienting themselves from hexagonal to lamellar conformation in each cell as its water content rises above the critical level, there is a short period when cell contents are becoming hydrated, but the membrane is not yet fully re-established as a semi-permeable barrier around the cell and solutes can leak out. This leakage is especially rapid in the first few minutes of imbibition, then waning in intensity although it continues slowly for many hours. Removing the seed coat allows faster imbibition of water and increases the initial rate of leakage.

There are several reasons for linking chilling injury with an enhancement of the leakage of solutes from seeds. First, chilling injury is induced by exposing seeds to cold water at the start of imbibition - two minutes in cold water is enough to injure soybean cotyledons (1); this early period of imbibition is just the time when membrane restitution is under way and leakage is most profuse. Second, leakage from imbibing seeds and embryos is itself intensified at low temperatures (14, 22, 27). Low temperature thus has a dual effect, enhancing leakage but restricting growth (1, 23).

The severity of chilling injury and leakage are both abated if the tissues are first allowed to imbibe in the warm (28). Finally it must be pointed out that rapid imbibition of warm water is itself enough to reduce seedling vigour. If isolated embryos are allowed to imbibe water and are then planted out in sand they grow poorly by comparison with intact seeds (which imbibe more slowly); pea embryos grew into plants which were only 70% as high as the controls after 18 days (15, 25).

In summary, chilling injury and leakage respond alike to changes of temperature and seed hydration; both reduce seedling vigour and both are manifestations of events in early imbibition. Furthermore chilling is known to increase leakage. These parallels between chilling and leakage lead to the hypothesis that chilling injury is a consequence of enhanced leakage (25), resulting from a slower restitution of membrane integrity in the cold. Although the loss of solutes from seeds may be quite extensive (29), leakage should perhaps be regarded as an extracellular indication of what is also happening within cells – that is to say, a temporary loss of compartmentation which may have serious repercussions for the subsequent operation of the cell.

Is it also possible that chilling injury could be due to a membrane phase shift causing mitochondrial dysfunction? This possibility must be viewed with some reserve in view of the evidence (29) that mitochondria are "immature" and inactive in the first few hours of imbibition, the time when seeds are most susceptible to chilling injury. Cohn and Obendorf (8) have examined the situation at a later stage when the integral membrane proteins would have returned to their rightful places alongside the phospholipids so re-establishing mitochondrial membrane integrity. The activity of mitochondria from corn with either 5 or 13% initial moisture content was examined after 12, 24 or 48 hours of imbibition at 5°C. The low moisture grains grew more slowly than the others in subsequent growth tests, showing that they had suffered chilling damage. However, the respiration rate, ATP content and energy charge were the same in high and low moisture grains after imbibition in the cold. Mitochondrial oxidase activity, ADP/O ratio and respiratory control ratio were also substantially independent of moisture level. Cohn and Obendorf came to the conclusion that "a disruption of energy metabolism is not a primary cause of imbibitional chilling injury."

II. THE LOW TEMPERATURE LIMIT FOR GERMINATION

The effect of temperature on the germination of seeds has been under investigation for more than a century. Already in 1874 Haberlandt recognized that barley, wheat and oats would still germinate when the temperature was as low as 2 or 3°C, but that a number of other crop plants such as corn, rice, cotton and tobacco failed to germinate below about $10-12^{o}$C. Taken in conjunction with the work on chilling injury discussed elsewhere in this Volume these early observations seem to imply that plants can be divided into two groups (the chill-resistant and the chill-sensitive) both as regards their behaviour as mature plants and also in respect of their power to germinate. Below about 10°C the plants are injured and their seeds fail to germinate. It would be a short step from this to suppose that comparable mechanisms underlie chilling in the two situations.

Recent work casts doubt on this view. First it should be said that species like *Agrostemma githago* (Fig. 1), lettuce (33) and mustard (30) are able to germinate readily down to temperatures near freezing. On an Arrhenius plot such data yield a single line (Fig. 2) corresponding to an activation energy of 25.6 kcals *(Agrostemma)* and 17.1 kcals (mustard).

However, many species produce curves of a different kind, 50% of the seeds only germinating over a limited range of temperature. For *Ajuga reptans* (Fig. 1) this extends from 35°C down to about 16°C where the curve appears to become vertical indicating a very sharp cut-off, suggesting that 50% germination would not be attained even after prolonged periods at 15°C or less. Cucumber seeds have a sharp cut-off at 11.5°C, and seeds maintained at 10°C achieved no more than 14% germination in 100 days (30). The transition from tempertures at which seeds germinate readily, to temperatures at which few if any seeds will ever germinate, is surprisingly abrupt (32).

The temperature at which this cut-off occurs shows a wide variation from species to species. Mason (17) has recorded data for 115 native British species of which 74 had a sharp cut-off, at a temperature ranging from 6°C for *Trifolium pratense* up to 30°C for *Polygonum persicaria* (Fig. 1). It must be emphasized that this last is no rare exception. Seeds just emerging from dormancy under the influence of after-ripening (dry storage at room temperature) or stratification (storage at 5°C in damp sand or soil) can at first only germinate over a narrow temperature range; for some seeds this initial range is positioned at a surprisingly high point on the temperature scale. Thus Vegis (34) has a diagram for the germination of five weed species before after-ripening from which it can be seen that the low temperature limit for three species was 33, and for two 36°C.

The low temperature limit may also vary within a single species, between different cultivars or different seed lots. The minima for a series of cultivars of carrot, leek and *Brassica* each vary over a range of 2-3°C, while the minimum for *Petrorhagia prolifera* in Germany was 5°C, and in Hungary 9°C (33). A survey of 12 species examined by Thompson indicates that on average the low temperature limit may shift in this way by about 3-4°C.

Much more remarkable is the shift that may occur as seeds lose their dormancy. Thompson (32) found that seeds of *Silene conoidea* would germinate down to 8°C immediately after harvest but the minimum then fell to 3°C after one year's storage at room temperature. *Hemerocallis* seed would only give 50% germination between 20 and 25°C after one week of stratification, but the lower limit fell in 8 weeks to less than 10°C (10). According to Vegis (34) fresh birch seeds only germinate at about 30°C but acquire the ability to germinate down to temperatures little more than 0°C after several months of stratification. These considerable shifts in the low temperature limit during emergence from dormancy, and the opposite movement said to occur as seeds enter dormancy (34, 35), must clearly be considered in any discussion of the mechanism that sets the low temperature limit for germination (19).

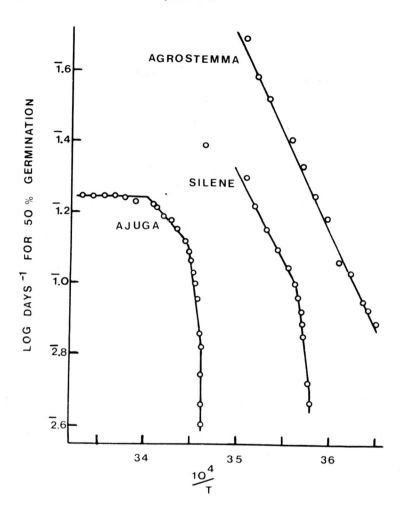

FIGURE 1. The effect of temperature on seed germination. Following the practice introduced (32) the time required for 50% germination is shown as a function of temperature. The lines representing the high temperature limits for the germination of each species are interrupted to draw attention to the position of the low temperature limits. A, Agrostemma githago; B, Ajuga reptans; C, Silene nutans; D, Carex nigra; E, Polygonum persicaria. A, B, C are from (32); D and E from (17).

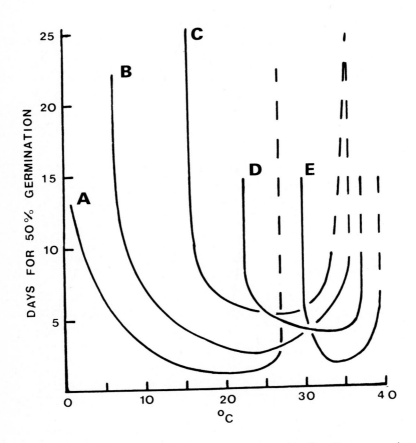

FIGURE 2. Arrhenius plots of germination for three species shown in Figure 1.

The range of temperatures over which the lower limit for germination may occur therefore extends from near freezing right up to about 35°C, quite different from the range up to 10 or 12°C that is sometimes regarded as typical for chilling injury. The range for the low temperature limit is also wider than that of the discontinuity in Arrhenius plots for membrane-bound enzymes in chill-sensitive plants, which seems to run from about 4°C in apple (18) up to 15°C in mung bean (26) and 20°C in postclimacteric avocado (13). Moreover there does not seem to be any counterpart in the mitochondrial work to the shift that occurs as seeds emerge from dormancy during dry storage.

When the germination data are expressed as Arrhenius plots (Fig. 2) it becomes apparent that they differ from mitochondrial oxidase plots in two additional respects. In *Ajuga* there is a quite extended range of

TABLE I.   Activation Energies (kcal) for Germination Above and Below the Break in the Arrhenius Plot.

Species	Above break	Below break	Reference
Cucumber	12	95	(30)
Mung bean	12	77	(30)
Fragaria vesca	19	61	(32)
Ajuga reptans	12	92	(32)
Primula farinosa	14	170	(32)
Silene nutans	23	89	(32)
Cistus creticus	25	210	(32)
Silene otites from Germany	32	170	(32)

temperatures over which germination proceeds at a rate little influenced by temperature (presumably because the physical process of imbibition is then rate-limiting). This is followed by a range of temperature encompassing the break in the plot and finally the curve becomes vertical at the sharp cut-off temperature. The activation energy for germination at temperatures below the discontinuity (Table 1) is much higher than the figure of 23.0 kcal quoted by Simon et al., (30) as an average from 35 Arrhenius plots for the activity of sub-cellular fractions from chill-sensitive plants. Furthermore there is only a small temperature span between the discontinuity and the cut-off point for germination, amounting to 1.5°C in Ajuga and Silene (Fig. 2) as compared to a mean of 7.5°C in 20 published Arrhenius plots for membrane-bound enzyme systems from plants.

These differences between the Arrhenius plots for germination and for enzyme systems raise serious doubts as to whether the same underlying mechanism is responsible for each. It would be interesting to know whether Arrhenius plots for the activity of mitochondria isolated from each of the seeds shown in Figs. 1 and 2 would be similar in shape to the germination data. However, information of this sort is not yet available, but Arrhenius plots for the respiration of 16 hour imbibed seeds of cucumber and mung bean show no discontinuity, although the seeds fail to germinate the cold (30), an observation which seems contrary to the thesis of a common underlying mechanism.

Moreover if the block to germination were due to mitochondrial dysfunction there should be an accumulation of toxic products such as ethanol. Subsequent transfer of the seeds into warm conditions would allow them to germinate after a lag period while such products were metabolized away. No such delay is in fact observed, the seeds germinating a little sooner after such a temperature step-up than if they had been at 20°C from the start (30).

The evidence presented above does little to sustain the view that the inability of seeds to germinate in the cold is mediated by a dysfunction of mitochondria or other membrane-bound enzyme systems. There seems on the contrary to be a difference between the response to cold by seeds and by mature plants. Thus although the respiration of imbibed cucumber seeds yields an Arrhenius plot without discontinuity (30), there are clear breaks at $12^{\circ}C$ in plots of cucumber leaf respiration (20) and cucumber fruit succinoxidase (16). Again plants such as *Carex nigra* and *Polygonum persicaria* which can germinate when exposed to temperatures of 22 and $29^{\circ}C$ respectively (Fig. 1), could hardly survive for long in a British summer if the adult plants became chilled as soon as the temperature dropped any lower. Temperatures below the lower limit for germination evidently do not injure the seeds or plants of these species; they merely prevent germination. In brief, germination is prevented in these native species at temperatures far above those responsible for chilling injury. Why then, do these seeds fail to germinate at such temperatures?

Failure of cell division or a requirement for some simple hormone or metabolite do not seem to be responsible. Nor do enhancement of membrane permeability and loss of compartmentation seem to be important consequences of imbibition in the cold, because cucumber seeds can still germinate readily after 350 hours imbibition at $5^{\circ}C$ (30). Inhibitor experiments and work on the incorporation of ^{14}C-leucine into imbibing cucumber seed provide no support for the idea that there is a complete breakdown of protein synthesis in the cold, although inability to synthesize certain particular proteins vital to the process of germination remains a possibility (19).

It is also possible that protein denaturation underlies the failure of germination in the cold. This hypothesis is in accord with the wide range of temperatures at which the lower limit for germination may occur, with the very high activation energy observed in the cold, with the sharp cut-off just below the temperature of the break, and also with the temperature step-up data. In addition it has interesting consequences for our understanding of dormancy, which I hope to develop elsewhere.

III. REFERENCES

1.   Bramlage, W. J., Leopold, A. C. and Parish, D. J. *Plant Physiol 61,* 525-529 (1978).
2.   Cal, J. P. and Obendorf, R. L. *Crop Science, 12,* 369-373 (1972).
3.   Christiansen, M. N. *Plant Physiol. 38,* 520-522 (1963).
4.   Christiansen, M. N. *Crop Science, 4,* 584-586 (1964).
5.   Christiansen, M. N. *Plant Physiol. 42,* 431-433 (1967).
6.   Christiansen, M. N. *Plant Physiol. 43,* 743-746 (1968).
7.   Christiansen, M. N.   Beltwide Cotton Production Research Conference, 50-51 (1969).

8. Cohn, M. A. and Obendorf, R. L. *Crop Science, 16,* 449-452 (1976).
9. Cohn, M. A. and Obendorf, R. L. *Amer. J. Bot. 65,* 50-56 (1978).
10. Griesbach, R. A. and Voth, P. D. *Bot. Gaz. 118,* 223-237 (1958).
11. Highkin, H. R. and Lang, A. *Planta (Berl.), 68,* 94-98 (1966).
12. Hobbs, P. R. and Obendorf, R. L. *Crop Science, 12,* 664-667 (1972).
13. Kosiyachinda, S. and Young, R. E. *Plant Physiol. 60,* 470-474 (1977).
14. Kraft, J. M. and Erwin, D. C. *Phytopath. 57,* 866-868 (1967).
15. Larson, L. A. *Plant Physiol. 43,* 255-259 (1968).
16. Lyons, J. M. and Raison, J. K. *Plant Physiol. 45,* 386-389 (1970).
17. Mason, G. Effects of Temperature on the Germination and Growth of Native Species, Using Temperature-Gradient Techniques. Ph.D. Thesis. University of Sheffield (1976).
18. McGlasson, W. B. and Raison, J. K. *Plant Physiol. 52* 390-392 (1973).
19. McMenamin, M. M. Temperature Limits for Seed Germination. Ph.D. Thesis. Queen's University, Belfast (1978).
20. Minchin, A. and Simon, E. W. *J. Exp. Bot. 24,* 1231-1235 (1973).
21. Obendorf, R. L. and Hobbs, P. R. *Crop Science, 10,* 563-566 (1970).
22. Perry, D. A. and Harrison, J. G. *J. Exp. Bot. 21,* 504-512 (1970).
23. Pollock, B. M. *Plant Physiol. 44,* 907-911 (1969).
24. Pollock, B. M. and Toole, V. K. *Plant Physiol. 41,* 221-229 (1966).
25. Powell, A. A. and Matthews, S. *J. Exp. Bot. 29,* 1215-1229 (1978).
26. Raison, J. K. and Chapman, E. A. *Aust. J. Plant Physiol. 3,* 291-299 (1976).
27. Short, G. E. and Lacy, M. L. *Phytopath. 66,* 182-187 (1976).
28. Simon, E. W. *New Phytol. 73,* 377-420 (1974).
29. Simon, E. W. *In* "Dry Biological Systems" (J. H. Crowe and J. S. Clegg, eds.), pp. 205-224. Academic Press, New York (1978).
30. Simon, E. W., Minchin, A., McMenamin, M. M. and Smith, J. M. *New Phytol. 77,* 301-311 (1976).
31. Simon, E. W. and Wiebe, H. H. *New Phytol. 74,* 407-411 (1975).
32. Thompson, P. A. *Nature (Lond.) 225,* 827-831 (1970).
33. Thompson, P. A. *In* "Seed Ecology" (W. Heydecker, ed.), pp. 31-58. Butterworths, London (1973).
34. Vegis, A. *In* "Environmental Control of Plant Growth" (L. T. Evans, ed.), pp. 265-287. Academic Press, New York (1963).
35. Vegis, A. *Ann. Rev. Plant Physiol. 15,* 185-224 (1964).

DROUGHT RESISTANCE AS RELATED
TO LOW TEMPERATURE STRESS

J. M. Wilson

School of Plant Biology
University College of North Wales
Bangor, Gwynedd LL57 2UW

I.   INTRODUCTION

The first symptoms of chilling-injury to the leaves of many agriculturally important tropical and sub-tropical species are rapid leaf wilting and the development of sunken, necrotic patches within a few hours of the start of chilling (e.g. *Phaseolus vulgaris* and *Gossypium hirsutum)*. On return of the plants to the warmth, the leaf margins and necrotic patches usually dry out giving the leaf a mottled and brittle appearance. In the extremely chill-sensitive ornamental species *Episcia reptans* a chilling treatment of only 2 hrs at $5^{\circ}C$ results in the loss of leaf turgor and the development of water soaked patches on the leaf surface which become necrotic if chilling is prolonged or the plants are returned to the warmth. These observations have led to investigations on whether the permeability of the plasmalemma to water and ions increases as a result of a phase change in the membrane lipids.

In many crop plants, such as *Phaseolus vulgaris,* leaf wilting and injury at $5^{\circ}C$ can be prevented for up to 9 days by chill-hardening the plants at $12^{\circ}C$, 85% R.H. (relative humidity) for 4 days before chilling (23, 24). Injury to these species can also be prevented by drought-hardening the plants at $25^{\circ}C$, 40% R.H., by withholding water from the roots so that the leaves wilt over a 4-day period. Drought-hardening has been shown to be as effective as chill-hardening in preventing chilling-injury (25). Chilling-injury at $5^{\circ}C$ can also be prevented in many crop species simply by maintaining a saturated (100% R.H.) atmosphere around the leaf by enclosing the plant inside a polythene bag before transfer from 25 to $5^{\circ}C$ (25).

Broadly speaking, chill-sensitive plants can be divided into two categories based on A) the sensitivity of the species to chilling-injury, B) the ability to harden against chilling-injury and C) whether chilling-injury can be prevented on direct transfer from 25 to $5^{\circ}C$ by maintaining

47

TABLE I.   The Division of Chill-Sensitive Species into Two
     Categories Based on their Sensitivity to Chilling, their Ability to
     Chill and Drought-Harden Against Chilling Injury at $5^{O}C$, 85% R.H.,
     and on Whether Chilling Injury can be Delayed on Direct Transfer
     from 25 to $5^{O}C$ by Maintaining a Saturated (100% R.H.) Atmosphere.

Category 1. e.g.   Episcia reptans, Episcia cupreata, Nautilocalyx
     lynchii
   (a)   Extremely chill-sensitive species which show injured spots
         after only 2 h at $5^{O}C$.
   (b)   These plants cannot be chill-hardened at $12^{O}C$, 85% R.H. or
         drought-hardened at $25^{O}C$, 40% R.H. to withstand chilling
         injury at $5^{O}C$. Even prolonged periods of acclimatization at
         $15^{O}C$ result in little increase in chill-tolerance.
   (c)   Maintaining a saturated atmosphere at $5^{O}C$ does not delay the
         onset of injury.

Category 2. e.g.   Phaseolus vulgaris, Cucumis sativus, Gossypium
     hirsutum
   (a)   Less chill-sensitive species usually incurring severe leaf injury
         after 24 h at $5^{O}C$, 85% R.H.
   (b)   Chill-hardening and drought-hardening can protect the leaves
         against chilling injury at $5^{O}C$, 85% R.H. for up to 9 days in P.
         vulgaris.
   (c)   Maintaining a saturated atmosphere at $5^{O}C$ can prevent
         chilling-injury for up to 9 days on direct transfer of Phaseolus
         vulgaris leaves from 25 to $5^{O}C$. Chilling-injury can also be
         prevented for up to 3 days in Cucumis sativus leaves and 2 days
         in Gossypium hirsutum leaves by maintaining 100% R.H. at $5^{O}C$.

a saturated atmosphere around the leaf. Table I shows that species such
as Phaseolus vulgaris, Gossypium hirsutum and Cucumis sativus are
placed in category 2 as these plants are less chill-sensitive and usually
only incur 50% leaf injury after 24 hrs at $5^{O}C$, 85% R.H. In addition these
plants can be chill and drought-hardened against chilling-injury and
injury can be prevented for up to 9 days in Phaseolus vulgaris by maintaining a saturated atmosphere around the leaf on direct transfer from 25
to $5^{O}C$. Therefore, chilling-injury to these species is primarily due to
water loss. However, when chilling is prolonged for several days by
maintaining 100% R.H., metabolic changes must eventually lead to cell
death.
      In the extremely chill-sensitive category 1 species (Episcia rep-
tans), water loss is less important in the development of chilling-injury as
the rate of injury cannot be significantly reduced by maintaining a
saturated atmosphere around the leaves. In addition it is not possible to
chill or drought-harden the leaves of this species to withstand chilling-

injury. Even a prolonged period of acclimatization in a cool, well ventilated greenhouse at 15°C resulted in little increase in chill-tolerance. Therefore, chilling-injury to the leaves of category 1 species is primarily metabolic. Tropical fruits also possess little ability to harden against chilling-injury. Attempts at hardening sweet potatoes have not been successful in reducing chilling-injury (21) and hardening cucumbers is only effective against slight chilling (1).

II. FATTY-ACID CHANGES DURING CHILL-HARDENING

A role for lipids and fatty-acids in the prevention of chilling-injury is suggested by increases in the degree of unsaturation and, often, weight of lipid during the acclimatization of plants as well as poikilothermic and homoeothermic animals to low temperatures (24). Increases in the degree of unsaturation of the membrane fatty-acids of 5 to 12% may prevent chilling-injury by lowering the phase transition temperature to below 5°C. Plant cell membranes usually contain at least 70% of their fatty-acids in the unsaturated form and Lyons and Asmundson (14) showed that, in artificial mixtures of unsaturated and saturated fatty-acids at this concentration, a 10% increase in unsaturation could lower the solidification temperature by as much as 20°C. In agreement with this hypothesis Wilson and Crawford (24) reported increases of 5 to 12% in the degree of unsaturation of the fatty-acids associated with the phospholipids of *P. vulgaris* and *G. hirsutum* leaves during chill-hardening at 12°C (Table II). These increases in total percentage unsaturated fatty-acid were mainly due to an increase in the percentage of linoleic acid (24). No increase in the degree of unsaturation of the glycolipids was detected. Table 2 also shows that no increase in the degree of unsaturation of the phospholipids occurred during the ineffective hardening of *Episcia reptans* at 15°C except for a slight increase in phosphatidylcholine.

Phase changes in the leaves of chill-sensitive plants at 12°C and below may result in chilling-injury by increasing the activation energy of membrane bound enzymes thus causing metabolic imbalances with non-membrane bound systems (13). In addition phase transitions may also increase the permeability of membranes to water and electrolytes leading to a loss of cell compartmentation and eventually death of the leaves. The following sections examine the evidence for metabolic imbalances and increased membrane permeability in tropical leaves at chilling temperatures.

III. INCREASED ACTIVATION ENERGIES

The most frequently quoted effect of a reduction in the rate of tri-carboxylic acid cycle activity below 12°C due to a lipid phase transition,

TABLE II.   Changes in Total Percent Unsaturated Fatty-Acid
              Associated with the Phospholipids During the Chill-
              Hardening of Phaseolus vulgaris and Gossypium hirsutum
              Leaves at 12^OC, 85% R.H. Changes in Unsaturation During
              the Ineffective Hardening of Episcia reptans at 15^OC are
              Included for Comparison.

Lipid*	Total percent unsaturated fatty-acid					
	Phaseolus vulgaris		Gossypium hirsutum		Episcia reptans	
	25^OC	12^OC	25^OC	12^OC	25^OC	12^OC
PC	69.3	80.8	70.6	81.3	73.8	74.8
PE	63.2	66.8	60.9	67.7	72.8	71.1
PI	69.4	64.7	52.6	60.6	58.3	57.6
PA	66.9	80.8	62.6	72.4	60.7	52.0
PG	54.4	59.7	45.2	55.3	66.8	61.8

*PC - Phosphatidylcholine: PE - Phosphatidyl-ethanolamine; PI -
Phosphatidylinositol; PA - Phosphatidic acid; PA - Phosphatidylglycerol.

without a similar reduction in the rate of glycolysis, is the accumulation
of the end products of glycolysis (e.g. ethanol and acetaldehyde) to toxic
levels, resulting in cell death. However, in tropical plant leaves and
fruits evidence for an increase in ethanol or acetaldehyde concentration
as a result of chilling is scarce. In E. reptans leaves Wilson (22) detected an increase in the ethanol content after 6 hrs chilling at 5^OC. In
banana pulp chilled at 6^OC for 15 days Murata (15) has shown an increase
in the levels of ethanol, acetaldehyde, pyruvate and α-ketoglutarate.
Whether the levels of these metabolites are sufficiently high to cause
injury to the cells has not been determined.
     Increases in the activation energy of NAD reduction by the
chloroplasts of P. vulgaris leaves (18) and impaired phosphorylation in
G. hirsutum leaves (19) have also been attributed to lipid phase
transitions.   It has been suggested that these changes lead to the
degeneration of the chloroplast structure and a level of ATP which would
be insufficient to maintain the metabolic integrity of the cytoplasm. The
significance of ATP changes during chilling is discussed in a later section
of this chapter.

IV. INCREASED MEMBRANE PERMEABILITY

Physical changes in membrane structure such as a phase transition in the membrane lipids can be expected to alter membrane permeability. Träuble and Haynes (20) suggested that an increase in membrane permeability would be expected to accompany the phase transition due to a) a decrease in membrane thickness, b) changes in the structure of the hydrocarbon chains important for diffusion across the membrane, or, c) changes in the arrangement of the polar head groups important for the entry of permeants into the membrane. In addition it has been speculated that 'cracks' or 'channels' may appear in the membrane at low temperatures due to the solidification of the lipid, thereby increasing membrane permeability. In agreement with these hypotheses the majority of studies on membrane permeability at chilling temperatures have shown an increase in the rate of electrolyte leakage from chill-sensitive tissues. Liebermann et al. (12) were the first to show that the rate of leakage of ions, mainly potassium, from sweet potato discs was increased at 7.5°C. In addition Christiansen et al. (4) and Guinn (19) detected an accelerated rate of leakage of electrolytes, proteins and carbohydrates from chilled cotton roots and cotyledons. Enhanced leakage of electrolytes from chilled leaf tissue has been reported by Wright and Simon (27); Creencia and Bram'age (5) and Patterson et al. (16). However, in all these studies the rate of leakage only became rapid after many hours or days at the chilling temperature and this argues against any rapid rise in permeability which can be attributed to lipid phase transitions.

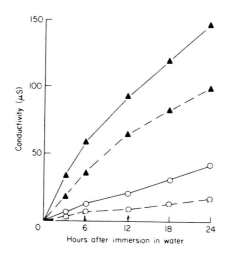

FIGURE 1. Leakage of electrolytes from leaves of Phaseolus vulgaris chilled at 5°C, 85% R.H. for 24 h (Δ) in comparison to unchilled leaves (O), placed either in water at 25°C (——), or 5°C (---).

Wilson (25) has compared the rates of ion leakage from *P. vulgaris* leaves after chilling in air at 5°C, 85% R.H., with the rate of leakage on direct transfer from air at 25°C to water at 5°C. Fig. 1 shows that leaves of *P. vulgaris* leak electrolytes if they are chilled in air for 24 hrs at 5°C, 85% R.H., so that the leaves are badly injured and wilted before they are transferred to distilled water at either 25 or 5°C. The rate of leakage from the air chilled leaves in water at 25°C is almost twice as fast as the leaves placed in water at 5°C. This may be due to a heat shock on transfer from 5 to 25°C. In contrast, the control leaves transferred directly from 25°C to water at either 25 or 5°C show a very slow rate of electrolyte leakage. The behaviour of the control leaves appears to be contrary to the phase transition hypothesis. If the phase transition in the membrane lipids causes an increase in membrane permeability then we would expect the leaves to leak electrolytes when transferred directly from 25°C to water at 5°C (i.e. without chilling in air). However, Fig. 1 shows that the leakage of electrolytes from unchilled leaves of *P. vulgaris* transferred directly to water at 5°C is very slow and that leakage is faster from the control leaves transferred directly to water at 25°C. This result suggests that initially a phase transition in the membrane lipids of the plasmalemma leads to a decrease in its permeability to ions, such as K^+, at low temperature. In support of this argument Blok *et al.* (2) have shown that the water permeability of liposomes prepared from synthetic lecithin decreases drastically below the transition

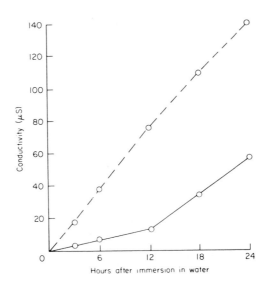

FIGURE 2. *Leakage of electrolytes from leaves of Episcia reptans transferred directly from 25°C, 85% R.H. to water at 25°C (———), or 5°C (---). Total quantity of electrolytes in leaves = 175 μ S.*

temperature. Furthermore, vesicles of *Escherichia coli* preloaded with labelled proline did not lose their radioactivity when incubated at temperatures below the transition (7). Increased rates of ion leakage from leaf tissues may therefore only occur after prolonged chilling which has resulted in the damage or death of the cell membranes. Fig. 1 shows that, in leaves, considerable water loss and cell damage must occur during chilling in air at 5°C, 85% R.H., for the leaves to lose their electrolytes rapidly when transferred to distilled water.

The leaves of some chill-sensitive species leak electrolytes more rapidly than *P. vulgaris* leaves on transfer from air at 25°C to water at 5°C. For example, Fig. 2 shows that leaves of the extremely chill-sensitive category 1 species, *Episcia reptans,* lose approximately 25% of their total electrolytes within 5 hrs of transfer from 25°C to water at 5°C. In this species visible signs of cell death accompany the rise in the conductivity of the water so that it is not possible to differentiate between whether leakage is an immediate effect due to a phase transition in the membrane lipids, or, whether leakage is a secondary effect due to cell death. At the present time efflux experiments using radioactive tracers are in progress to try to resolve whether there is any rapid change in the permeability of the plasmalemma of the above species at low temperature which can be attributed to a lipid phase transition. Changes in the permeability of leaf cells of *Phaseolus vulgaris* during prolonged chilling at 5°C, 100% R.H., are also being investigated.

V. THE CAUSE OF LEAF WILTING IN *PHASEOLUS VULGARIS* LEAVES

To investigate the significance of the increase in the degree of unsaturation of the fatty-acids during chill-hardening of *P. vulgaris* leaves (Table II), the changes in the fatty-acid composition of the phospholipids during drought-hardening were followed. Although drought-hardening was as effective as chill-hardening in preventing chilling-injury at 5°C, 85% R.H., no increase in the degree of unsaturation of the phospholipids or glycolipids was detected. In general the degree of unsaturation of the fatty-acids associated with the phospholipids decreased during drought-hardening. For example, Table III shows that there is a decrease in the total proportion of unsaturated fatty-acid associated with phosphatidylcholine during drought-hardening so that, according to the phase transition hypothesis, we would expect the drought-hardened plants to be more chill-sensitive and not chill-resistant. Chill-hardening of *P. vulgaris* leaves at 12°C is not effective if the plants are maintained at 100% R.H. (25). Although the degree of unsaturation of the phospholipids increased during an ineffective hardening treatment of 4 days at 12°C, 100% R.H., there was no increase in chill-tolerance, indicating that chill-hardening is not dependent on a highly unsaturated fatty-acid composition. The phase transition

TABLE III.   Changes in % Fatty Acid Composition of Phosphatidyl
Choline from Leaves of Phaseolus vulgaris during Chill-
Hardening at $12^{o}C$, 85% R.H. and Drought-Hardening at
$25^{o}C$, 40% R.H.

Fatty acid[a]	Control	Fatty acid composition of phosphadidylcholine (%)	
		Chill hardened at $12^{o}C$ 85% R.H. for 4 days	Drought-hardened at $25^{o}C$, 40% R.H. for 4 days
14:0	3.8	2.1	5.3
16:0	20.4	12.8	24.3
16:1	0.9	1.0	0.7
16:2	0.9	1.1	0.4
18:0	6.5	4.3	6.2
18:1	4.0	3.5	2.0
18:2	27.5	40.0	24.1
18:3	36.0	35.2	37.0
Total % unsaturated fatty acid	69.3	80.8	64.2

[a]The numbers shown are the ratios of the number of carbon atoms to
the number of double bonds in the molecule.

hypothesis is also unable to account for the prevention of chilling-injury
to leaves of P. vulgaris by enclosing the plant inside a polythene bag
before transfer to $5^{o}C$. Plants maintained in a saturated atmosphere for
7 days do not wilt at $5^{o}C$. If lipid phase transitions resulted in an increase
in the permeability of the plasmalemma of the leaf cells at $5^{o}C$ then we
would expect the leaves to wilt on transfer to $5^{o}C$, 100% R.H., as the
turgor pressure of the cell would facilitate the loss of water and
electrolytes.

It has been demonstrated that the primary cause of chilling-injury
to P. vulgaris leaves at $5^{o}C$, 85% R.H., is water loss due to the opening of
the stomata at a time when the permeability of the roots to water is low
(25). The opening of the stomata after 2 hrs at $5^{o}C$, 85% R.H., (Fig. 3), is
surprising as the leaf is wilted and in most plants the stomata close in the
early stages of water stress before visible wilting occurs.   The
replacement of the water lost by evapotranspiration from the leaf is
prevented by the low permeability of the roots to water at $5^{o}C$ (Fig. 4),

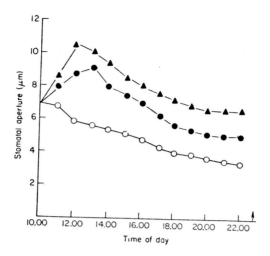

FIGURE 3. Changes in stomatal aperture on transferring entire plants of Phaseolus vulgaris directly from 25°C, 85% R.H. to (a) 5°C, 85% R.H. (△), (b) 12%, 85% R.H. (●), compared to the controls maintained at 25°C, 85% R.H. (O). Arrow shows start of night period.

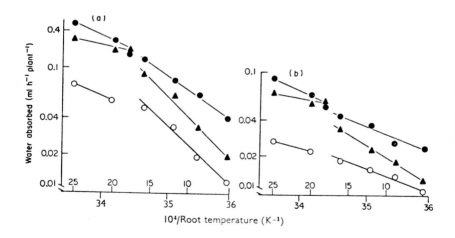

FIGURE 4. Arrhenius plots of the effect of root temperature on the rate of water absorption by Phaseolus vulgaris plants grown at 25°C, 85% R.H. (△), chill-hardened at 12°C, 85% R.H. (●), and drought-hardened at 25°C, 40% R.H. (O). (a) Shows the rate of water uptake plus exudation under 50 cm Hg vacuum and (b) the rate of exudation alone. Each point represents the average value from at least five plants.

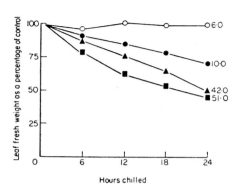

FIGURE 5. Changes in leaf fresh weight on chilling either the leaves alone (O), roots alone (●), or the whole plant of Phaseolus vulgaris (△), for 24 h at 5°C, 85% R.H. ◢, Denotes plants grown at 12°C, 100% R.H. for 4 days before whole plant chilled at 5°C, 85% R.H. Figures beside the points show the percentage of the leaf which became necrotic after 24 h chilling and 2 days recovery at 25°C, 85% R.H.

resulting in rapid leaf dehydration and injury. Hence the severity of chilling-injury depends on a synergistic effect between stomatal opening and reduced permeability of the roots to water at 5°C. Fig. 5 shows that when only the leaves of *P. vulgaris* were chilled at 5°C, 85% R.H., there was no detectable fresh weight loss after 24 hrs, except for a slight wilt after 3 to 4 hrs chilling. Chilling the leaves alone for 24 hours resulted in only 6% injury to the leaf after 2 days recovery at 25°C, 85% R.H. (Fig. 5). In the reverse experiment, when roots alone were chilled and the leaves held at 25°C, 85% R.H., there was a 30% decrease in fresh weight after 24 hrs but this resulted in only 10% injury. Chilling the entire plant of *P. vulgaris* at 5°C, 85% R.H., produced a more rapid fresh weight loss than chilling either the roots or leaves alone. Fig. 5 shows that chilling the whole plant for 24 hrs resulted in a 50% decrease in fresh weight and 42% injury on return to 25°C. The cause of stomatal opening in chill-sensitive species at 5°C is unknown. Perhaps a phase transition in the membranes of the guard cells alters their permeability properties or, alternatively, K^+, H^+-ATPase pumps may be inactivated at low temperatures.

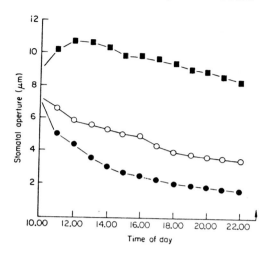

Figure 6. Changes in stomatal aperture of Phaseolus vulgaris plants on chilling at 5°C, 85 % R.H. after being hardened at 12°C, 85% R.H. (●), or ineffectively hardened at 12°C, 100% R.H. (by enclosure in a polythene bag) (■), compared to the controls maintained at 25°C, 85% R.H. (O). Arrow shows start of night period.

VI. THE MECHANISM OF CHILL- AND DROUGHT-HARDENING

Chill-hardening at 12°C, 85% R.H., prevents leaf dehydration by conditioning the stomata so that they close on transfer to 5°C, 85% R.H. (Fig. 6). Similarly, drought-hardening causes stomatal closure and the stomata remain closed on transfer to 5°C, 85% R.H. Although chill-hardening resulted in an increase in the permeability of the roots to water at low temperature, drought-hardening produced a large decrease in root permeability (Fig. 4) and yet drought-hardening was as effective as chill-hardening in preventing leaf injury. Therefore, the most important factor in the prevention of chilling-injury to *P. vulgaris* during chill- and drought-hardening is the closure of the stomata. This can be demonstrated by spraying the leaves of plants grown at 25°C, 85% R.H., with 100 μm abscisic acid (ABA) which causes stomatal closure within 24 hrs. On transfer to 5°C, 85% R.H., the sprayed leaves do not wilt as the stomata remain closed and injury is prevented for approximately 2 days by which time the effectiveness of the ABA has decreased. This decrease in the effectiveness of applied ABA is surprising as one might expect the plant to synthesize ABA during this 2-day period. Perhaps 5°C is too low a temperature for the synthesis of ABA in chill-sensitive plants.

During chill-hardening at 12°C, 85% R.H., the plant experiences a water stress (as shown by the temporary wilting of the leaves) due to the opening of the stomata (Fig. 3) and the decrease in the permeability of the roots to water (Fig. 4). However, at the intermediate temperature of 12°C the stress is not severe enough to result in damage and the wilting vanishes after 12 hrs. Similarly, during drought-hardening the water stress is imposed simply by withholding water from the roots under conditions of high evapotranspiration so that the leaves wilt. Plants maintained at 5° or 12°C, 100% R.H., do not harden because they experience no water stress. Even though the stomata are open under these conditions no water can be lost from the leaf so that the stomata remain fully open on transfer to 5°C, 85% R.H., (Fig. 6). Plants held at 12°C, 100% R.H., for 4 days and then chilled for 24 hrs at 5°C, 85% R.H., suffer a 55% decrease in fresh weight and approximately 50% leaf injury, after 2 days recovery at 25°C (Fig. 5). Therefore, enclosure in a polythene bag is not a method which can be used to lower the hardening temperature below 12°C. Although the above experiments indicate that ABA is not synthesized in *P. vulgaris* leaves at 12°C, 100% R.H., it is not known whether low temperature alone, in the absence of water stress, can induce ABA synthesis. However, the correlation between chill and drought-hardening has shown that an intermediate temperature of 12°C is not essential for hardening. Therefore, water stress and not low temperature *per se* is the primary factor inducing hardening against chilling-injury in *P. vulgaris* leaves.

VII. CAUSES OF CHILLING-INJURY TO *PHASEOLUS VULGARIS* LEAVES AT 5°C, 100% R.H.

Although changes in stomatal aperture and root permeability to water are able to explain the rapid wilting, dehydration and ultimately injury to the leaves during chilling at 5°C, 85% R.H., an alternative explanation must be sought for the death of the leaves after 9 days at 5°C, 100% R.H. The loss of turgor, bleaching and necrosis of *P. vulgaris* leaves after 9 days at 5°C, 100% R.H., suggests that injury is probably due to a combination of factors which may develop from a phase transition event in the membrane lipids.

A. Lipid and Fatty-Acid Changes

Preliminary experiments on the changes in membrane lipid composition during prolonged chilling have revealed no significant decrease in the weights of phospholipids or glycolipids during the first 4 days at 5°C, 100% R.H. This indicates that up to this time there is no increase in phospholipase activity. However, during this period there was a decrease in the percentage of linoleic acid associated with the phospholipids but no change in the fatty-acid composition of the glycolipids. Changes in the fatty-acid and lipid composition during prolonged chilling may lead to changes in membrane permeability and function and severely affect the reversibility of the phase change on return of the plants to the warmth.

B. Photo-oxidation of Plant Pigments

The bleaching of some *P. vulgaris* leaves after 7 days chilling at 100% R.H. indicates photo-oxidative degradation of the leaf pigments. Hasselt and Strikwerda (10) have shown that severe photooxidation of the leaf pigments and membrane lipids occurs in cucumber leaves after 2 to 3 days chilling at 1°C. At temperatures higher than 1°C this type of damage develops more slowly.

C. Photosynthesis and Translocation

It is well known that photosynthesis is more sensitive to low temperature than respiration so that starvation of plant tissue may occur during prolonged chilling. Translocation is inhibited in chill-sensitive species at 5°C which may lead to the starvation of non-photosynthetic parts of the plant, and the accumulation of starch in the chloroplast may further inhibit photosynthesis. Giaquinta and Geiger (8) suggested that the cessation of translocation in chill-sensitive species at 5°C is due to a phase change in the membrane lipids of the plasmalemma of the sieve tube resulting in the collapse of the material lining the cell and the blockage of the sieve plate by the flow of cytoplasm, organelles, P-protein and membranes into it.

D. ATP Supply

Fig. 7a shows that the ATP and ADP levels in *P. vulgaris* leaves chilled at 5°C, 100% R.H., increased over the first 24 hrs and remained high during the following seven days. Even after 8 or 9 days chilling in a saturated atmosphere there was only a slight fall in the ATP and ADP levels and this coincided with the development of visible signs of chilling-injury to the leaf. A decrease in ATP supply below that necessary to maintain the metabolic integrity of the cytoplasm cannot therefore be considered to be the cause of chilling-injury to leaves at 5°C, 100% R.H. In agreement with this result Jones (11) has also reported an increase in the ATP level of *P. vulgaris* leaves at low temperatures. The increases in ATP level of *P. vulgaris* leaves at 5°C may be due to the cold sensitivity of ATPase which is readily inactivated at low temperatures (17).

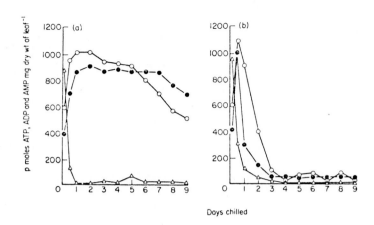

FIGURE 7. Changes in the levels of ATP (●), ADP (O), and AMP (Δ) in the leaves of Phaseolus vulgaris, (a) during the prevention of chilling-injury and water loss by maintaining a saturated (100% R.H.) atmosphere around the leaves on transfer from 25°C to 5°C and (b) during the development of chilling-injury and water loss on direct transfer from 25°C to 5°C, 85% R.H. Each point is the mean of three replicates.

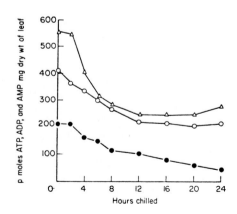

FIGURE 8. Changes in the levels of ATP (●), ADP (O) and AMP (Δ) in the leaves of Episcia reptans during chilling in the dark at 5°C. Each point is the mean of three replicates.

In contrast, chilling leaves of *P. vulgaris* at $5^{o}C$, 85% R.H., resulted in rapid leaf wilting and injury and the ATP and ADP levels decreased after 12 hours chilling (Fig. 7b). Although the leaves chilled at $5^{o}C$, 85% R.H., were approximately 50% injured after 24 hrs, the level of ATP had decreased by less than 33%, indicating that leaf dehydration and not a fall in ATP supply is the cause of cell' death.   The impaired phosphorylation of cotton leaves at $5^{o}C$ reported by Stewart and Guinn (19) can be attributed to the effects of water stress and not low temperature *per se.*

VIII. CAUSES OF CHILLING-INJURY TO *EPISCIA REPTANS* LEAVES

A.   ATP Supply

Although chilling leaves of *Episcia reptans* for 5 hr at $5^{o}C$ produced severe leaf injury there was no rapid decrease in ATP level during the first 5 hr of chilling. Fig. 8 shows that the ATP and ADP levels fell by only 25% after 5 hr of chilling. Prolonged chilling between 5 and 24 hr resulted in a further gradual decline in ATP level and a stabilization of ADP and AMP levels. A reduced ATP supply is therefore unlikely to be the cause of chilling-injury to *E. reptans* (26).

B.   Leaf Respiration

Although changes in respiratory behaviour have been widely investigated in relation to chilling-injury in fruits there have been comparatively few studies of changes in the respiration rate of chill-sensitive leaves at $5^{o}C$, especially in very sensitive leaves such as *Episcia reptans.* In cucumber fruits Eaks and Morris (6) detected a doubling of the respiration rate after 8 days at $5^{o}C$ and this increase coincided with the onset and development of chilling-injury. This was followed by a decline in the respiration rate at the time of the general death of the tissue. Although Eaks and Morris were unable to explain the cause of the respiratory increase Creencia and Bramlage (5) have shown, using the uncoupling agent 2, 4-dinitrophenol (DNP), that the respiratory increase in *Zea mays* leaves after prolonged chilling for 24 to 48 hrs at $0.3^{o}C$ is partly due to the' uncoupling of oxidative phosphorylation at chilling temperatures.

In *E. reptans* leaves Fig. 9a shows that after only 80 min at $5^{o}C$ the oxygen uptake rate is three times higher than that of the controls maintained at $25^{o}C$.   The rate of carbon dioxide production also increased to a maximum after 80 min at $5^{o}C$ but only to a maximum value of one-third of the rate of oxygen uptake (Fig. 9b). The respiratory peak after 80 min at $5^{o}C$ coincided with the development of dark water soaked patched on the leaf and the loss of leaf turgor. Plants chilled for 80 min

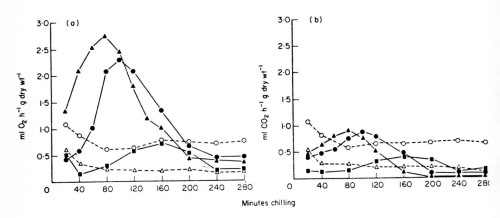

FIGURE 9. (a) Oxygen uptake and (b) carbon dioxide evolution in the dark by leaf discs of Episcia reptans chilled (———) at 5°C (△), 10°C (●) and 11°C (■) in comparison to unchilled (----) leaf discs at 12.5°C (△) and 25°C (O).

at 5°C showed few signs of injury on return to 25°C. However, more prolonged chilling resulted in a gradual decline in respiration rate and an increase in leaf necrosis until after 5 hrs at 5°C none of the leaves recovered on subsequent transfer to 25°C.

Chilling at temperatures higher than 5°C delayed the onset of the respiratory rise and reduced its height (Fig. 9a). In leaves chilled at 10°C the respiration rate was maximal after 100 min and peak height was reduced by one-fifth in comparison to leaves chilled at 5°C. The first visible signs of chilling-injury at 10°C also occurred at approximately the same time as maximum respiration rate. The extreme chill-sensitivity of E. reptans is demonstrated by the 90% injury incurred after 5 hrs at the relatively high chilling temperature of 10°C followed by 2 days at 25°C (26). At 11°C there was an initial decrease in respiration rate followed by a gradual increase to the level of the control leaves maintained at 25°C after 160 min (Fig. 9a). The development of chilling-injury was also slow at 11°C, the leaves incurring only 20% injury after 5 hrs chilling and 2 days recovery at 25°C, which indicates that 11°C is near the upper temperature limit for chilling-injury in this species. An upper temperature limit of 11 to 12°C for chilling-injury in E. reptans is supported by the absence of any increase in the respiration rate of the leaves held for 5 hrs at 12.5°C (Fig. 9a) and the development of only 5% injury on transfer to 25°C (26). At 12.5°C the respiration rate decreased to approximately one-third of the rate of the control leaves at 25°C and remained at this level over the 5 hr period. However, prolonged chilling over several days at 12 to 15°C can cause chilling-injury to *E. Reptans* (24). In contrast to *E. reptans* there was no increase in the respiration rate of *Phaseolus vulgaris* leaves during the first 12 hrs of chilling at 5°C (26).

Experiments on the cause of this respiratory increase in *E. reptans* leaves at 5°C, have been made using 2,4-dinitrophenol (DNP) and potassium cyanide (KCN). Fig. 10 shows that treating leaf discs of *E. reptans* with 1 mM DNP doubled the oxygen uptake rate at 25°C. However, transferring the DNP-treated discs to 5°C did not lessen the effect of DNP, as would be expected if chilling caused the uncoupling of oxidative phosphorylation. Fig. 10 shows that chilling the DNP-treated leaf discs resulted in an extremely rapid three-fold increase in oxygen uptake within 40 min of the start of chilling, whilst in the untreated controls at 5°C, the oxygen uptake did not attain its maximum value until 80 min after the start of chilling. The rapid acceleration of respiration by the addition of DNP showed that this method could not be used to determine whether any part of the respiratory increase was due to the uncoupling of oxidative phosphorylation. Fig. 10 also shows that the respiration rate of leaf discs treated with 10 mM KCN at 5°C was accelerated in the same manner as the DNP-treated discs. A seven-fold

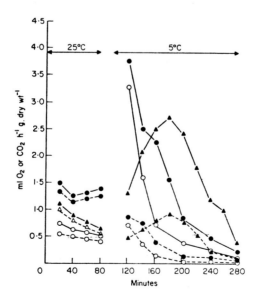

FIGURE 10. The effects of 1 mM DNP (●) and 10 mM KCN (O) on the oxygen uptake (——) and carbon dioxide evolution (----) rates of Episcia reptans leaf discs at 5°C in comparison to untreated, control discs at 5°C (Δ).

increase in the oxygen uptake rate of KCN treated leaves occurred within 40 min of the start of chilling. The rapid increase in the rate of oxygen uptake by the DNP- and KCN-treated leaf discs at 5°C is not accompanied by a higher rate of carbon dioxide production (Fig. 10). The addition of DNP and KCN to *E. reptans* leaves on chilling is thought to increase the rate of oxygen uptake by decreasing membrane stability at low temperatures.

It is suggested that the cause of this rapid wound-induced respiration in *E. reptans* leaves at 5°C is an increase in the permeability of the cell membranes at low temperature due to either lipid phase transitions (13), protein denaturation (3), changes in lipid-protein interaction (28), or a combination of these events. The magnitude of the phase change in the membranes of *E. reptans* leaves may be far greater than in *P. vulgaris* and this may account for the more rapid development of chilling-injury in *E. reptans*. A rapid change in the permeability of the plasmalemma and tonoplast would account for the speed of injury to *E. reptans* and the rapid rise in respiration rate due to the loss of cell compartmentation which would allow enzymes increased access to substrates. It is speculated that the oxidation of phenols released from the vacuole as a result of a change in tonoplast permeability may account for the rapid rise in wound-induced respiration in this species.

Experiments are in progress on the changes in membrane permeability and lipid composition of *E. reptans* leaves during chilling. An electron microscope study of the development of chilling-injury in *Episcia reptans* leaves has revealed unusual changes in cytoplasmic structure.

IX. REFERENCES

1. Apeland, J. *Bull. Int. Inst. Refrig. 46,* 325 (1966).
2. Blok, M. C., van Deenen, L. L. M. and De Gier, *Biophys. Acta. 433,* 1 (1976).
3. Brandts, J. F. "Heat Effects on Proteins and Enzymes." In *Thermobiology* (Ed. by A. H. Rose). Academic Press, London (1967).
4. Christiansen, M. N., Carns, H. R. and Slyter, D. J. *Pl. Physiol. 46,* 53 (1970).
5. Creencia, R. P. and Bramlage, W. J. *Pl. Physiol. 47,* 389 (1971).
6. Eaks, I. L. and Morris, L. L. *Pl. Physiol. 31,* 308 (1956).
7. Esfahan, M., Limbrick, A. R., Knutton, S., Oka, T. and Wakil, S. J. *Proc. Nat. Acad. Sci. U.S.A. 68,* 3180 (1971).
8. Giaquinta, R. T. and Geiger, D. R. *Pl. Physiol. 51,* 372 (1973).
9. Guinn, G. *Crop Sci. 11,* 101 (1971).
10. Hasselt, Ph.R. van and Strikwerda, J. T. *Physiol. Plant. 37,* 253 (1976).
11. Jones, P. C. T. *J. Exp. Bot. 21,* 58 (1970).
12. Liebermann, M., Craft, C. C., Audia, W. V. and Wilcox, M. S. *Pl. Physiol. 33,* 307 (1958).

13. Lyons, J. M. *Ann. Rev. Plant Physiol. 24,* 445 (1973).
14. Lyons, J. M. and Asmundson, C. M. *J. Am. Oil Chem. Soc. 42,* 1056 (1965).
15. Murata, T. *Physiol. Plantarum, 22,* 401 (1969).
16. Patterson, B. D., Murata, T. and Graham, D. *Aust. J. Plant Physiol. 3,* 435 (1976).
17. Penefsky, H. S. and Warner, R. C. *J. Biol. Chem. 250,* 4694 (1965).
18. Shneyour, A., Raison, J. K. and Smillie, R. M. *Biochem.   Biophys. Acta. 292,* 152 (1973).
19. Stewart, J. McD. and Guinn, G. *Pl. Physiol. 44,*605 (1969).
20. Träuble, H. and Haynes, D. H. *Chem. Phys. Lipids, 7,* 324 (1971).
21. Wheaton, T. A. and Morris, L. L. *Proc. Am. Soc. Hort. Sci. 91,* 529 (1967).
22. Wilson, J. M. Ph.D. Thesis, University of St. Andrews (1974).
23. Wilson, J. M. and Crawford, R. M. M. *J. Exp. Bot. 25,* 121 (1974a).
24. Wilson, J. M. and Crawford, R. M. M. *New Phytol. 73,*805 1974b).
25. Wilson, J. M. *New Phytol. 76,* 257 (1976).
26. Wilson, J. M. *New Phytol. 80,* 325 (1978).
27. Wright, M. and Simon, E. W. *J. Exp. Bot. 24,* 400 (1973).
28. Yamaki, S. and Uritani, I. *Pl. Cell Physiol. 15,* 385 (1974).

Q

LOW TEMPERATURE RESPONSES
OF THREE *SORGHUM* SPECIES

David Bagnall

CSIRO Division of Plant Industry
Canberra City, 2601, Australia

I.   INTRODUCTION

Plants possessing the C_4 pathway of photosynthesis (13) are often susceptible to injury at temperatures above $0°C$ (32), with growth and development usually very slow within the chilling range of 0 to $15°C$ (6, 15, 33). The grain cereal *Sorghum bicolor* L. (Moench) is a C_4 plant that is particularly chilling sensitive (8, 10, 20, 29), requiring not only frost free periods for growth but temperatures above $10°C$ for germination and emergence (25). Specific stages of development can be particularly temperature sensitive; for example night temperatures of $13°C$ or less at the time of pollen meiosis may induce male sterility (3, 10).

Other species of sorghum are less susceptible to chilling injury than *S. bicolor* (9, 30). Johnson grass *(S. halepense)* and an Australian wild sorghum *(S. leiocladum)* are two species able to withstand the severe climate of Canberra, Australia, having in one year survived 40 severe frosts and mean daily temperatures of $6°C$ in the months of July and August. *S. leiocladum* is a native grass of eastern Australia and is the only native sorghum found in the colder southern tablelands region of Australia (4). Johnson grass, a native of the Mediterranean region, is a weed in Australian pastures. It has a significantly higher growth rate than *S. bicolor* at low day/night temperatures – $15°/10°C$ (30). In the experiments of Taylor *et al.*(30) leaves of *S. bicolor* were chlorotic while those of Johnson grass and its hybrids were greener than other chilling resistant species and lines. The ability of Johnson grass to survive frosting temperatures may however be due to the existence of extensive creeping rhizomes which allow the plant to generate new shoots in spring (9), and not to some characteristic of the plant's leaf tissue which mainly dies back in winter.

In this paper the growth responses of vegetative plants of these three species are examined at different temperatures to determine the relative importance of night temperature, photosynthesis, chlorophyll formation and degradation, and water stress to their growth and survival at low temperatures. Within this study the critical temperatures is defined as that temperature at which a plant fails to accumulate dry matter over one week (27). This temperature corresponds closely to the temperature at which the plant does not survive in the long term and at which chilling lesions appear as described by Taylor and Rowley (32).

A. Experimental

1. Relative Growth Rates at Different Constant Temperatures.
Plants of the species S. bicolor L. (Moench) cv Texas 610, S. leiocladum (Hack) G. E. Hubbard and S. halepense (L.) Pers. were germinated and grown in 13 cm pots containing equal proportions of perlite and vermiculite in a $21^{\circ}/16^{\circ}C$ glasshouse of the Canberra phytotron (23). The plants were transferred to naturally lit glasshouse C cabinets (23) held continuously at 5, 10, 15, 20 and $25^{\circ}C$ for 10 days and relative growth rates computed. The ages of the plants at the time of transfer were 72 days for S. leiocladum, 36 days for S. halepense and 21 days for S. bicolor. These differences in age were used to obtain plants of similar size. Maximum radiation in the cabinets reached 1,550 μE $m^{-2}s^{-1}$ at midday.

2. Relative Growth Rates at Various Night Temperatures. S.
bicolor plants that had been grown in a $21^{\circ}/16^{\circ}C$ glasshouse were transferred to a range of night temperatures 18 days from planting. Control plants continued to receive 8 h (day) at $21^{\circ}C$ and 16 h at $16^{\circ}C$ in the naturally lit glasshouse. Between 0800 h and 1600 h both experimental and control plants were held at $21^{\circ}C$ in the glasshouse. At 1600 h the experimental plants were transferred to artificially lit L.B. cabinets. Between 1600 and 1800 h and 0600 and 0800 h the experimental plants were held at $13^{\circ}C$ in the dark, while between 1800 and 0600 h they were subjected to temperatures of 1.5°, 4°, 7°, 10° or $13^{\circ}C$ in the dark.

3. Photosynthetic Response to Low Temperature. Carbon dioxide
exchange rate was measured in a water-cooled assimilation chamber (14) under normal L.B. cabinet lighting (23) supplemented with a 1000W Phillips HPLR Mercury Vapour Lamp. The irradiance at the leaf was 950 $\mu E m^{-2}s^{-1}$. Air flow rates varied from 1.5 to 2.0 litres min^{-1} and the cabinet temperature was held at $20^{\circ}C$. The temperature of the leaf zone within the jacket was lowered from 20° to $5^{\circ}C$ over a period of about 2 h, the leaf zone was then kept below $5^{\circ}C$ for times ranging from 10 min to 3 h (average about 75 min) and then subsequently reheating from 5 to $20^{\circ}C$ over about 2 h. Variation of time at low temperature was used to assess whether the time factor was important in the pattern of recovery. The temperature of the leaf zone in the chamber was measured with a copper-constantan thermocouple placed against the underside of the leaf.

4. *Relative Growth Rates and Light Intensity.* Plants of the three species that had been grown in the $21^\circ/16^\circ$C glasshouse were transferred to naturally lit glasshouse B cabinets (23) held at constant temperatures of 8.5°, 11.0° and 13.0°C. In each cabinet one-half of the seedlings were kept under shade cloth that reduced radiation to 40% of that within the cabinet. The seedlings were 35 day old for *S. bicolor,* 40 day old for *S. halepense* and 115 day old for *S. leiocladum* at the time of transfer and were held at the chilling temperatures for 40 days. Radiation levels within the cabinets were monitored with a Li-cor quantum meter and Rimco integrating pyranometers under shaded and unshaded conditions.

5. *Low Temperature, Water Stress and Chlorophyll Destruction.* *S. bicolor* plants that had been grown to the 10th leaf stage at $30^\circ/25^\circ$C were transferred to a glasshouse B cabinet held at 7°C constant. Chlorophyll content and relative water content (RWC) were monitored over the subsequent 10 days using the methods of Arnon (2) and Weatherley (35) respectively. Chlorophyll was sampled from the last fully expanded leaf from the top (usually the third leaf) while RWC was determined in the 5th leaf from the top. Water potential was measured using a model 1000 P.M.S. pressure chamber (16). A control plant was kept in the $30^\circ/25^\circ$C glasshouse.

II.   RESULTS AND DISCUSSION

A.   *Relative Growth Rates at Constant Temperatures*

The response of plants to low temperature can be measured in several ways but relative growth rate is the fundamental measure of organic growth and it is a very sensitive yardstick (36). Over the 10 day experimental period (Figure 1) Johnson grass and *S. bicolor* had similar growth rates throughout the temperature range, while *S. leiocladum* had significantly slower growth at the higher temperatures. At a constant 5°C, however, all three species were suffering from severe chilling lesions. The inability of these three species to grow at 5°C appears to be at variance with field data that show that these species can withstand mild frosting temperatures [*S. bicolor* (20)] or severe frosts [Johnson grass – (9)]. The failure of the three species to grow at a constant 5°C while being able to survive frosts at night raises the possibility that a different critical temperature exists at night from that during the day.

B.   *Effect of night temperature on relative growth rate*

The growth response of *S. bicolor* to different night temperatures is shown in Figure 2. *S. bicolor* is the most chilling sensitive of the three species in this series of experiments. All of the plants were grown

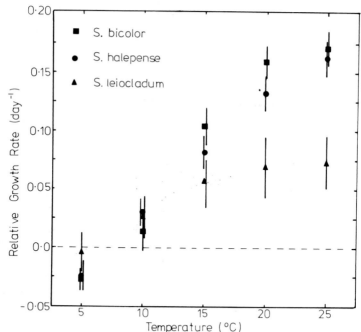

FIGURE 1. *Relative growth rates of three Sorghum species over 10 days at 5 constant temperatures. Each point is the mean value of 15 plants ± 1 s.e.*

at $21^{\circ}C$ in the day while night temperature was varied between $16^{\circ}C$ (control) and $1.5^{\circ}C$. Although growth tended to decrease with decreasing night temperature, this was only a small drop and it continued at a rate about equal to growth at a constant $15^{\circ}C$. All of the experimental plants showed reduced growth and development, but none showed any sign of chilling damage. Tissue formed during the seven days when night temperature fell to $1.5^{\circ}C$ had similar chlorophyll content (1.6 mg/g fresh wt) to control plants (night temperature equal to $16^{\circ}C$). The CO_2 exchange rates of leaves from the control and $1.5^{\circ}C$ night treatments were also similar (27.3 mg $CO_2 dm^{-2} hour^{-1}$.

The absence of a critical night temperature for *S. bicolor* cv 610 above $1.5^{\circ}C$ was confirmed by field trial data. Begg and Turner (unpublished data) found that night temperatures between 0° and $10^{\circ}C$ did not stop crop growth and that only frosts caused damage to mature sorghum plants.

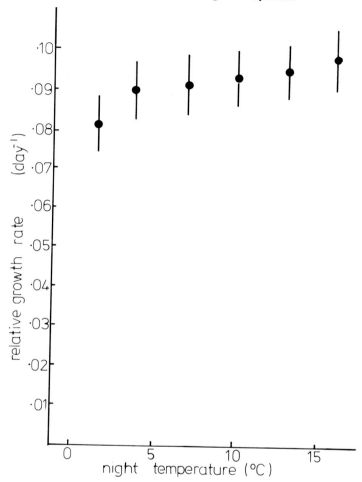

FIGURE 2.   *Relative growth rates of S. bicolor over 7 days at different night temperatures. Day temperature was 21°C in all cases. Each point is the mean value of 25 plants ± 1 s.e.*

C.   *Photosynthetic Response to Low Temperature*

The difference between critical night temperature as found in the field and phytotron and the critical temperature when the day as well as the night temperature is low (Fig. 2 vs Fig. 1) might be attributable to the photosynthetic response of the plants to low temperature.   Figure 3 shows the CO_2 Exchange Rate (CER) in a leaf zone (5 cm length) of *S. bicolor* and *S. leiocladum* as temperature was lowered from 20° to 3°C, and then subsequently returned to 20°C.   The leaf zone was kept below 5°C for variable amounts of time from 10 min to 3 h although on an average this time was 75 min.

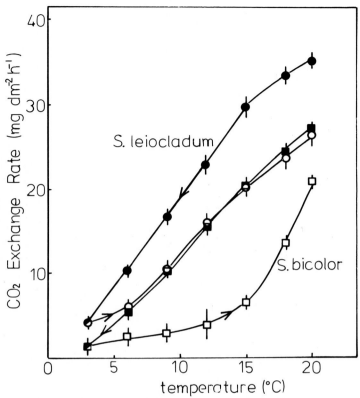

FIGURE 3. CO_2 exchange rate of S. leiocladum and S. bicolor under decreasing (solid symbols) and then increasing (hollow symbols) temperature. Both species were grown at $21°/16°C$ in a glasshouse (S. bicolor; square symbols, S. leiocladum: circular symbols). Each loop is a mean of 5 plants $\pm$ 1 s.e.

Plots of the CER as a function of temperature take the form of hysteresis loops. During the cooling phase CER drops rapidly with temperature down to $3°C$, but during the warming phase a lag occurs. This lag was evident in both species irrespective of how long the plants were kept below $5°C$ (from 10 min to 3 h). The lag was, however, greater in S. bicolor than in wild sorghum and this may be an adaptive mechanism that allows S. leiocladum to respond to warmer temperatures after cold nights faster than S. bicolor.

After the leaf zone had been rewarmed to $20°C$, complete recovery of CER did not occur even after 12 h. This photosynthetic reduction could be either a stomatal response (24) or photoinhibition (26), but this aspect was not pursued in these experiments.

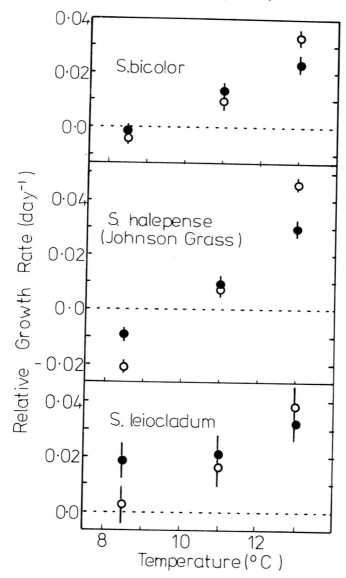

FIGURE 4. *Relative growth rates of three Sorghum species at 3 constant temperatures and two irradiance levels over 40 days. Hollow symbols represent unshaded plants while solid symbols represent the shaded plants. Each point represents the mean value of 10 plants ± 1 s.e.*

Wild sorghum had a higher photosynthetic rate throughout the range $3°$ to $20°C$ than *S. bicolor* (Fig. 3) although generally having a lower growth rate over most of this temperature range (Fig. 1). The higher CER of wild sorghum at low temperatures may provide a reason for its ability to grow at low temperatures than *S. bicolor* but it does not resolve why both species suffer severe chilling damage at a constant $5^{\circ}C$, when both have positive CERs.

D.   Effect of Light Intensity on Relative Growth Rate

Severe lesions occur in *S. bicolor* plants subjected to high light intensity and $10^{\circ}C$ constant, within 2 days after transfer to the cold (32). Chlorophyll levels dropped rapidly within these lesions. Under low light and $10^{\circ}C$ the chilling effects were much less severe. Light intensity is therefore important to chilling injury. Figure 4 shows the relative growth rates of the three sorghum species at three temperatures and under two light intensities during 40 days at low temperature. The highest irradiances reached in the cabinets on clear sunny days at midday were 1480 $\mu E\ m^{-2}s^{-1}$ (unshaded) and 605 $\mu E\ m^{-2}s^{-1}$ (under shade cloth) although on average the plants only received 470 g cal cm^{-2} day (unshaded) and 165 g cal cm^{-2} (shaded, over the 40 days at low temperature.

All of the plants from the three species survived and put on dry weight in both shaded and unshaded treatments at 11° and $13^{\circ}C$, but at $8.5^{\circ}C$ Johnson grass and *Sorghum bicolor* lost dry weight under shaded and unshaded conditions while in *S. leiocladum* 2 of the 10 plants survived at $8.5^{\circ}C$ in the open and all grew successfully under the shaded conditions. Therefore, under light intensities typical of winter the perennial *S. leiocladum* can still grow at a constant $8.5^{\circ}C$, although only very slowly.

TABLE I.   *Effect of Low Temperature and Shade on Chlorophyll Content of Three Sorghum Species*

		Species		
Temperature	Leaf age	S. bicolor	S. leiocladum	S. halepense
$8.5^{\circ}C$	old	15.3 ± 1.9	61.8 ± 5.2	12.6 ± 3.2
(shaded)	new	*	39.5	*
$21/16^{\circ}C$	old+new	49.6 ± 4.0	74.0 ± 6.0	19.7 ± 6.4

Chlorophyll was measured in $\mu g\ cm^{-2}$ or leaf area. Each value is mean of 6 plants $\pm$ s.e. Old leaves were formed before transfer to $8.5^{\circ}C$, new leaves were formed during 40 d. at low temperature.
*No new leaf tissue was formed at $8.5^{\circ}C$ in these species.

TABLE II.   *Effect of Temperature on Chlorophyll Content of Old and New Leaves of Three Sorghum Species*

		Species		
Temperature	Leaf age	S. bicolor	S. leiocladum	S. halepense
11^OC	old new	22.9 + 5.8 0.8 + 0.0	60.4 + 1.5 56.0 + 14.5	20.8 + 2.5 0.2 + 0.0
13^OC	old new	31.6 + 1.8 6.5 + 3.4	63.2 + 1.7 50.6 + 4.5	25.1 + 6.0 7.4 + 5.8
$21/16^O$C	old+new	49.6 + 4.0	74.0 + 6.0	19.7 + 6.4

Chlorophyll was measured in $\mu g\ cm^{-2}$ of leaf area. There were not significant differences in the chlorophyll contents of shaded and unshaded plants. Each value is the mean of 12 plants $\pm$ s.e.

TABLE III.   *Effect of Temperature on the Chlorophyll Content of Week Old Seedlings of Three Sorghum Species*

	Species		
Temperature	S. bicolor	S. leiocladum	S. halepense
15.5^OC 16.5^OC 17.5^OC	0.04 + 0.00 0.37 + 0.04 1.00 + 0.02	0.76 + 0.04 - -	0.87 + 0.04 - -

Chlorophyll was measured in $mg\ g^{-1}$ fresh weight. Each value is the mean of 3 samples of 3-6 seedlings that were germinated and grown on moist blotting paper in glasshouse cabinets.

The appearance of the plants undergoing the three temperature treatments varied greatly. None of the unshaded plants at 8.5°C had significant chlorophyll levels in their leaves at the end of 40 days. At 8.5°C, under the shade, the Johnson grass and *S. bicolor* plants survived but both species had suffered chilling injury of the type described by Taylor and Rowley (32) with significantly reduced chlorophyll content (Table I). Wild sorghum, however, produced additional dry matter (Fig. 4), retained chlorophyll in its old tissue and formed chlorophyll in its new leaves. At 11°C and 13°C, the *S. bicolor* and Johnson grass plants retained chlorophyll in the tissue formed before transfer to the cold, but the new leaves formed in the cold lacked chlorophyll (Table II). This occurred under both shaded and unshaded treatments for these two species. Wild sorghum was able to manufacture chlorophyll at both of these temperatures as well as under shade at 8.5°C. Rhykerd *et al.*(29) found that chlorophyll formation limited growth below 16°C in *S. bicolor* while Alberda (1) found that maize seedlings scarcely grew at 15°C because of their failure to "green up" but, despite the failure to produce chlorophyll in new leaf tissue of older plants, the growth rate of these was largely maintained by the pre-existing green tissue.

Seedlings of *S. bicolor* were germinated and grown at 15.5°C, 16.5°C and 17.5°C for 1 week and chlorophyll contents compared with Johnson grass and wild sorghum grown at 15.5°C (Table III). The reduced chlorophyll formation in *S. bicolor* observed below 16°C must limit growth, particularly in the seedling stage. Wild sorghum and Johnson grass formed significant levels of chlorophyll at 15.5°C and are likely to survive earlier in spring than crop sorghum because of this.

E.   Water Relations

It has been suggested that plant water relations play a part in chilling injury (5, 7, 37). When shoots and roots were chilled in C_3 plants, both photosynthesis and stomatal opening subsequently decreased due to temporary conditions of water stress (7). Wright (37) also working with a C_3 plant concluded that a partial water deficit was a prerequisite for chilling injury and that water deficits in the shoots arise when the roots alone are chilled, probably due to reduced water uptake.

Changes in relative water content (RWC) and chlorophyll over 10 days of chilling at 7°C in *S. bicolor* plants are shown in Figure 5. The fall-off in RWC was preceded by chlorophyll destruction in this species.

The leaf water potential up to the morning of day 5 varied between -1.1×10^{5} Pa and -1.5×10^{5} Pa (1 bar = 10^{5} Pascal) for all of the low temperature plants. At day 9 when RWC had dropped to 60%, the mean leaf water potential for all tissue was -5.1×10^{5} Pa while the green, healthy tissue had a mean value of -3.3×10^{5} Pa and the bleached and damaged leaves gave a mean value of -6.1×10^{5} Pa.

Plant water potentials were in fact quite high when compared with those of plants of a similar age and variety undergoing water stress.

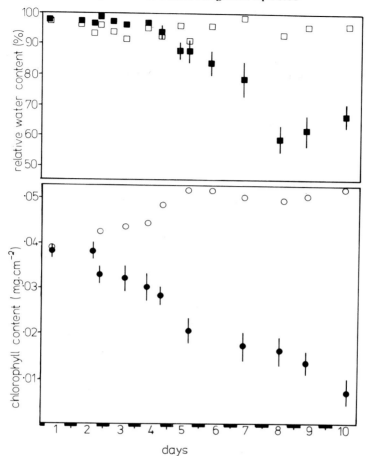

FIGURE 5. *Changes over a 10 day period in relative water content (upper graph) and chlorophyll content (lower graph) of S. bicolor plants transferred to 7°C. Plants were transferred from a 30°/25°C glasshouse at 0900 h on day 1. Hollow symbols represent control plant held in a 30°/25°C glasshouse, solid symbols are experimental plants at a constant 7°C. Each experimental point is a mean of 10 plants.*

Water stressed *S. bicolor* cv 610 plants with RWCs of 60% have leaf water potentials lower than -20×10^5 (Turner and Long, unpublished). Water stressed plants and chilling injured plants were totally different in appearance. Water-stressed plants had senescent lower leaves with generally healthy upper leaves showing some browning around the edges of the younger leaves. Chilling injured plants had lost all chlorophyll from leaf tissue exposed perpendicularly to sunlight. Upper leaves were most affected while tissue that was shaded or close to the leaf sheath was less affected.

To assess the importance of root water uptake in chilling injury in this species the response of detached leaves in flasks of water was compared with whole plants in well watered pots at $7^{\circ}C$ constant. The leaves and the whole plants looked qualitatively similar throughout: chilling lesions appeared on the second day of low temperature, while wilting did not occur in either set of plants.

These results suggest that water stress was not the initial cause of low temperature stress in *S. bicolor*. The loss in water content then is symptomatic of chilling injury rather than a causal factor. It probably reflects breakdown of membranes, leakage of cell contents and drying of the leaf. It appears likely that reduced water uptake may be a factor in chilling stress in C_3 plants but not in C_4 plants (7).

F.   Formation and Destruction of Chlorophyll and its Importance to Low Temperature Injury

Chlorophyll formation and destruction are major factors in chilling injury in plants. Reduced chlorophyll formation has been found to limit the growth of maize at low temperature (1) and chlorophyll development was shown to be temperature and light intensity dependent in this species (21). It appears that a photosensitized oxidative destruction of chlorophyll can occur under high light conditions at low temperature (12, 17, 32). It has been suggested that the rates of CO_2 fixation are insufficient to utilize the phosphorylative and reducing capacity of the chloroplast under low temperature condtions (31), an effect possibly related to stomatal closure at low temperature (26, 28).

Millerd and McWilliam (22) found that the minimum temperature for chlorophyll formation could be lowered using flashing light, and suggested that the intermittent dark period avoided photooxidation of pigments following the initial photoreduction. Recent experiments have shown that the antioxidant tocopherol reduces the chilling breakdown in chloroplasts and results in protection of unsaturated lipids (18).

G.   Plant Critical Temperatures

Raison and Chapman (27) have postulated that a single critical temperature may exist for plant growth, but an examination of the literature suggests that this is an oversimplified view. Martin (20) for example found different temperatures limiting growth at different stages of the life cycle of sorghum. Ivory and Whiteman (15), when measuring the effect of temperature on the growth of several C_4 grasses, found that "critical" night temperature appeared to vary depending on the accompanying day temperature and that critical day temperature also varied with accompanying night temperature. These changes can be explained by changes in net photosynthesis and may be due to stomatal response, starch accumulation or dark respiration (19, 24, 34).

The results reported here do not support the concept of a single critical temperature for plant growth since the effect depends on the stage of growth and the level of irradiance.

ACKNOWLEDGMENT

I gratefully acknowledge advice from Ian Wardlaw, Ross Downes and Neil Turner and superb assistance from the staff of the phytotron and Mike Moncur.

III. REFERENCES

1.   Alberda, T. *Acta. Bot. Neerl. 18*, 39-50 (1969).
2.   Arnon, D. I. *Pl. Physiol. 24*, 1-15 (1949).
3.   Brooking, I. R. *Aust. J. Pl. Physiol. 3*, 589-596 (1976).
4.   Burbidge, N. T. *Australian Grasses Vol. 1, A.C.T. and Southern Tablelands of N.S.W.* Angus and Robertson, Sydney (1966).
5.   Chu, T. M., Jusaitis, M., Aspinall, D., and Paleg, L. G. *Physiol. Plant. 43*, 254-260 (1978).
6.   Cooper, J. P. and Tainton, N. M. *Herb. Ab. 38*, 167-176 (1968).
7.   Crookston, R. K., O'Toole, J., Lee, R., Ozbun, J. L., and Wallace, D. H. *Crop Sci. 14*, 457-464 (1974).
8.   Doggett, H. "Sorghum" Longmans, London (1970).
9.   Downes, R. W. *In* "Sorghum in Seventies" (N.G.P. Rao, and L. R. House, eds.), pp. 265-274. Oxford and IBH, New Delhi (1972).
10.   Downes, R. W. and Marshall, D. R. *Aust. J. Exp. Agric. Anim. Husb. 11*, 352-356 (1971).
11.   Hasselt, P. R. van *Acta Bot. Neerl. 21*, 539-543 (1972).
12.   Hasselt, P. R. van *Acta Bot. Neerl. 23*, 159-169 (1974).
13.   Hatch, M. D., Slack, C. R., and Johnson, H. S. *Biochem. J. 102*, 417-422 (1967).
14.   Helms, K., and Wardlaw, I. F. *Phytopathology 67*, 344-350 (1977).
15.   Ivory, D. A., and Whiteman, P. C. *Aust. J. Plant Physiol. 5*, 138-148 (1978).
16.   Jones, M. M. and Turner, N. C. *Pl. Physiol. 61*, 122-126 (1978).
17.   Kandler, O. and Sironval, C. *Biochim. Biophys. Acta 33*, 207-214 (1959).
18.   Kok, L. J. de, van Hasselt, P. R. and Kuiper, P. J. C. *Physiol. Plant. 43*, 7-12 (1978).
19.   Ludlow, M. M. and Wilson, G. L. *Aust. J. Biol. Sc. 24*, 1065-1075 (1971).
20.   Martin, J. H. *U.S. Dept. of Ag. Yearbook*, 343-347 (1941).
21.   McWilliam, J. R. and Naylor, A. W. *Pl. Physiol. 42*, 1711-1715 (1967).

22. Millerd, A. and McWilliam, J. R. *Pl. Physiol. 43*, 1967-1972 (1969).
23. Morse, R. N. and Evans, L. T. *J. Agric. Eng. Res. 7*, 128-140 (1962).
24. Pasternak, D. and Wilson, G. L. *New Pytol. 71*, 682-689 (1972).
25. Pinthus, M. J. and Rosenblum, J. *Crop Sci. 1*, 293-296 (1961).
26. Powles, S. B. and Osmond, C. B. *Aust. J. Plant Physiol. 5*, 619-629 (1978).
27. Raison, J. K. and Chapman, E. A. *Aust. J. Plant Physiol. 3*, 291-299 (1976).
28. Raschke, K. *Planta 91*, 336-363 (1970).
29. Rhykerd, C. L., Cross, C. F. and Sullivan, E. F. *Crops and Soils 12/9*, 24 (1960).
30. Taylor, A. O., Halligan, G. and Rowley, J. A. *Aust. J. Plant Physiol. 2*, 247-257 (1975).
31. Taylor, A. O., Slack, C. R. and McPherson, H. G. *In* "Mechanisms of Regulation of Plant Growth". (R. L. Bieleski, A. R. Ferguson and M. M. Cresswell, eds.) *Bull. 12, Royal Soc. N.Z.*, 519-524 (1974).
32. Taylor, A. O. and Rowley, J. A. *Pl. Physiol. 47*, 713-778 (1971).
33. Teeri, J. A. and Stowe, L. G. *Oecologia 23*, 1-12 (1976).
34. West, S. H. *Proc. XI Int. Grassl. Congr.* 514-517 (1970).
35. Weatherley, P. E. *New Phytol. 49*, 81-87 (1950).
36. Williams, R. F. "The Shoot Apex and Leaf Growth". Cambridge University Press, London (1974).
37. Wright, M. *Planta 120*, 63-69 (1974).

PHYSIOLOGY OF COOL-STORAGE DISORDERS OF FRUIT AND VEGETABLES

N. L. Wade

N.S.W. Department of Agriculture
located at CSIRO Division of Food Research
P.O. Box 52
North Ryde, N.S.W. 2113, Australia

I. INTRODUCTION

Cool-storage is an important technique for preserving fresh plant produce. Deterioration due to ripening, senescence and disease can be retarded by storage at reduced temperature, with the optimum temperature often being slightly above the freezing point of the produce. Many commodities, however, suffer injury when cool-stored at temperatures above their freezing point. By definition, such commodities might be classified as "chilling sensitive" (19), but this designation may sometimes be ambiguous.

Low temperature has been ascribed as the incitant of many storage disorders without critical appraisal. When produce is cool-stored its life is usually increased, permitting senescent disorders to develop which are never observed at higher temperatures due to rapid tissue disintegration or invasion by disease. Low temperature can also aggravate an injury incited by some other cause, such as heat or mechanical injury.

The critical threshold temperature below which injury develops ranges from less than $5^{\circ}C$ for apple and orange fruit to $10\text{-}13^{\circ}C$ for mango and $12\text{-}13^{\circ}C$ for banana fruit (18). Cucumber, eggplant, papaya and peach fruit suffer injury below about $7^{\circ}C$, while the lime, muskmelon and pineapple have thresholds of $7\text{-}10^{\circ}C$. Temperature and time interact in the development of cool-storage injury. As the temperature at which produce is stored decreases below the critical threshold temperature, the severity of ultimate injury increases, although the rate of development of injury during cool-storage decreases (36).

In tropical fruit such as the banana, severe injury is expressed after exposures to chilling temperatures of only a few hours, and each fruit in a particular sample suffers injury (9). In other commodities such as apple and grapefruit, injury can take weeks or months of storage to develop,

81

and only a proportion of the fruit in a particular sample may manifest injury. Biological variation in the critical threshold temperature for chilling within samples of similar material was postulated by van der Plank and Davies (36). These authors observed that only a proportion of grapefruit in a sample suffered chilling injury and that this proportion reached a plateau which did not change further with time. They proposed that when a sample of fruit is cool-stored, a portion of the sample is "out of equilibrium" (i.e. suffers a metabolic imbalance) from the "very instant of cooling, and is predestined to subsequent injury". Time does not cause this imbalance in the susceptible portion but merely enables the development of visible lesions. The rest of the sample does not suffer injury because these fruit have a critical threshold temperature which is below the storage temperature.

This paper discusses evidence which has accrued about the mechanisms by which cool-storage temperatures can injure fresh produce. The plant organs which we harvest for food usually contain no dividing cells and have often ceased expansion growth. These detached organs are no longer dependent on the parent plant for water, minerals and photosynthate, and their metabolism is directed towards maintenance of existing tissues. In the cool-storage environment produce is subject to a constant temperature, unlike plants growing in the field, where diurnal temperature fluctuations may relieve chilling conditions for part of every day.

II.   ROLE OF THE MEMBRANE IN COOL-STORAGE DISORDERS

The injury response observed when tropical commodities are stored below the threshold temperature for chilling is consistent with the hypothesis of Lyons and Raison (20) that an instantaneous temperature-induced change in membrane structure causes the injury. Where, however, the symptoms of injury develop after months of cool-storage and susceptibility to injury varies amongst organs from the same clone and the same trees, a simple cause-and-effect relation between a change in membrane structure and injury is obscured. There is very little evidence which enables us to relate disorders of this type with the critical temperature, T_s (24) below which membrane structure and function become abnormal. Few storage experiments have been conducted in which T_s of cell membranes was monitored throughout the storage period. Studies with cool-stored artichoke tuber tissue showed that T_s was not constant but changed with time (2). There is both direct and presumptive evidence that the value of T_s changes as a fruit ripens (see section IV A). It is not sufficient to determine the value of T_s on a few samples of a particular commodity and assume that this value applies to other samples from different harvests before and during storage.

An attempt has been made to correlate the low temperature breakdown disorder of apple fruit with the T_s of mitochondrial membranes from pulp tissue sampled before storage (21). The apparent T_s, estimated using both succinoxidase activity and spin-labels, was 4-10°C, depending on variety and locality. There was no correlation between apparent T_s and susceptibility to breakdown, since some of the varieties tested were completely resistant to the disorder whilst others showed symptoms when stored below 5°C. Some reservations must, however, be held about the reliability of the T_s estimates because of the subsequent work of Raison et al. (25), who demonstrated how erroneous high estimates of T_s can be derived.

Until reliable estimates of T_s are available, critical appraisal of the role of membrane structure in many cool-storage disorders of fresh produce will not be possible. The inherent variability of horticultural produce and the effects of ripening and storage history on T_s dictate that T_s should be monitored throughout each storage experiment. A simple method is needed to achieve this aim. Assays of succinoxidase activity or of spin-label motion are neither simple nor routine.

III. CHILLING INJURY AND THE ACCUMULATION OF METABOLITES DURING COOL STORAGE

A. Glycolytic Products

Inhibition or disruption of mitochondrial function may lead to an accumulation of pyruvate, which may in turn be metabolized to such compounds as acetaldehyde, ethanol and acetate. Lyons (19) observes that such a metabolic imbalance would result from a temperature-induced change in membrane structure which inhibits membrane-bound enzymes, such as those of the tricarboxylic acid cycle, whilst not affecting the glycolytic enzymes of the cytoplasm.

Pyruvate, acetaldehyde, ethanol and a-keto acids accumulate in chilled banana fruit tissues (22). Acetate (43) and acetaldehyde (31) accumulate in cool-stored apple fruits.

The occurrence of abnormal concentrations of various metabolites in cold-stressed tissues is unequivocal. It is difficult, however, to interpret the significance of such observations. Changes in metabolite concentrations may occur as a direct result of a primary effect of chilling stress, as proposed by Lyons and Raison (20), or as a terminal event in tissue injury. The presence in visibly injured tissues of high concentrations of substances such as those arising from the incomplete oxidation of pyruvate may be commonplace in morbid tissues. Changes in metabolite concentrations which precede visible injury are more informative, since confusion with immediate pre- and post-mortem events is avoided.

A further difficulty arises in deciding if a metabolite which accumulates in response to chilling stress exerts a toxic effect on the tissue. In a review of the role of acetaldehyde in fruit disorders, Smagula and Bramlage (31) concluded that it was not possible from existing information to distinguish if acetaldehyde accumulation is a cause or an effect of tissue disorganization. A similar conclusion can probably be drawn for related metabolites such as ethanol. Proof of toxicity at physiological concentrations is not always easy to obtain. When banana fruit slices were treated with ethanol solutions and then induced to ripen with ethylene, the slices showed a normal respiratory rise and ripened normally, as judged by peel colour, aroma, softening and soluble solids accumulation (Table I). Assuming complete dilution of the applied ethanol with tissue water, infiltration with 1.0 M ethanol gives a tissue concentration of 100 mM ethanol. The highest endogenous ethanol concentrations measured in the pulp of chilled, green banana fruit was about 15 mM, and in chilled, yellow fruit was about 50 mM (22). Regardless of the rate at which applied ethanol volatilizes or is detoxified, injury sustained by the tissue at the time of application should prevent a normal response to applied ethylene.

B.   Oxaloacetate versus Acetate

Hulme *et al.* (14) observed a strong positive correlation between the accumulation of oxaloacetate in cool-stored apples and the subsequent development of low temperature breakdown. A short interim period of storage at higher temperature reduced both the accumulation of oxaloacetate and the severity of breakdown. Oxaloacetate did not accumulate in some varieties of apple which do not suffer from low temperature breakdown. This promising correlation has not been sustained. In another study no differences were found in the levels of tricarboxylic acid cycle intermediates which could be associated with differences in breakdown (39).

The incidence of low temperature breakdown in apples is correlated with the concentration of acetate in the fruit (40, 43). The severity of the disorder can be exacerbated by injecting acetate into the fruit and alleviated by treatments which decrease the acetate concentration in the tissues by increasing the rate of evolution of acetate esters. The long time lapse observed between injection of acetate and appearance of the disorder suggest that a metabolite of acetate is the toxin (44). Indeed, several compounds including the acetate derivative, mevalonate, are potent inducers of breakdown (41).

TABLE I.   Effect of Ethanol on the Response of Banana Fruit Slices to Applied Ethylene

	No Ethylene	Ethylene Applied					
	Untreated control	Untreated control	Water treated	0.10	0.25 Methanol	0.50	1.00
Soluble solids content (%)*	2.1±0.2	17.7±0.7	16.0±0.4	15.3±0.6	16.0±1.2	15.3±1.3	15.7±2.7

Transverse slices of preclimacteric banana fruit (6mm thick) were vacuum-infiltrated with water or ethanol solutions and incubated in air at 20^0C for 7 days. Ethylene (10 μ l/l) was then applied and the soluble solids content of juice expressed from the pulp was determined 14 days after ethanol application.
* g sucrose equivalents/100 g fresh weight.

C.   α-farnesene

Farnesene has been implicated in the superficial scald disorder of cool-stored apples (11), and the skin injury sustained by chilled banana fruits (44). Although the concentration of this compound can increase somewhat in response to chilling stress prior to the onset of tissue morbidity, a better correlation exists between the concentration of farnesene in the peel tissues of apples and bananas at harvest and the subsequent severity of injury after cool-storage. Farnesene can be regarded as a pre-existing potential toxin, which becomes harmful only after a chilling stress has been applied. Farnesene itself does not appear to be toxic to peel tissues, but it produces toxic hydroperoxides when it oxidizes during cool-storage (12). Chilling may promote oxidation of farnesene by damaging cellular compartmentation.

D.   Sorbitol

Sorbitol can accumulate in apple fruit during cool-storage, and this accumulation is accompanied by low temperature breakdown (9). A causal connection was not established in this work. Infiltration of apple fruit with sorbitol solutions can also increase the incidence of breakdown (1). The involvement of sorbitol in cool-storage injury is discussed in section IV.E.5.

IV.   STORAGE TREATMENTS WHICH ALLEVIATE COOL-STORAGE INJURY

A number of empirical storage methods have been discovered which reduce the occurrence or severity of injury during cool-storage. Some of these methods are of considerable practical commercial importance, and they are also a useful tool in the study of cool-storage disorders. Speculation about the mode of action of a particular storage method can lead to useful hypotheses, and the evaluation of these hypotheses is aided by the ability to induce or control the disorder at will.

A.   Delayed Cool-Storage

A remarkable reduction in the subsequent development of injury has been observed in certain instances where fruit are incubated at room temperature for a short time before cool-storage. Davies et al. (3, 4) observed that peach fruit benefited if storage was delayed for about two days. All reported instances of a beneficial response to delayed storage appear to involve fruit which were ripening at harvest or began to ripen immediately after harvest. The crucial effect of delaying storage is, therefore, to alter the stage of ripening at which the fruit enter storage. The maximum benefits of delayed storage have been obtained when rapid, uniform ripening was ensured by treatment with a hydrocarbon gas (5).

One consequence of the stage of ripeness at which a fruit enters storage is obvious. The ripening mechanism is acutely sensitive to chilling stress and an inevitable symptom of chilling in unripe fruits is the subsequent failure of the fruit to ripen (8). Avocado fruit are most sensitive to chilling stress during the climacteric rise and at the climacteric peak (16). The fruit is markedly less sensitive when pre- or post-climacteric, with post-climacteric fruit being least sensitive. The possible involvement of membrane structure in this phenomenon has been studied, by obtaining Arrhenius plots for succinoxidase activity of mitochondria isolated from avocado fruit at different stages of ripeness (17). These authors concluded that the apparent T_s was about $9^{\circ}C$ in both the preclimacteric stage and the climacteric rise. At the climacteric peak, apparent T_s rose to $11\text{-}12^{\circ}C$, whilst in the post-climacteric phase T_s fell to $2\text{-}5^{\circ}C$. These results agree qualitatively with the relative susceptibilities of fruit to chilling injury at each ripeness stage, but they do not explain why, for example, preclimacteric (but not climacteric rise) fruit could be stored at $2^{\circ}C$ for about 30 days before incipient chilling injury appeared. Kosiyachinda and Young (17) proposed that a change in membrane lipid composition during ripening could account for the changes in apparent T_s.

Studies with the ripening banana fruit have shown that appreciable changes in membrane lipid composition occur in this fruit during ripening (37). The change is greatest in the phospholipid fraction and entails an increase in the proportion of esterified linolenic acid and in the total unsaturation of the phospholipids (Table II). The fluidity of liposomes prepared from the extracted phospholipids increased as the proportion of linolenic acid increased (Table III) and it can be predicted that T_s would decrease concomitantly. This does not imply, however, that the banana fruit should become less sensitive to chilling as it ripens. The threshold temperature for chilling may change during ripening of the banana and avocado, but if either fruit is stored well below the threshold, then the observed sensitivity to chilling will depend on the metabolic state of the fruit at the time. The disruptive effect of chilling a ripening fruit may be particularly severe due to the intense metabolic activity which characterizes ripening.

B.   Intermittent Warming

Interruption of cool storage by periods of exposure to warm temperatures postpones the development of injury. The beneficial effect of warming after a period of cool storage appears to reside in recovery from the harmful effects of the stress, although the recovery effect can be confounded with the ripening which also occurs at the higher temperature.

TABLE II.   Changes in the Fatty Acid Composition of Banana Pulp
              Phospholipids during Ripening (37).

Fatty acid	Days after ethylene applied		
	0	3	6
Palmitic	42.0	31.2	35.1
Oleic	6.5	7.3	6.2
Linoleic	31.6	34.7	26.4
Linolenic	6.6	13.5	25.5
TOTAL UNSATURATED (includes 16:1, 16:2)	53.3	65.2	63.2

Banana fruit were induced to ripen with $10\mu\ell/\ell$ ethylene and
sampled at the times shown for extraction, isolation and analysis of fatty
acids esterified to phospholipids. Results for the four major compoents
only are shown.

TABLE III.   Changes in the Physical Properties of Banana Pulp
              Phospholipids during Ripening (37)

Temperature of measurement $(0^{o}C)$	Days after ethylene applied		
	0	3	6
	S_n*		
10	0.765	0.764	0.742
20	0.699	0.690	0.671
30	0.628	0.624	0.608
40	0.579	0.573	0.561

* S_n is the order parameter (13).
Liposomes were prepared from the phospholipid samples analyzed in
Table II and the motion of infused 5-nitroxide stearic acid spin label
monitored. The order parameter, S_n, decreased both as the measurement
temperature increased and as the time of ripening increased, showing
that the liposomal membranes from the lipids of ripe tissue were more
flexible or fluid than those from unripe tissue.

Intermittent warming has received particular attention in the storage of stone fruit, such as the Victoria plum (33), although apples also respond to this treatment (34). It has been argued that transfer to a warm temperature permits the fruit to metabolize a toxic metabolite which has accumulated in cool storage (32). Unless special precautions are taken, increased water loss will occur when produce is transferred to a higher temperature. The consequence of this is discussed in IV.C.

C. Humidity Control

Chilling injury can be reduced if the storage humidity is either high or low, depending upon the particular commodity. High humidity may simply suppress the expression of symptoms, by reducing desiccation of necrotic tissues (19). Low humidity enhances the loss of volatile esters of acetate, a suspected toxin (III.B) in cool-stored apples (38). Simon (29) has suggested that an early stage in the etiology of low temperature breakdown of apples is water-soaking of the intercellular spaces, followed by hydrostatic rupture of the protoplasts. Evaporation of intercellular water would raise the tonicity of the remaining solution and reduce this postulated rupture of protoplasts.

D. Controlled Atmospheres

Controlled atmospheres which are depleted in oxygen and/or enriched in carbon dioxide relative to air are used to delay ripening and senescence in fresh produce, usually in conjunction with cool storage. In many instances controlled atmospheres have exacerbated cool-storage disorders (7), although there are examples where controlled atmospheres have proven beneficial. A harmful effect of controlled atmospheres can be readily explained in produce which is suffering a severe metabolic imbalance due to chilling. The inhibition of vital metabolic pathways by cool-storage will probably be exacerbated by reduced oxygen and increased carbon dioxide tensions. For example, inhibition of succinoxidase activity by high concentrations of carbon dioxide may be highly injurious in a tissue where mitochondrial function is also being inhibited by low temperature.

The beneficial effects which are sometimes observed during controlled atmosphere treatment are less amenable to explanation. An example of such a response is given by the peach fruit. Addition of carbon dioxide to the storage atmosphere reduces injury in cool-stored peaches (Fig. 1). In an atmosphere of 20% v/v carbon dioxide (balance air), no injury (mealy texture or flesh browning) was detected after storage for 30 days at 1°C, followed by 7 days at 20°C, whereas the storage life in air was only 14-21 days.

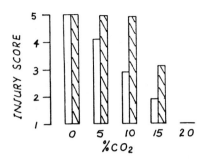

FIGURE 1. Peach fruit (cv. J. H. Hale) were stored at $1^{\circ}C$ for 30 days in air containing the amounts of added carbon dioxide indicated. After a further 7 day ripening period at $20^{\circ}C$, the fruit were subjectively assessed for the presence of flesh browning (▢) and mealy texture(▨). A score of 1 denotes no flesh browning or mealy texture and 5 denotes a severe level of each attribute.

Cool-storage disrupts normal ripening in the peach fruit. When peaches were stored at $1^{\circ}C$, fruit transferred to $20^{\circ}C$ after 14 days or less evolved ethylene at rapid rates after transfer (Fig. 2) and underwent a respiratory rise (Fig. 3). After 21 days at $1.0^{\circ}C$ ethylene evolution was greatly reduced upon removal and after 28 days was very low (Fig. 2). The respiration rates of fruit transferred at 21 to 28 days were initially abnormally high and then declined, showing no evidence of a normal respiratory rise. Incipient symptoms of flesh mealiness and retarded carotenoid pigmentation were observed in fruit transferred at 21 days, and severe symptoms of flesh mealiness and browning were present at 28 days.

When 20% v/v carbon dioxide was added to the storage atmosphere and fruit were removed to air at $20^{\circ}C$ after 30 days, a large increase in ethylene evolution preceded by a lag phase of six days was observed (Fig. 4). These post-storage changes in gas exchange resembled those of air-stored fruit removed within the first fourteen days of storage. Peaches cool-stored in air lose the ability to ripen normally upon removal from storage, whilst peaches stored in 20% v/v carbon dioxide retain the ability to ripen.

The mechanism by which carbon dioxide exerts this protective effect on peach fruit is unknown. Control experiments using low oxygen atmospheres verify that a true effect of carbon dioxide is being observed. The possible sites of carbon dioxide action in the cell are numerous. In the context of the role which membranes play in chilling stress, it is interesting that a direct effect of carbon dioxide on the hydration and permeability of membranes has been proposed (26).

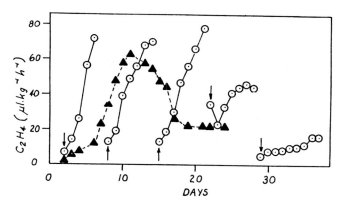

FIGURE 2. Peach fruit (cv. J. H. Hale) were stored in air at 1 or 20°C. Ethylene evolution was measured during continuous storage at 20°C (▲--- ▲),while ethylene production at 1°C was barely detectable. At the times indicated by the arrows, samples of fruit were removed from 1 to 20°C and their ethylene production at 20°C (O——O) was measured for several days.

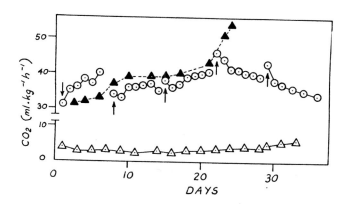

FIGURE 3. Peach fruit (cv. J. H. Hale) were stored in air at 1 or 20°C. Carbon dioxide evolution was measured during continuous storage at 20°C (Δ--- Δ) and 1°C (Δ——Δ). At the times indicated by the arrows, samples of fruit were removed from 1 to 20°C and their respiration rates at 20°C (O——O) measured for several days.

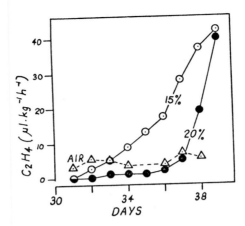

FIGURE 4. Peach fruit (cv. J. H. Hale) were stored for 30 days at $1^{o}C$ in air containing 0% (△ --- △), 15% (O——O) or 20% (●——●) of added carbon dioxide. The ethylene evolution of the fruit upon transfer to air at $20^{o}C$ was measured.

E.   Chemical Treatments

1.   Antioxidants. Diphenylamine, ethoxyquin and butylated hydroxytoluene (10, 11) reduce the severity of superficial scald in apple fruit. These three compounds are antioxidants and may act by reducing the oxidation of farnesene (III.C). Diphenylamine, however, is also known to inhibit carotenoid biosynthesis (6) and butylated hydroxytoluene can perturb membrane structure (35).

2.   Oils. A treatment for the control of superficial scald in apple fruit which preceded the use of antioxidants entailed wrapping each fruit in paper impregnated with mineral or vegetable oils. The mineral oil wraps absorbed farnesene from the peel of the fruit (11), but oil which contaminated the peel may have also affected the oxidation of the farnesene and its derivatives which remained in the peel (III.C). Recently, Jones et al. (15) have found that treatment of banana fruits with dimethylpolysiloxane, safflower oil and mineral oil prevents peel-discoloration after a brief chilling treatment. It was suggested that these materials may perturb membrane structure, but another valid hypothesis is that the oxidation of farnesene or its derivatives is affected.

3.   *Gibberellic acid and phorone (2,6-dimethyl-2,5-heptadien-4-one).* Gibberellic acid (41) and phorone (42) reduce subsequent wastage from low temperature breakdown when applied to apple fruit. The mode of action of these compounds is unknown.

4.   *Benzimidazole derivatives.* The fungicide thiabendazole (2-(4-thiazolyl) benzimidazole) reduces surface pitting in cool-stored grapefruit (26). Benzimidazole and its derivatives display weak cytokinin activity (30). Thiabendazole may inhibit senescent changes which lead to surface pitting.

5.   *Calcium.* Calcium accumulates in growing fruits with difficulty, and mature fruits often suffer from calcium-related disorders (29). Good correlations have been found between tissue calcium content and the susceptibility of the produce to cool-storage disorders (23). Produce with relatively high calcium content is less likely to suffer from cool-storage disorders than produce with low calcium. The application of calcium to fruit after harvest reduces the incidence of disorders such as low temperature breakdown of apples (1, 27). Variations in tissue calcium may explain why commodities such as apple fruit do not respond uniformly to chilling, so that only a proportion of fruit in a particular sample manifest injury.

Sorbitol has been implicated in the breakdown of apples in cool-storage (III.D). Bangerth *et al.* (1) found that infiltration of fruit with sorbitol increased breakdown, whilst addition of calcium to the infiltration solution prevented this adverse effect. It was suggested that calcium enhances the uptake and compartmentation of substrates such as sorbitol. Simon (29) suggests that apples are predisposed to breakdown by permeation of the intercellular spaces with phloem sap. The affected tissues become incidentally high in sorbitol because phloem sap contains sorbitol. Osmotic uptake of this intercellular water causes individual pulp cells to swell. In calcium-deficient tissues the cell walls are unable to withstand the turgor pressure and the cells rupture, causing the necrotic lesions characteristic of breakdown. The role of temperature in this hypothesis is unclear, since the sorbitol is externally-derived and not synthesized by the chilled tissues. Perhaps apple fruits are a specific example of the situation referred to in Section I, where temperature is not a primary incitant of injury but merely facilitates the expression of symptoms. Contrary to this proposition is the evidence that apples are normally cool-stored below the critical temperature T_s (Section II).

V.   CONCLUSION

The etiologies of the cool-storage disorders of fresh produce are at best poorly understood. A unifying concept has been provided by the hypothesis that a temperature-induced change in membrane structure causes changes in cell compartmentation and imbalances in metabolism. There is, however, a dearth of experimental evidence which either supports or refutes this hypothesis for many storage disorders, since in many instances it is not known if produce is being stored above or below the critical temperature at which the postulated change in membrane structure begins. A problem also exists in linking the postulated change in membrane structure with the various abnormalities in metabolism which have been found to follow chilling stress. In no case does it seem possible to set out in detail a sequence of events starting with a change in membrane structure and ending in cell death. Regardless of what progress is made in our understanding of the effect of temperature on membrane architecture, the conclusion appears inescapable that the effects which temperature has on the cells and tissues of stored produce will remain enigmatic.

VI. REFERENCES

1.   Bangerth, F., Dilley, D. R., and Dewey, D. H. *J. Amer. Soc. Hort. Sci. 97,* 679-682 (1972).
2.   Chapman, E. A., Wright, L., and Raison, J. K. *Plant Physiol, 63,* 363-366 (1979).
3.   Davies, R., Boyes, W. W., and de Villiers, D. J. R. *Union of South Africa Department of Agriculture and Forestry Low Temperature Research Laboratory, Capetown. Annual Report.* 1936-1937, pp. 53-67 (1938).
4.   Davies, R., Boyes, W. W., and de Villiers, D. J. R. *Union of South Africa Department of Agriculture and Forestry Low Temperature Research Laboratory, Capetown. Annual Report* 1937-1938, pp. 51-53 (1939).
5.   Davies, R., and Boyes, W. W. *Union of South Africa Department of Agriculture and Forestry Low Temperature Research Laboratory, Capetown. Annual Report.* 1938-1939, pp. 41-43 (1940).
6.   Davies, D. D., Giovanelli, J., and Ap. Rees, T. *"Plant Biochemistry"*p. 340. Blackwell Oxford (1964).
7.   Eaks, I. L. *Proc. Amer. Soc. Hort. Sci. 67,* 473-478 (1956).
8.   Fidler, J. C., and Coursey, D. G. *In* "Proceedings of the Conference on Tropical and Subtropical Fruits, 1969" pp. 103-110. Tropical Products Institute, London (1970).
9.   Fidler, J. C., and North, C. J. *J. Hort. Sci. 45,* 197-204 (1970).
10.   Gough, R. E., Shutak, V. G., Olney, C. E., and Day, H. *J. Amer. Soc. Hort. Sci. 98,* 14-15 (1973).
11.   Huelin, F. E., and Coggiola, I. M. *J. Sci. Food Agric. 19,* 397-301 (1968).
12.   Huelin, F. E., and Coggiola, I. M. *J. Sci. Food Agric. 21,* 44-48 (1970).

13. Huestis, W. H., and McConnell, H. M. *Biochem.   Biophys.   Res. Commun. 57,* 726-732 (1974).
14. Hulme, A. C., Smith, W. H., and Wooltorton, L. S. C. *J. Sci.   Food Agric. 15,* 303-307 (1964).
15. Jones, R. L., Freebairn, H. T., and McDonnell, J. F. *J. Amer. Soc. Hort. Sci. 103,* 219-221 (1978).
16. Kosiyachinda, S., and Young, R. E. *J. Amer. Soc. Hort. Sci. 101,* 665-667 (1976).
17. Kosiyachinda, S., and Young, R. E. *Plant Physiol. 60,* 470-474 (1977).
18. Lutz, J. M., and Hardenburg, R. E. "The Commercial Storage of Fruits, Vegetables, and Florist and Nursery Stocks." Agriculture Handbook No. 66 pp. 19 and 34.   United States Department of Agriculture, Washington D.C. (1968).
19. Lyons, J. M. *Annu. Rev. Plant Physiol. 24,* 445-466 (1973).
20. Lyons, J. M., and Raison, J. K. *Plant Physiol. 45,* 386-389 (1970).
21. McGlasson, W. B., and Raison, J. K. *Plant Physiol. 52,* 390-392 (1973).
22. Murata, T. *Physiol. Plant 22,* 401-411 (1969).
23. Perring, M. A. *J. Sci. Food Agric. 19,* 640-645 (1968).
24. Raison, J. K. and Chapman, E. A. *Aust. J. Plant Physiol. 3,* 291-299 (1976).
25. Raison, J. K., Chapman, E. A., and White, P. Y. *Plant Physiol. 59,* 623-627 (1977).
26. Schiffmann-Nadel, M., Chalutz, E., Waks, J., and Lattar, F. S. *HortScience 7,* 394-395 (1972).
27. Scott, K. J., and Wills, R. B. H. *HortScience 10,* 75-76 (1975).
28. Sears, D. F., and Eisenberg, R. M. *J. Gen. Physiol. 44,* 869-887 (1961).
29. Simon, E. W. *New Phytol. 80,* 1-15 (1978).
30. Skene, K. G. M. *J. Hort. Sci. 47,* 179-182 (1972).
31. Smagula, J. M., and Bramlage, W. J. *HortScience 12,* 200-203 (1977).
32. Smith, W. H. *Nature (London) 159,* 541-542 (1947a).
33. Smith, W. H. J. Pomol. Hort. Sci. 23, *92-98 (1947b).*
34. Smith, W. H. *Nature (London) 181,* 275-276 (1958).
35. Snipes, W., Person, S., Keith, A., and Cupp, J. *Science 188,* 64-66 (1975).
36. Van der Plank, J. E., and Davies, R. *J. Pomol. Hort. Sci. 15,* 226-247 (1937).
37. Wade, N. L., and Bishop, D. G. *Biochim. Biophys. Acta 529,* 454-464 (1978).
38. Wills, R. B. H. *J. Sci. Food Agric. 19,* 354-356 (1968).
39. Wills, R. B. H., and McGlasson, W. B. *Phytochemistry 7,* 733- 739 (1968).
40. Wills, R. B. H., and McGlasson, W. B. *J. Hort. Sci. 46,* 115-120 (1971).
41. Wills, R. B. H., and Patterson, B. D. *Phytochemistry 10,* 2983-2986 (1971).

42. Wills, R. B. H., and Scott, K. J. *J. Hort. Sci. 49,* 199–202 (1974).
43. Wills, R. B. H., Scott, K. J., and McGlasson, W. B. *J.   Sci.   Food Agric. 21,* 42–44 (1970).
44. Wills, R. B. H., Bailey, W. McC., and Scott, K. J. *Plant Physiol. 56,* 550–551 (1975).

SEQUENCE OF ULTRASTRUCTURAL CHANGES
IN TOMATO COTYLEDONS DURING SHORT
PERIODS OF CHILLING

Reinfriede Ilker

General Foods Corporation
Technical Center
Tarrytown, New York

R. W. Breidenbach

Plant Growth Laboratory
Department of Agronomy & Range Science
University of California
Davis, California

J. M. Lyons

Department of Vegetable Crops
University of California
Davis, California

I.   INTRODUCTION

When mature grapefruits (9) or tomato fruits (8) were exposed to chilling temperatures for extended periods, subcellular fine structure was extensively modified. Chloroplast conversion to chromoplasts was inhibited, mitochondria were swollen and more opaque, and the nucleoplasm often appeared more transparent, with accumulations of condensed heterochromatin. Electron micrographs of cotyledons of tomato seedlings exposed to 5°C for 3 days also showed severe damage at every level of cellular organization. The cytoplasmic volume appeared diminished, the membrane systems were generally disorganized, there was a loss of cytoplasmic ultrastructure. Cell death occurred in a ratio of one in ten cells (4).

Evidence also suggests that different membrane systems within the cell may be differentially sensitive to chilling. Minchin and Simon (7) reported that aerobic respiration was impaired at 12°C in expanding cucumber leaves, but that water and solutes underwent leakage at 8°C, implying that the tonoplast and plasmalemma are less sensitive than the mitochondrial membranes. So-called "lipid clustering" (16) seen as discrete areas with smooth fracture faces and other areas of aggregated particles in freeze-fracture electron-microscope studies of mammalian lymphocytes and *Tetrahymena pyriformis,* also suggested that different membrane systems in the cell undergo thermotropic transitions at different temperatures (5).

The intensity of chilling injury depends on the temperature and on the duration of exposure. This report on tomato cotyledons used time increments to assess when and where the first effects on cellular ultrastructure were seen and how they progressed during exposure for periods up to 24 h.

Figures 1, 2, 6, 7, 8, 12, and 18 illustrate the ultrastructure of mesophyll cells from 8-day-old unchilled tomato cotyledons. Organelle envelopes were distinct, well defined boundaries. Cisternae of the endoplasmic reticulum and dictyosomes were relatively infrequent. Microtubules with interconnecting bridges and connections to the plasma membrane were numerous. Vacuolar storage protein was abundant and the general cellular ultrastructure was typical for plant cells, especially those of storage organs.

The 2 h of chilling at 5°C may have induced small changes in ultrastructure in some cells, but not in others. These changes may be seen as a slight decrease in the definition of membrane profiles, or as discontinuities in the thylakoids and envelopes of plastids and the outer membrane of mitochondria. At 4 h (Figs. 3, 9, 13, 19), definite changes due to chilling exposure were observed, including vacuolization of the cytoplasm (Fig. 2), a loss of definition of the thylakoids and prolammelar bodies (Fig. 8), and the development of discontinuities in the envelopes of mitochondria (Figs. 2, 12), peroxisomes (Fig. 12), and nuclei (Fig. 18). The plasmalemma and ground plasm (Fig. 2), including the cortical microtubules, appeared unchanged. At 8 h, the ground plasm appeared generally less structured with a loss of order and definition of ribosomes. The plasmalemma still appeared intact (Fig. 13). The plastid envelopes, prolamellar bodies, and thylakoids (Fig. 9), as well as the mitochondrial (Fig. 13) and the nuclear membranes (Fig. 19), became less definite. An increased electron opacity of the organelle matrices was common. Microtubules were now rarely seen. Swelling of the endoplasmic reticulum and dictyosome cisternae and loss of both the peroxisomal envelope and matrix had already occurred at 4 h (Fig. 12). In many instances, the plastid membranes were barely resolved at 12 h (Fig. 4), while staining of the plastid matrix was intense. Changes in the cell walls consisted of changes in opacity and occasional swelling of the middle lamella. At 16 h (Fig. 10, 14) all the chilling symptoms appeared intensified. In addition, the protein precipitates within the vacuoles

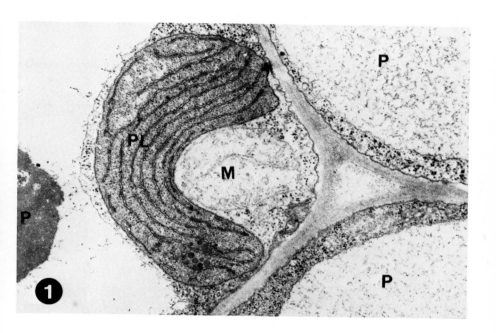

FIGURE 1. Tomato (Lycopersicon esculentum cv. VF145) mesophyll cells from cotyledons of non-chilled seedlings. Seedlings used for all treatments were germinated and grown in vermiculite in the dark for 6 days at 20°C. The seedlings were then transferred to slant boards as described by Waring et al. (1976) after carefully washing them free of vermiculite. The seedlings were established on the slant boards for 1 day at 20°C. Exposure to the chilling temperature of 5° was initiated beginning with those seedlings receiving the longest exposure so that all seedlings were the same age when killed and fixed. Ultrathin sections were prepared as described by Ilker et al (4) and observed with a JEOL 100S electron microscope. Plastid, PL; mitochondria, M; storage protein, P; vacuole, V; nucleus, N; spherosome, G; microbody, M.B. dictysome, D; middle lamella, L.

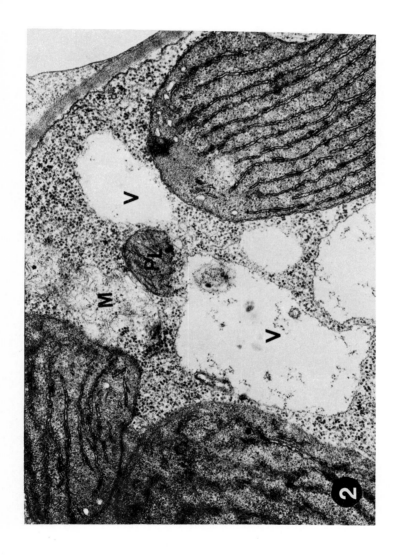

FIGURE 2. Part of a mesophyll cell after 4 h of chilling.

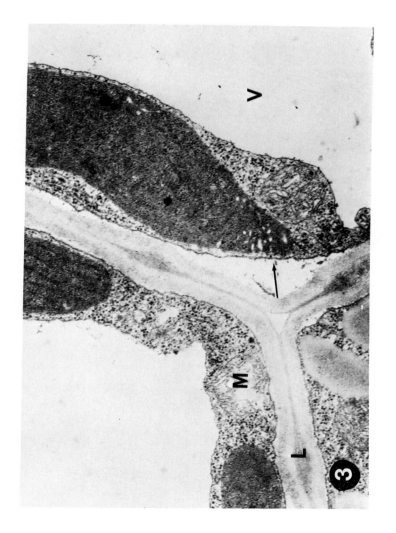

FIGURE 3. *View of mesophyll cells after 12 h of chilling. One cell is slightly plasmolyzed (arrow) cell walls are swollen at the region of middle lamella.*

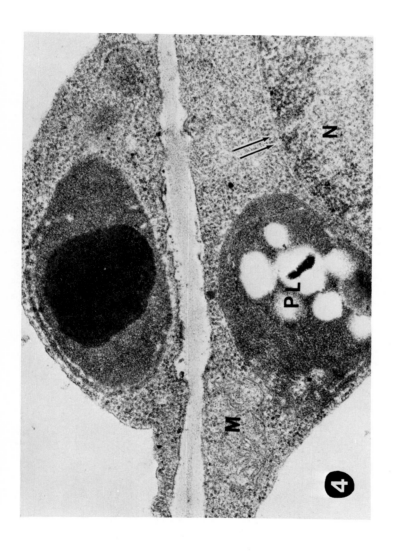

FIGURE 4. *Appearance of mesophyll cells after 24 h of chilling. Remnant of what could be microtubules are seen at the arrow. Double arrows indicate the nuclear envelope.*

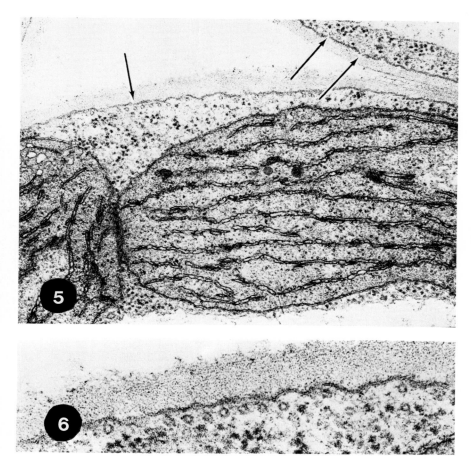

FIGURE 5. A view of an unchilled mesophyll cell to illustrate the distribution of microtubules (arrows). Note the sharpness of all cellular details.

FIGURE 6. An enlargement of Figure 5 to show that microtubules are bridged between themselves and to the plasmalemma.

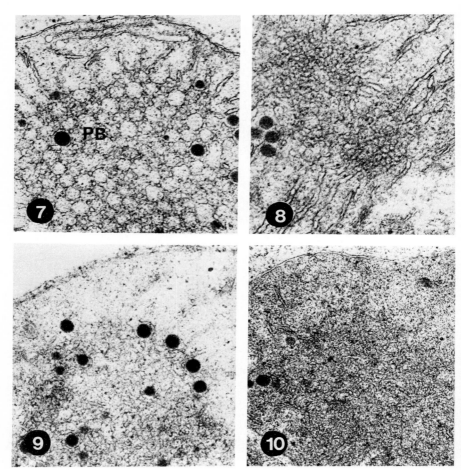

FIGURE 7. Appearance of prolammelar body (PB) from a non-chilled cell. A fairly regular hexagonal pattern is seen (x 58,000).

FIGURE 8. After 4 h of chilling, the prolammelar body appears collapsed. Thylakoids are intact.

FIGURE 9. After 12 h of chilling, most membraneous parts of plastids are indistinct.

FIGURE 10. After 16 h of chilling, the membrane changes are still progressing; the entire plastid stains darkly.

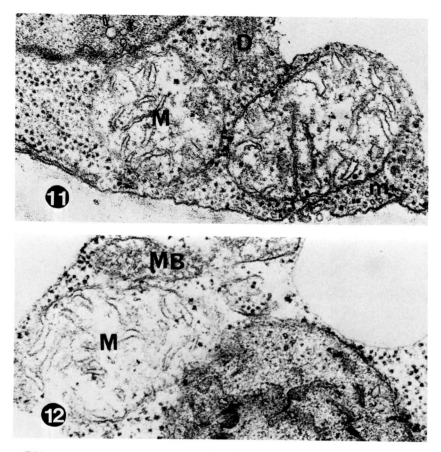

FIGURE 11. The appearance of mitochondria from unchilled tissue.

FIGURE 12. Mitochondrian (M) from tissue chilled for 4 h. Envelope and cristae are indistinct. Peroxisome (MB) and plastid membrane are indistinct (x 42,000).

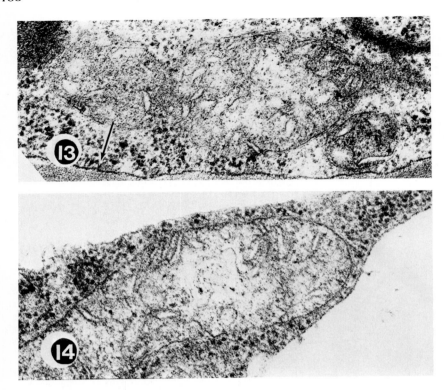

FIGURE 13. Mitochondria after 8 h of chilling. Envelopes appear faint or disassembled. Matrix is darkened (x 42,000). Arrow points to microtubular remnants.

FIGURE 14. Mitochondrian after 16 h of chilling. Mitochondrial structure appears to have stabilized because injury appears about the same as at 8 h (x 42,000).

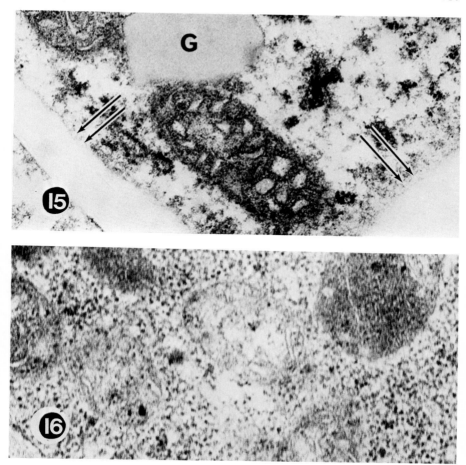

FIGURE 15.   Mitochondrian from a vascular cell after 20 hours. Cristae are distended, envelope is missing. Fat body (G) and cytoplasmic details, including plasma membrane (double arrows) are highly damaged (x 42,000).

FIGURE 16.   Mitochondria after 24 h of chilling. Most membranes are indistinct (x 42,000).

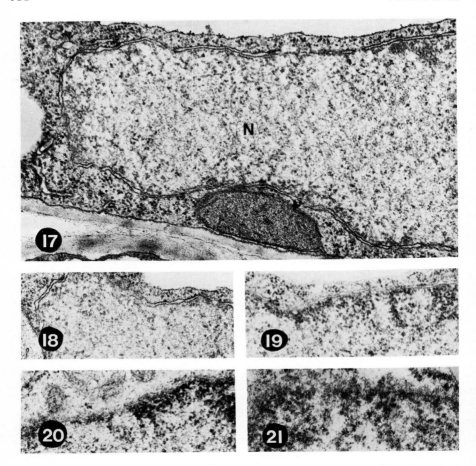

FIGURE 17. Nucleus (N) from an unchilled mesophyll cell. Both nuclear envelopes appear distinct. The nucleoplasm is evenly dispersed.

FIGURE 18. The nucleus after 4 h of chilling envelope has small discontinuities.

FIGURE 19. The nucleus after 12 h of chilling, large parts of the envelope may be missing.

FIGURE 20. Nuclear envelope after 20 h of chilling seems "smeared-out". A small amount of heterochromatin has accumulated.

FIGURE 21. Nuclear envelope after 24 h of chilling. In this instance a filamentous "beard" is associated with the nuclear surface.

appeared to have diminished. Twenty and 24 h of chilling often resulted in an almost complete loss of membranes, as illustrated in Fig. 5. Plastid, endoplasmic reticulum, and nuclear and mitochondrial membranes were absent or nearly absent. The tonoplast and plasmalemma (see Fig. 15), as well as any remnants of microtubules, were poorly resolved. Both nuclear membranes often appeared to be completely disassembled (Fig. 20), with filamentous perinuclear membrane associations (Fig. 21). However, despite the apparent absence of membranes, cellular organization remained (its loss was observed only rarely in this study). Furthermore, cytoplasmic dehydration and cell-volume changes, which might be expected from the apparent loss of structure and function of the plasmalemma, were uncommon. At this time, the cell wall exhibited marked irregularities. It appeared thinner in some areas, while in others it was expanded, with increased material in the middle lamella.

II.   DISCUSSION

Electron micrographs from tomato cotyledon cells show that changes from chilling exposures occur very quickly at the membrane level. Slight discontinuities in some of the membrane profiles of the plastids and mitochondria after very brief exposure to chilling temperatures give evidence that the tissue has been subjected to a stress. Damage in membranes is seen several hours before the appearance of any plasmolytic phenomena or changes in staining.

After prolonged chilling (4, 9), the mitochondria generally appeared less severely altered than the other organelles. However, the relatively short exposures of 2, 4, and 8 h used in the present study resulted in discontinuities in the mitochondrial envelope, followed by swelling of the cristae and increased opacity of the matrix. It appears that the structure of the mitochondria may somehow stabilize during the longer chilling exposures, whereas that of the other organelles continues to change.

The plastids of tomato mesophyll are highly susceptible to chilling, and visible disorganization progresses more quickly and completely in the plastids than in the mitochondria or other organelles. An important cause of this differing behavior could be the unique lipid composition of plastids, with a large amount of galacto and sulfo lipids and relatively small amounts of phospholipids. This may relate to the report of Wilson and Crawford (13), who found that the degree of saturation could be altered in the phospholipids, but not in the glycolipids during the hardening of cucumber seedlings. Shneyour *et al.* (11) found a discontinuity in the plots of $1/T$ vs the log of the rate of photoreduction of $NADP^+$ by the chloroplasts at around $12^\circ C$, about the same temperature where discontinuities are observed for membrane-bound mitochondrial enzymes(6).

The nuclear envelope went through a progressive loss of definition, culminating in what often appeared to be filamentous perinuclear associations. Franke (2) has described such perinuclear "beards" in HeLa and other cells. He believes that they occur at times of general membrane disassembly, as for example, just before mitosis.

Despite the apparent loss of the nuclear envelope (Fig. 5), the nucleoplasm did not undergo the extensive changes observed with 3 days of chilling (4). Only occasionally were there small accumulations of heterochromatin. Most of the time the granularity of the nuclear matrix remained unchanged.

Cortical microtubules are believed to serve as a cytoskeleton in expanding plant cells. In animal cells, there is evidence that interactions between microtubules, microfilaments, and the plasmalemma aid protein mobility at the cell surface (1, 10). Our electron micrographs indicate that microtubules may disassemble or collapse after short periods of chilling. Since these structures are known to depolymerize at low temperature *in vitro,* their almost complete disappearance by 8 h of exposure is not surprising.

The swelling and discoloration of the cell walls starting at 12 h may indicate a leakage and/or accumulation of solutes from the cells at that time. The premature loss of storage protein may not be associated with the normal metabolism of actively growing seedlings. The storage protein could be dissolved and dispersed or hydrolyzed without further metabolism during the chilling period.

Ultrastructural changes in the plasmalemma occurred at 20 and 24 h, much later than in the other membranes. We believe that the absence of visible cytoplasmic dehydration in this study is due primarily to the higher resistance of this membrane to chilling. In animal cells, a loss of cell-volume regulation initiates a sequence of ultrastructural changes that can lead to death (12). In plants, Guinn (3) has pointed out that dehydration is a prerequisite to macroscopic chilling injury because cotton leaf-discs floated on 0.2 M mannitol remained free of symptoms during chilling, whereas exposure of whole plants injured or killed the leaves. Wright and Simon (14) found that phospholipids decline during chilling only after water loss is considerable. According to Minchin and Simon (7), solutes leak from cucumber leaves at temperatures below $12^{\circ}C$, the temperature where discontinuities occur in the plot for the uptake of oxygen.

In lymph cells (15, 16) and in *Tetrahymena* (5), the plasma membrane is highly resistant to chilling-induced freeze-fracture patterns, a property associated, at least in part, to the strong damping effect of sterols (cholesterol and tetrahymenol, respectively) upon lipid fluidity of these membranes. Even though little is known about the lipid composition in plant plasma membranes, similar principles may act to make them more resistant to chilling.

Each membrane type in freeze-fractured *Tetrahymena* initiates smooth membrane regions or "lipid clustering" at specific temperatures. While electron micrographs of thin sections cannot reveal such precise relationships, they did illustrate a progression or continuum of chilling injury, and the differential sensitivity to temperatures of individual membrane types.

TABLE I. Dependency of Chilling Symptoms on Time in Various Compartments of Tomato Cotyledon Chilled at 5°C.[a]

Duration of Chilling (h)	Plasma-lemma	Tono-plast	Mito-chondria	Plastids	Nuclear envelope	Peroxi-somes	Micro-tubules
2	--	--	+	+	--	--	--
4	--	+/-	+	+	+	+/-	--
8	--	+/-	+	++	+	+	absent
12	--	+	++	+++	+	+	absent
16	+/-	+	+++	++++	+	++	absent
20	+	++	+++	++++	+++	++	absent
24	++	++	+++	++++	++++	++++	absent

[a] About six samples of each treatment were observed.

-- = no injury
+ = slight
++ = moderate
+++ = severe
++++ = extreme

In summary, the ultrastructural chilling symptoms of tomato-seedling cotyledons (held at 5°C from 2 to 24 h) manifested themselves primarily as a progression of membrane deteriorations (see Table I). We have found that: 1) membrane alterations and the appearance of small vacuoles preceded other cellular changes (these were followed by more severe alterations of the membranes of plastids, mitochondria, endoplasmic reticulum, peroxisomes, and nuclei, often proceeding to complete loss of ultrastructure); 2) severe cytoplasmic dehydration, new cell-wall deposits, and accumulation of osmiophilic material, previously observed after 3 days of chilling at 5°C (4), was uncommon with the shorter exposures used here; 3) different lengths of exposure were required to damage different types of organelles; and 4) the plasmalemma appeared relatively more resistant to chilling than the other membranes.

ACKNOWLEDGMENTS

We thank Dr. E. M. Gifford, Department of Botany, University of California, Davis, for kindly allowing us to use his laboratory. We thank Dr. C. Rick, Department of Vegetable Crops, University of California, Davis, for providing us with tomato seeds.

III.   REFERENCES

1.   Berlin, R. D., Fera, J. P. *Proc. Natl. Acad. Sci. USA 74,* 1072-1076 (1977).
2.   Franke, W. W. *Protoplasma 73,* 263-292 (1971).
3.   Guinn, G. *Crop Sci. 11,* 11-12 (1971).
4.   Ilker, R., Waring, A. J., Lyons, J. M., Breidenbach, R. W. *Protoplasma 90,* 229-252 (1976).
5.   Kitajima, Y., Thompson, G. A. *J. Cell Biol. 72,* 744-755 (1977).
6.   Lyons, J. M., Raison, J. K. *Plant Physiol. 45, 386-389 (1970).*
7.   Minchin, A., Simon, E. W. *J. Exp. Bot. 24,* 1231-1235 (1974).
8.   Moline, H. E. *Phytopathology 66,* 617-624 (1976).
9.   Platt-Aloia, K. A., Thomson, W. W. *Cryobiology 13,* 95-106 (1976).
10.   Poste, G., Papahadjopoulos, D. *Proc. Natl. Acad. Sci. USA 73,* 1603-1607 (1976).
11.   Shneyour, A., Raison, J. K., Smillie, R. M. *Biochem. Biophys. Acta 292,* 152-161 (1975).
12.   Trump, B. F., Croker, B. P., Jr., Mergher, W. J. *In* Cell Membranes (Richter, G. W., Scaprelli, D. G., eds.) The Williams and Wilkins Co., Baltimore (1971).
13.   Wilson, J. M., Crawford, R. M. M. *New Phytol. 73,* 805-820 (1974).

14. Wright, J., Simon, E. W. *J. Exp. Bot. 23*, 400-411 (1973).
15. Wunderlich, F., Hoelzl Wallach, D. F., Speth, V., Fisher, H. *Biochim. Biophys. Acta 373*, 34-43 (1974).
16. Wunderlich, F., Ronai, A., Speth, V., Seelig, J., Blume, A. *Biochem. 14*, 3730-3735 (1975).

MOVEMENT AND LOSS OF ORGANIC SUBSTANCES FROM ROOTS AT LOW TEMPERATURE

M. N. *Christiansen*

U.S. Department of Agriculture
Science and Education Administration
Agricultural Research
Plant Physiology Institute
Plant Stress Laboratory
Beltsville, Maryland

Substances released from roots are commonly called exudates, be they passively or actively transported to the root rhizosphere. Root exudates include sugars, amino acids, protein, and most other entities common to plant root cells. Several environmental stresses including water deficiency, anaerobiosis, low pH, reduced light intensity, and temperature extremes, are known to increase loss of organic and inorganic substances from roots. Alteration of the chemical quality of exudates is also reported as a consequence of stress. The loss of substances to the rhizosphere has been related to subsequent increased root disease incidence and has, therefore, been of considerable interest to plant pathologists. Allelopathic effects or the creation of unfavorable conditions for neighboring plants or subsequent plantings may be due to exudates (13). The earliest reports of the nature of root exudation were by Knudson (10). Roveria (14, 15) has presented several reviews of the subject. Our interest in the subject stems from a conviction that stress-induced exudation results from membrane malfunction, and that exudation is an easily measured and reliable *in vivo* indicator that adverse treatments have induced injury.

The documentation that stresses induce greater exudation generally supports the concept that the mechanism relates directly to the physical nature of membranes or to metabolic events that support membrane function and integrity. From this basis, one can extrapolate to the thesis that temperature extremes are affecting the control systems that enclose, compartmentalize and transport organic and inorganic substances.

It is well established that most organic substances found in plant root cells also occur in exudates [see reviews by Roveria (14; Schroth and Hildebrand (16)]. Other evidence has been presented that topically applied non-endogenous chemicals can also be transported within the plant and exudated from roots (9). As one might expect, exudates from vacuolate cells may vary from those of non-vacuolate cells (1). Garrad and Humphreys (6) showed variation in quality of saccharides from corn scutellar tissue in varying conditions.

There is much evidence that membranes under cold stress fail in their role as metabolic barriers. Banana *(Musa sp.)* is an excellent example; only brief chilling is necessary to induce marked fruit discoloration (20). The color change is a consequence of oxidation of phenols by release of membrane-bound polyphenol oxidase. Other examples of leaf (8) and root exudation (2) show a rapid alteration of membranes and reduced control of cell contents. Exudation, therefore, provides *in vivo* evidence that temperature extremes alter membranes.

One of the questions presented by chilling-induced exudation is whether it is a consequence of cold inhibition of metabolic systems that are essential for maintenance of membrane integrity or if it is due to a simple biophysical disordering of the lipid bilayer. The question can perhaps in part be answered by observation of response time. A rapid chilling induction of leakage might well indicate physical membrane "channelization" alteration, while a response that requires hours of even days may be a consequence of metabolic blockage, energy unbalance or toxin accumulation.

Respiration inhibition by cold can be duplicated by chemical inhibitors or O_2 deficiency and can induce exudation rather rapidly which somewhat nullifies an argument for categorizing stress effects within a short or long time frame as related to physical or metabolic effects on membranes. Rapid low temperature induced exudative response of leaves or roots (Fig. 1) as contrasted to fleshy organs such as fruits or tubers may be only a consequence of tissue mass and heat loss. Sweet potato *(Ipomoea batatas, (L.) Lam.),* for example, only develops leakage and tissue browning after a prolonged exposure of 2-4 weeks at 10° (12). The same situation occurs in internal browning of apple.

The intervention of divalent cations such as calcium or magnesium in cold-induced exudation provides evidence that the process is physical in nature. Calcium is known to rapidly stabilize root cell membrane function (3). Much of the published information concerning Ca and membranes does not relate to cold stress but to the role of Ca in mediating membrane ion discrimination or uptake (5, 18). The need for Ca for normal root formation and function was noted by Sorokin and Sommer (17). The presence of Ca can lead to a general enhancement of ion absorption commonly called the "Viets" effect (19) which is attributed to the effect of Ca on membrane permeability. The manner in which Ca functions in membranes is a point of considerable debate. Dodds and Ellis (4) hold that membrane-bound ATPase is activated by Ca. Removal of Ca with a chelating agent (EDTA) reduces respiration and results in loss of nucleotides from roots (7). This again implicates respiration as an important factor in maintenance of membrane integrity. It has also been established that Ca mediates membrane control of cell organic contents. Garrard and Humphreys (6) have shown that Ca or Mg prevents sucrose leakage from corn scutellar tissue, especially at low temperatures. They theorize that Ca binds anionic groups of the membrane structure to form cross bridges between membrane structural components, thereby maintaining a selective permeability by pore radius or surface charge. Other studies with Ca^{45} indicate that Ca localizes on the surface of root cells (11), thereby suggesting a plasmalemma membrane structural maintenance role in roots.

Christiansen *et al.* (2) noted that Ca exerts a marked effect on cold-stress-induced exudation (Table I). Cold or O_2 deficiency induced exudation is blocked by Ca or Mg; low pH (3.0) induced exudation could not be controlled by added Ca. The inhibition of exudation was effective as a pre-treatment (Treatment F, Table I), or by addition after induction of exudation (Treatment E). The blockage of ongoing exudation by Ca is quite rapid (within minutes) and strongly suggests a biophysical binding of membrane structure rather than a stimulation of ATPase function or an effect on other metabolic events. This also suggests that cold-induced exudation is due to a direct physical alteration of membranes rather than an impact on membrane-associated metabolism.

We have also performed experiments concerned with latent effects of chilling. The aim of these studies was to determine the chilling dosages required to induce exudation; to develop chemical amelioration techniques; and, to determine by chemical inhibitors if metabolism was involved in recovery or "mending the barriers." In general, four days at 10°C was the point of no return for recovery of cotton roots. Calcium applied after prolonged cold did little to hasten recovery of membrane control. We found no chemical methods of ameliorization, but did find many "metabolic inhibitors" that slow recovery as well as induce root exudation. In general, respiration inhibitors cause increases in exudation; protein synthesis inhibitors have little effect, and the uncouplers such as DNP induce exudation as well as prevent recovery after chilling (Tables II and III).

In using "specific" metabolic inhibitors, one learns that they are more often dull research tools, first because they are not specific in metabolic action site, secondly because little can be determined about rate of influx to action sites. The most interesting point in these data is that EDTA (Table I) exerted a much greater inductive effect than iodoacetate, fluoride, or fluorocitrate although EDTA is reputed to move into tissue much more slowly (21). One might surmise that EDTA acts at the membrane surface by removing Ca or other stabilizing cations rather than by exerting an effect upon metabolism, whereas the respiration inhibitors must penetrate the cell mitochondria to affect respiration.

M. N. Christiansen

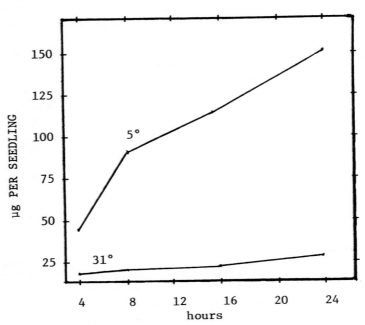

FIGURE 1. Cumulative radicle loss of carbohydrate at 31° and 5° by seedlings germinated 1 day at 31°.

TABLE I. Cotton Radicle Exudation at Low Temperature in N_2 or at Low pH

Treatment	Exudation µg/seedling/hour carbohydrate
A. Control 31°	4.5
B. Control 5°	14.4
C. Control + $CaSO_4$ $(10^{-6}M)$	2.0
D. 5° + $CaSO_4$ $(10^{-6}M)$	4.4
E. 5° for 2 hours → 5° $CaSO_4(10^{-6}M)$	2.1
F. 31° + $CaSO_4$ $(10^{-6}M)$ → 5° Water	4.4
G. N_2	22.0
H. N_2 + $CaSO_4$ $(10^{-6}M)$	1.7
I. pH 3-31°	121.0
J. pH 3 + $CaSO_4(10^{-6}M)$	106.0
K. EDTA $(2 \times 10^{-3}M)$	109.0

TABLE II.   Induction of Radicle Exudation by Chemical Agents

Treatment	Carbohydratee Exudation µg/seedling/hour	Function inhibited[a]
Sodium Azide 3.5×10^{-5} M	50	Respiration
Iodoacetate 10^{-5} M	48	Respiration glycolysis
Sodium Fluoride 10^{-5} M	24	Respiration enolase
Fluorocitrate 10^{-5} M	20	Krebs cycle
2,4DNP 10^{-4} M	66	Oxidative phosphorylation
Sodium Arsenate 10^{-5} M	5	Respiration
Chloroamphenicol 10^{-4} M	20	Protein synthesis
Control	4.5	

[a]Davenports Law: "The specificity of an inhibitor is inversely proportional to how much is known about it" (21).

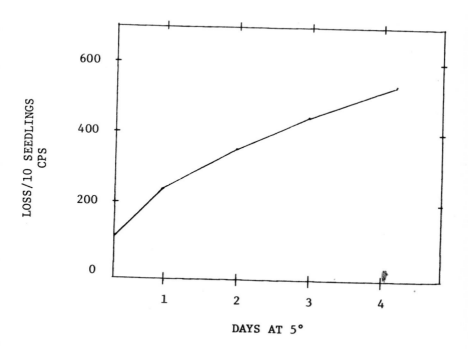

FIGURE 2. Loss of C_{14} Glycine at 30° after chilling at 5°.

TABLE III. Carbohydrate Exudation after Cold as Affected by Metabolic Inhibitors, Calcium and ATP.

Treatment	Hours of treatment 31°			
	1	2	3	4
	μ g/hour/seedling			
Control H_2O 3 days 5°	73	27	20	14
$CaSO_4$ ($10^{-5}M$)	65	31	22	14
Sodium Azide ($3.5 \times 10^{-4}M$)	67	77	91	73
DNP ($10^{-3}M$)	53	126	143	110
Chloroamphenicol ($10^{-4}M$)	38	18	10	9
Cyclohexamide ($10^{-4}M$)	53	37	27	27
DNP ($10^{-3}M$) + ATP ($10^{-4}M$)	60	133	131	117
Sodium Azide ($10^{-4}M$) + ATP ($10^{-4}M$)	38	29	24	37

Treatment	Hours at 31° after 4 days cold			
	1	2	3	4
Control 31° H_2O	81	37	36	26
$CaSO_4$ ($10^{-5}M$)	71	34	23	16
Sodium Azide ($3.5 \times 10^{-4}M$)	65	70	59	48
DNP ($10^{-3}M$)	100	124	141	105
Chloroamphenicol ($10^{-4}M$)	63	25	23	20
Cyclohexamide ($10^{-4}M$)	60	35	32	32
DNP ($10^{-3}M$) + ATP ($10^{-4}M$)	50	98	138	114
Sodium Azide ($10^{-4}M$) + ATP ($10^{-4}M$)	32	19	29	30
Control-No Cold	20	12	9	6
Sodium Azide ($10^{-4}M$)-No Cold	38	49	49	52

INTERNAL SOLUTE MOVEMENT AND ROOT EXUDATION

In an effort to determine organic movement within seedling tissue and the effect of chilling on these processes, a method involving topical applications of C^{14}-labeled glycine was used. The labeled substance was applied quantitatively in 0.5% agar to cut surfaces of cotyledons of seedlings which had been germinated 30 hours at 31° (2 cm uniform radicle length). The seedlings were placed upright in holders with 1 cm of the radicle tip emersed in distilled water. Root loss of topically applied glycine increased at a constant rate over 4 days' treatment at 5^oC (Fig.

TABLE IV. Calcium Effect on Loss of Topically Applied C^{14} Glycine from Radicles of Chilled and Nonchilled Cotton Seedlings

Temperature Regime	Glycine loss/hr from 10 seedlings as counts per second			
	3 hr reading		6 hr reading	
	Ca	No Ca	Ca	No Ca
30° Control	10	8	11	10
5° after 8 hours at 30°	47	35	63	50
30° after 8 hours at 5°	144	350	48	108
5° continuous	102	166	44	102

2). The response is comparable to previously reported exudation of sugars and amino acids from roots (2). The results indicate that chilling influences membranes of the entire seedling more or less equally with that of the root cell plasmalemma. Presumably the movement of cotyledonary applied C^{14} glycine would be through cells of the cotyledonary mesophyl, the phloem cells of the axis and root, through cortex cells of the root to the rhizosphere, or in other words, the pathway of organic nutrient flow from cotyledons to the root.

We also attempted to further elucidate the role of calcium by incorporating $CaSO_4$ into the topically applied agar-glycine C^{14} mixture. It had no effect on movement at 30^oC. In seedlings transferred from 30° to 5^oC, little effect was noted until after 6 hours at 5^oC, glycine movement through the seedling axis and loss from the root was reduced 1/3 to 1/2 (Table IV). The site of the calcium effect in the seedling has not been determined. Possibly it localized near the cotyledon site of application and exerted a major limiting influence at that point. Alternatively, it could diffuse throughout the seedling. Calcium is normally rather immobile in tissue due to its reactivity within the plant. The answer lies in use of Ca^{45} tracing which we have not yet done.

The present information indicates that chilling-sensitive cotton seedlings suffer membrane alteration throughout the plant which can affect solute as well as water movement. The rapidity of induction of root exudation suggests physical alteration of membranes; the reversibility by calcium and the action of EDTA likewise support alteration of structure rather than metabolic events.

REFERENCES

1. Burstrom, H. *Biol. Rev. 43,* 287-316 (1968).
2. Christiansen, M. N., Carns, H. R., and Slyter, D. J. *Plant Physiol. 46,* 53-56 (1971).
3. Christiansen, M. N., and Foy, C. D. *Comm. in Soil Sci. and Plt. Analysis 10,* 427-442 (1979).
4. Dodd, J. A. A., and Ellis, R. J. *Biochem. J. 101,* 31 (1966).
5. Epstein, E. *Plant Physiol. 36,* 437-444 (1961).
6. Garrard, L. A., and Humphreys, T. E. *Phytochem. 6,* 1085-1095 (1967).
7. Hanson, J. B. *Plant Physiol. 35,* 372-379 (1960).
8. Guinn, G. *Crop Science 11,* 101-102 (1971).
9. Hurtt, W., and Foy, C. L. *Plant Physiol. (Sup.) 40,* 58 (1965).
10. Knudson, L. *Am. J. Bot. 7,* 371-379 (1920).
11. Leggett, J. E., and Gilbert, W. A. *Plant Physiol. 42,* 1658-1664 (1967).
12. Lieberman, M., Craft, C. C., Audia, W. V., and Wilcox, M. S. *Plant Physiol. 33,* 307-311 (1958).
13. Risser, P. G. *Bot. Rev. 35,* 251-284 (1969).
14. Roveria, A. D. *In* "Ecology of Soil-Borne Pathogens." K. Baker and W. Snyder (eds.). pp. 170-186. University of California Press, Los Angeles (1965).
15. Roveria, A. D. *Bot. Rev. 35,* 35-57 (1969).
16. Schroth, M. N., and Hildebrand, D. C. *Ann. Rev. Plant Phytopathol. 2,* 101-132 (1964).
17. Sorokin, H., and Sommer, A. L. *Am. J. Bot. 27,* 308-318 (1940).
18. True, R. H. *Am. J. Bot. 1,* 255-273 (1914).
19. Viets, F. G., Jr. *Plant Physiol. 19,* 466-480 (1944).
20. Wardlaw, C. W. In "Banana Diseases Including Plantains and Abaca." Pub. Longmans, London (1961).
21. Webb, J. Leyden. *In* "Enzyme and Metabolic Inhibitors," Vol. 1, Academic Press, N.Y. and London.

COLD-SHOCK INJURY AND ITS RELATION
TO ION TRANSPORT BY ROOTS

F. Zsoldos

Department of Plant Physiology
Attila Jozsef University
Szeged, Hungary

B. Karvaly[1]

Institute of Biophysics
Biological Research Center
Hungarian Academy of Sciences
Szeged, Hungary

I.   INTRODUCTION

The overwhelming majority of experimental studies on the effects of low-temperature stress on various life processes of plants have indicated that membrane-linked events, including ion transport, are primarily damaged when plants are subjected to a sudden fall in temperature. Owing to the temperature-balancing properties of soil, roots are usually somewhat less exposed to sudden variations in temperature. In spite of this, their scanty adaptiveness to extreme temperatures and their higher sensitiveness to rapid temperature fluctuations make it understandable why the root temperature is critical as regards the surviving of either high- or low-temperature stress and shock (21).

The key-role of potassium in maintaining structures and physiological functions at a cellular level is generally acknowledged (6). This, as well as a low-temperature anomaly observed solely for K^+ (and Rb^+), initiated comparative investigations on the low-temperature K^+-transport of roots of cold-resistant and cold-sensitive (thermophilic)

[1] *Present address: Chemistry Department, University of California, Los Angeles, 405 Hilgard Avenue, Los Angeles, California 90024, U.S.A.*

123

plants (30, 31, 32, 34, 35, 37, 38, 39). In the following an account will be given for the most important results obtained with winter wheat *(Triticum aestivum* L. cv. Mironowskaya 808), as a model species of non-thermophilic character, and rice *(Oryza sativa* L. cv. Dunghan Shali), representing thermophilic species. Most experimental findings, to be summarized below will pertain to these related species (cereals); they are, however, typical of other non-thermophilic (e.g. winter barley, etc.) and thermophilic (e.g. cucumber, melon, sorghum, etc.) seedlings, respectively. Mostly low-temperature effects will be discussed.

Conditions of growth and handling of seedlings, the experimental procedure used, the reproducibility, reliability and accuracy of data have been described previously (34, 38). ^{86}Rb labelling was employed for K^+, the suitability of this being carefully checked and substantiated in experiments with double (^{86}Rb and ^{42}K) labelling.

II.   RESULTS AND DISCUSSION

A. K^+ *Uptake by Excised Roots Following Cold-Shock Treatment*

When 6-7 cm long, excised roots are suddenly immersed into a Ca^{2+}-free uptake solution of a given temperature, the initial (60 min) K^+ uptake exhibits a very characteristic temperature pattern, depending upon both variety and species (Fig. 1). Roots of non-thermophilic plants display a monotonously decreasing K^+ uptake when the temperature is lowered, as usually expected. For thermophilic species an anomalous K^+ uptake occurs, with negative temperature coefficient, which may exceed that in the physiologically optimum temperature range. The trough-like temperature pattern is typical only of seedlings of thermophilic species. As regards the positions of the minima of the trough-like curves, they lie at higher temperatures the more thermophilic the species and/or variety (muskmelon > cucumber > rice > sorghum).

Anomalous K^+ uptake arises solely when roots are suddenly exposed to and not gradually cooled down to low-temperature (Fig. 2). Another kind of anomaly occurs in the initial K^+ uptake when experiments are performed on suddenly precooled roots in Ca^{2+}-free absorption medium and at $25^{\circ}C$ (Fig. 3). The initial K^+ uptake measured at $25^{\circ}C$ exceeds the normal physiological value, "remembering" the precooling temperature much more than recognizing the actual, more favourable uptake temperature. As a function of the precooling temperature, the $25^{\circ}C$ K^+-uptake rate is very reminiscent of the curve given in Fig. 1. No such "memory-effect" can be found, however, if Ca^{2+} is present. These observations indicate that anomalous initial K^+ uptake is a dynamic property with a fairly large inertia and a potentiality for irreversibility. Moreover, these features cannot be ascribed to individual molecular processes, but rather to changes in molecular organization of the transport barrier, primarily the plasma membrane.   The striking differences between responses of initial K^+ uptakes of non-thermophilic

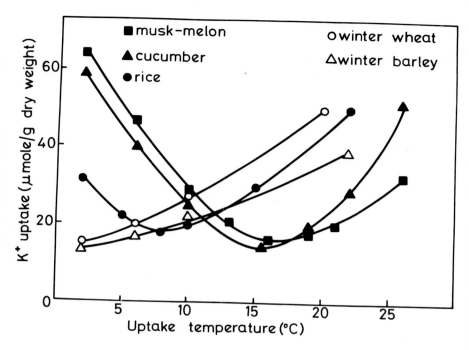

FIGURE 1. Temperature-dependence of initial $K^+(^{86}Rb)$ uptake by excised roots of various plants, after sudden cooling to the uptake temperature. Absorption solution: 0.5 mM KCl; uptake time: 50 min.

and thermophilic plants to cold-shock suggest that significant dissimilarities must exist in the chemical compositions and organization principles of the plasma membranes in the roots of thermophilic plants.

1. *Effect of pH.* As shown in Fig. 4, external pH influences the initial K^+ uptake only at temperatures out of the range of anomaly. Near $0°C$ profound changes occur, however, when the external pH is lowered: a sharp reduction of the anomalous K^+ uptake accompanied decreasing pH from 6.5 to 5.5, and the anomaly completely disappeared below 5.5 (33). The abrupt change in the anomalous K^+ uptake cannot be ascribed simply either to a pH induced change in membrane structure or to the reduction of a selective H^+-K^+ exchange diffusion due to a balancing pH gradient. Consequently, the observed titration-like behaviour must be connected with the pH-dependent affinity of K^+ for the root material.

2. *Effect of Ca^{2+}.* In the physiological range of temperature, Ca^{2+} usually exerts a stimulation on K^+ absorption (Fig. 5) (1, 4, 25, 36, 38) in both groups of plants. An inhibition of K^+ uptake, however, occurs after cold-shock treatment. The anomaly (at either $0°C$ or $25°C$) is

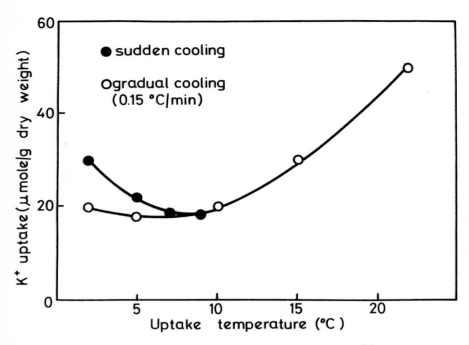

FIGURE 2. *Temperature-dependence of initial $K^+(^{86}Rb)$ uptake by excised roots of rice, after gradual cooling to the uptake temperature. Absorption solution: 0.5 mM KCl; uptake time: 50 min.*

gradually reduced when the Ca^{2+} is increased and is completely eliminated when the Ca^{2+} is present well in excess of a threshold concentration of about 10^{-5} M. In our case, Ca^{2+} apparently counteracts the spectacular effects of cold-shock on K^+ uptake, completely abolishing the anomaly, but it does not prevent the growth disturbances. Figure 6 shows that the initial anomalous K^+ uptake can be not only eliminated, but also reversed by adding Ca^{2+}, which is reminiscent of a cation-exchanger as well (4). This Ca^{2+}-produced reversal could be attributed to the loss of K^+ (^{86}Rb) absorbed, due to passive exchange gaining ascendancy over uptake. Such a mechanism alone, however, cannot account quantitatively for the observations. Therefore, it is compulsory to reason that the anomalous K^+ uptake may be closely related to the peculiar cation-exchange properties of the root material produced by cold-shock.

3.   *Interactions of K^+ with Other Electrolytes.* Cations, both mono- and divalent, greatly affect the anomaly, depending upon valency and concentration. Monopositive ions (Li^+, Na^+, Cs^+, Rb^+, NH_4^+) only reduce,

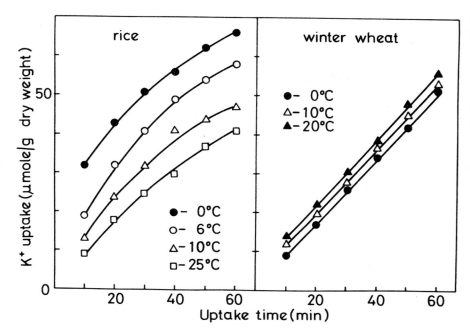

FIGURE 3. Time-dependence of $25^{\circ}C$ $K^{+}(^{86}Rb)$ uptake by excised rice and winter wheat roots after precooling. Absorption solution: 0.5 mM KCl; precooling time: 50 min.

but do not eliminate the anomaly, with varying effectiveness. As a tentative order, Li^{+} < Na^{+} < Cs^{+} < Rb^{+} < NH_{4}^{+}) can be given, which coincides quite well with the sequence of the surface charge density on the respective cation in hydration form. This finding is also in line with the assumption of a competitive translocation of monopositive ions through negatively-charged sites, e.g. in cation-exchange regimes.

Divalent cations such as Ca^{2+}, Mg^{2+}, Sr^{2+} and Mn^{2+} without exception abolish the anomaly, with practically the same efficiency. The presence of different anions (NO_{3}^{-}, Cl^{-}, SO_{4}^{2-}, $H_{2}PO_{4}^{-}$) does not exert any significant influence on the anomalous uptake process. The observed impacts of different ions on the K^{+} uptake anomaly do not parallel their ordering/disordering effects on lipids (8, 13). Hence, their interactions with K^{+} are not directly related to membranes.

4.   *Efflux and Exchange.* The efflux of ions from plant roots is usually regarded as a passive process with very moderate temperature-dependence, occurring via either simple or exchange diffusion along the electrochemical gradient (17, 18, 19). Figure 7 shows that the non-thermophilic wheat roots obey this quite general rule, even when they

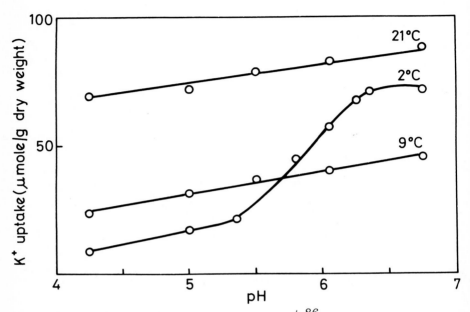

FIGURE 4. pH-dependence of initial $K^+(^{86}Rb)$ uptake by excised rice roots at different temperatures. Absorption solution: 0.5 mM KCl: uptake time: 60 min (33).

were exposed suddenly to cold, regardless of whether Ca^{2+} is present or not. In contrast, the K^+ efflux rate for rice, which exhibits a moderate Ca^{2+}-dependence at $20^\circ C$, responds dramatically to cold shock, increasing markedly to an extent independent of the presence of Ca^{2+}. No significant change was found in the time-course of K^+ efflux when the pH was lowered from 6.5 to 4.5 where the uptake anomaly is abolished (Fig. 4). Under exchange conditions the corresponding rates are influenced by Ca^{2+} similarly as for the efflux in both experimental systems.

A comparison of the characters of the K^+-uptake, -efflux and -exchange processes suggests that cold-treatment of thermophilic plant roots opens new, separate pathways for the passive inward and outward movements, respectively, of K^+, which are very different in nature, and that it transforms the overall passive K^+ transport of thermophilic roots in an irreversible way (see also the "memory effect" in Fig. 3).

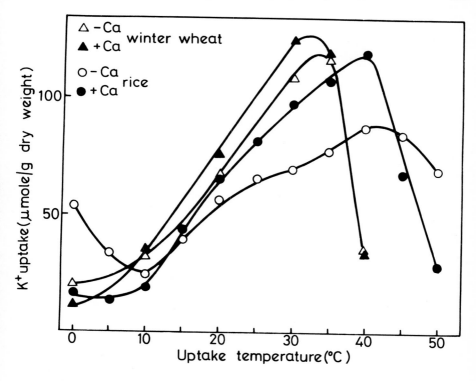

FIGURE 5.   Effect of Ca^{2+} upon the temperature-dependence of initial K^+ (^{86}Rb) uptake by excised roots of rice and wheat. Absorption solution: 1 mM KCl with or without 1 mM $CaCl_2$; uptake time: 60 min.

B.   Relations of Cold-Shock Effect to Root Zones

1. K^+ uptake distribution along roots. Figure 8   presents the zonal distributions of the K^+ uptake in rice and winter wheat roots and their responses to Ca^{2+}, at $0°C$ and $25°C$.   Indisputably, the extraordinarily large K^+ influx does not extend over the whole root; it is confined to the apical root portions, which are also the most responsive to Ca^{2+}. As regards the non-thermophilic winter wheat, the respective uptake patterns are similar to those published by others (2, 5, 16) and bear witness to completely different behaviour, as discussed in detail recently (38).

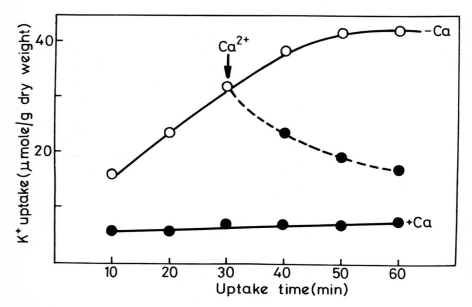

FIGURE 6. Reversal of low-temperature anomalous initial K^+ (^{86}Rb) uptake of rice roots by addition of 1 mM Ca^{2+} during uptake. Absorption solution: 1 mM KCl.

2. *Potassium content distribution along roots.* The overall potassium distribution patterns (Fig. 9) for untreated roots exhibit qualitatively rather similar appearances; a high potassium content was found in the first 1 cm section, gradually decreasing (39). The absence of Ca^{2+} from the uptake solution exercises, in general, only a moderate influence on the uptake patterns, with the exception of rice at $0°C$. In this latter case a marked loss of K^+ occurs especially in the first two root zones, which is irreversible and cannot be prevented by Ca^{2+}. In addition, the presence of Ca^{2+} further increases (and does not arrest) the loss of K^+. Similar observations on other thermophilic plant roots (39) enable us to conclude that such a behaviour of the apical meristematic region of roots is peculiar to thermophilic plants. It is important to emphasize that the highest potassium content exists in the apical meristematic region, where K^+ is taken up anomalously.

3. *Calcium content distribution along roots.* Figure 10 depicts the results obtained for Ca^{2+} distribution. A comparison of data for control, $25°C$ and $0°C$ experiments, respectively, in the absence of Ca^{2+},

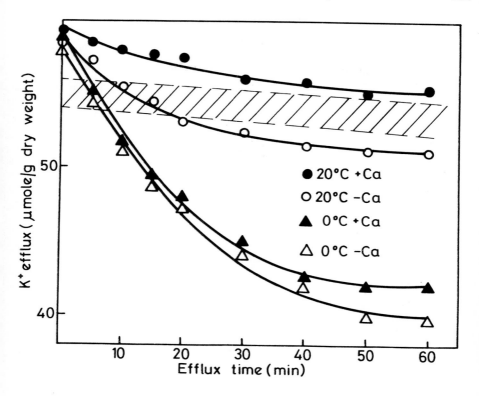

FIGURE 7. Effects of temperature and Ca^{2+} upon the time-course of efflux from excised roots of rice and wheat. Corresponding data for wheat fell within the ruled area, i.e. they coincided within experimental uncertainties. Labelling solution: 1 mM K^+ (^{86}Rb)Cl; efflux medium: distilled water with or without 1 mM $CaCl_2$; pre-incubation time: 50 min.

suggests that Ca^{2+} exists in roots in at least three different pools: (1) soluble Ca^{2+} removable in a Ca^{2+} free absorption solution at $25°C$; (2) strongly-bound Ca^{2+} remaining in the root tissue even after an incubation in Ca^{2+} free solution at $25°C$; and (3) Ca^{2+} retained in excess to the strongly-bound Ca^{2+} when cold-shock treatment is applied. Mobile Ca^{2+} is certainly situated in the water-free space (4); strongly-bound Ca^{2+} is thought to be adsorbed on the cell wall and the external surface of the plasmolemma, and located in various cell compartments (4), while Ca^{2+} retained after cold-shock is a proportion of the soluble mobile Ca^{2+} accessible for and adsorbed on cold-shock-produced, new

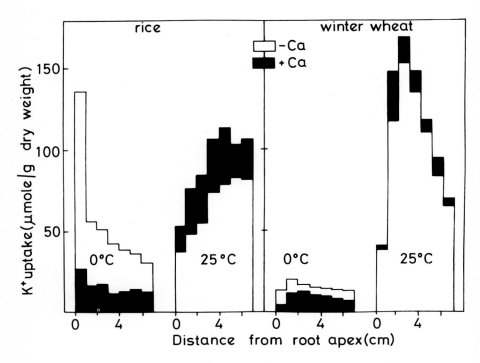

FIGURE 8. Effects of Ca^{2+} and temperature upon $K^{+}(^{86}Rb)$ uptake distribution patterns along primary roots of rice and winter wheat. Absorption solution: 1 mM KCl with or without 1 mM $CaCl_2$; uptake time: 60 min (38).

binding sites. If the external solution contains Ca^{2+}, then this ion is accumulated in a considerable excess, in the apical region for rice and in the more matured segments for winter wheat following cold-shock treatment.

It should be stressed that at $0^{\circ}C$ the entry of Ca^{2+} into the respective rice root segments takes place at the expense of K^{+} (Figs. 8, 9 and 10). In other words, Ca^{2+} is preferentially bound by root zones where the K^{+} uptake operates anomalously. This points again to the fact that, in the case of thermophilic plants, cold-shock treatment creates new cation-binding sites within apical root tissue, for which Ca^{2+} and K^{+} compete, Ca^{2+} having the greater affinity.

It has been demonstrated in the foregoing that the anomalous K^{+} uptake by thermophilic plant roots is restricted to the apical meristematic region, where intense loss of K^{+} and preferential Ca^{2+}

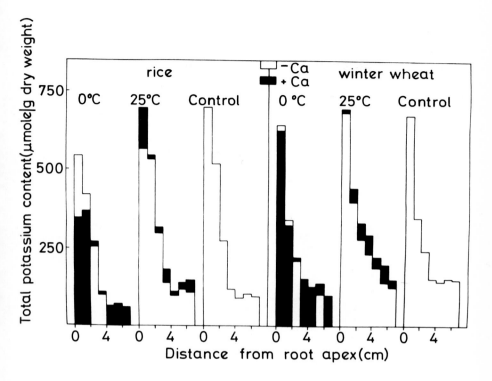

FIGURE 9. Effects of Ca^{2+} and temperature upon the distribution of potassium content along primary roots of rice and winter wheat. Otherwise as Figure 9 (39).

accumulation also occur following cold-shock. All these data underline that the observed anomaly in the K^+ uptake is related to, and may therefore be an indicator for cold-stress sensitivity (thermophily). This finding receives strong support from the well-visible chronic growth disturbances of roots: the arrestation of primary-root elongation and the intense side-root formation behind the root tip after chilling treatment (39). Preliminary morphological studies show that a cold-shock injury proceeds towards complete disorganization in the apical root section (40).

C. **Uptake of Other Ions by Excised Roots Following Cold-Shock Treatment**

As mentioned before, only potassium and its replacing ion, rubidium, exhibited the anomalous inward movement at $0°C$ and in the absence of Ca^{2+}. This is now demonstrated in Fig. 11 for the absorptions of NH_4^+,

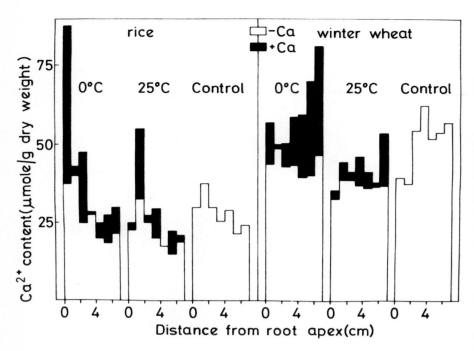

FIGURE 10. Distribution of calcium content along primary roots of rice and winter wheat. Otherwise as Figure 8.

NO_3^-, $H_2PO_4^-$ and I^-, respectively, by rice roots. The NH_4^+ and NO_3^- uptakes increase monotonously with rising temperature, either in the presence or in the absence of Ca^{2+}. A higher absorption can be experienced for NH_4^+, however, the most effective competitor of K^+ and Rb^+ at low temperatures and in the absence of Ca^{2+}. In contrast, Ca^{2+} considerably promotes the accumulations of $H_2PO_4^-$ and I^- ions at each temperature. In general, similar uptake patterns were found for the non-thermophilic plants (e.g. winter wheat). These results clearly document that the observed anomaly produced by cold-shock is a property of thermophilic species and peculiar to monovalent cations such as K^+ and Rb^+. Thus, of all the above effects, the K^+ uptake anomaly can provide the key to a better understanding of the mechanism of cold-shock injury.

III. PROPOSED MECHANISM FOR CHILLING-INJURY OF THERMO-
      PHILIC ROOTS

The indispensability of potassium in maintaining structural and functional organization in plants, in addition to very prompt reaction of

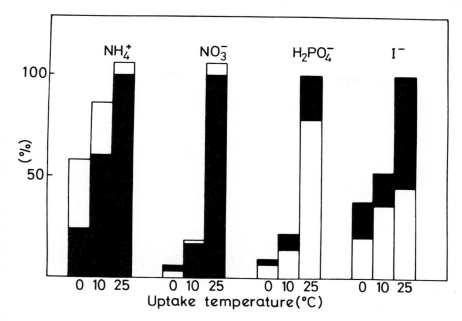

FIGURE 11. Temperature-dependence of initial uptake of different ions by excised rice roots after sudden cooling to the uptake temperature. The labelled absorption solution contained 0.5 mM of NH_4Cl, $NaNO_3$, NaJ each and 0.1 mM KH_2PO_4, respectively. Uptake time: 60 minutes (▫ -Ca, ◼ +Ca).

potassium movement and retention to, and the prolonged effects of cold-shock, suggest that the potassium economy and related events form key-elements of the cold-injury of thermophilic plant roots.

The above observations enable us to outline a tentative proposal for the possible mechanism of some primary processes involved in the cold-shock injury of roots of thermophilic plants. The distinctive features of thermophilic plants can be briefly summarized as follows:

1)   only a sudden drop in temperature brings about the anomalous inward movement of K^+ with negative temperature coefficient (Figs. 1 and 2);

2)   precooling results in an anomaly of similar character in the 25^oC uptake of K^+ (after- or "memory-effect", Fig. 3);

3)   the pH-dependence of anomalous K^+ uptake is very reminiscent of a titration curve (Fig. 4);

4) Ca^{2+} not only eliminates the K^+ uptake anomaly (Fig. 5) but even reverses the anomalous K^+ uptake (Fig. 6), which is typical of a cation-exchange regime;

5)   elevated low-temperature efflux elicited by cold-shock is practically insensitive to Ca^{2+} (Fig. 7);

6)   the anomalous K^+ uptake is restricted to the apical

meristematic zone (Fig. 8) where a marked Ca^{2+}-dependent potassium loss occurs (Fig. 9), and where Ca^{2+} accumulates preferentially as well (Fig. 10).

Bearing all this in mind, one has to assume that, as a first step, rapid cold-exposure very probably leads to the formation of randomly distributed pores and a prompt extrusion of water and solutes (preferentially K^+), especially from the cells in the apical region (3, 10, 11,12,20,39). Then, the cytoplasm, because of its abruptly decreasing volume, tends to carry and thus, to tear away the outer plasma membrane adhering to the supporting cell wall, the restricted plasticity of membrane in the more ordered state not allowing it to follow the volume shrinkage smoothly. The drastic removal of the plasmalemma from the primary cell wall (which predominantly exists in the apical zone) (22), certainly accompanies the breaking-up of ionic, protein-cell wall and lipid-cell wall linkages and the freeing of dissociable groups shielded originally from water by membrane constituents. Consequently, new, potent cation-binding sites are created on and in the cell wall, which serve as cation-exchange pathways for K^+. Such an interpretation is in line with the experimental data presented in Figs. 1-5. Since the low-temperature efflux of K^+ proved to be practically indifferent to Ca^{2+}, it is compelling to deduce that there evolve separate pathways for K^+ uptake and release in the (apical region of) roots of thermophilic plants during cold-shock treatment. It is easy to see that the newly-created cation-binding sites, which are prerequisites of anomalous K^+ uptake, operate as passive pathways peculiar to a K^+ (monovalent cation) proton antiport system, the K^+ movement being directed towards and protons leaving the root tissue.

As regards the origins of excess protons within the cell, they are certainly produced by the elevated respiration and the arrestation of the "proton-consuming" processes, including ATP synthesis. In the proposed model, the highly mobile K-ions accomplish their uphill movement at the expense of the electrochemical potential of protons produced inside the root cells, and K^+ and H^+ are selectively exchanged due to the preference of the cation-exchange regime for K^+. This latter must be connected with the peculiar cation-binding properties of the carboxylate chromophores of pectic acid, considered as principal constituent of primary cell wall (9). It is therefore suggested that this passive K^+/H^+ antiport manifests itself in the unidirectional, anomalous K^+ movement. Moreover, the parallel working of this passive K^+/H^+ antiport system with probably partially impaired K^+ uptake leads to the after- or "memory-effect" observed at $25^\circ C$ (Fig. 3). Accordingly, cold-shock on roots of thermophilic plants brings about initial events leading to structural and functional injury at the levels of both plasma membrane and intracellular (e.g. mitochondrial) membranes. This is visualized in a tentative network diagram in Fig. 12, which is an extension of Lyons' proposal (15), and applies to the apical region of roots of thermophilic plants.

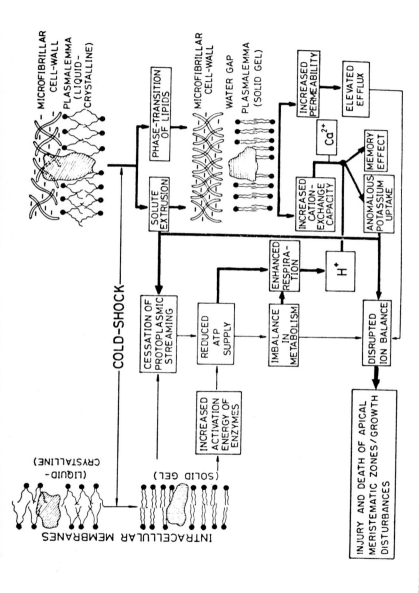

FIGURE 12. Suggested schematic pathway of membrane-linked events involved in cold-shock and chilling injuries of the apical meristematic zones os thermophilic plant roots. The model is an extension of that proposed originally by Lyons (15).

Finally, mention should be made about the possible relation between lipid and fatty acid compositions of membranes and cold-shock tolerance. This is especially important, because the thermotropic properties of membrane lipids and low-temperature responses of plants have appeared to correlate in a large number of studies (14, 26, 28, 29), although the lack of any correlation has also been reported (27). Very recent results obtained with shoots of a cold-sensitive wheat (Penjamo 62), when its frost-resistance was gradually improved by radiation treatment, fail to support a definite parallelism between unsaturation and frost-resistance (24). It seems very likely, however, that the "stoichiometry" of the C_{18} family and protoplasmic proteins jointly determine the cold-tolerance, both being decisive as regards water economy (permeability and retention) (7). Similar conclusions can be drawn from preliminary fatty acid analyses of root segments as well (Toth and Karvaly, unpublished). For these reasons, we venture to assume that the rapid, crystalline-to-solid gel transition of membrane lipids, caused by cold-shock, may result in non-equilibrium membrane structures: pores differing in either population density or size distribution, or both. (Beyond doubt, this is tightly and almost exclusively bound up with the fatty acid composition.) This can explain conclusively not only the results presented in this paper, but even the bases of dissimilarities in the behaviours of cold-resistant and cold-tolerant plants upon cold-shock.

IV. REFERENCES

1. Bowling, D. J. F. "Uptake of Ions by Plant Roots." Chapman and Hall, London (1976).
2. Canning, R. E. and Kramer, P. J. *Amer. J. Bot. 45,* 378-382. (1958).
3. Christiansen, M. N., Carns, H. R. and Slyter, D. J. *Plant Physiol. 46,* 53-56. (1970).
4. Epstein, E.   "Mineral Nutrition of Plants: Principles and Perspectives." Wiley and Sons, New York (1972).
5. Eshel, A. and Waisel, Y. *Plant Physiol. 49,* 585-589 (1972).
6. International Potash Institute. "Potassium in Biochemistry and Physiology." *Proc. 8th Coll. Intern. Potash Inst., Uppsala 1971.* Publ. IPI Berne, Switzerland (1971).
7. Kacperska-Palacz, A., Dlugokecka, E., Breitenwald, J. and Wcislinska, B. *Biol. Plant. (Praha) 19,* 10-17 (1977).
8. Karvaly, B. and Loshchilova, E. *Biochim. Biophys. Acta 514,* 274-285 (1977).
9. Kohn, R. and Sticzay, T. *Collection Czechoslov. Chem. Commun. 42,* 2372-2378 (1977).
10. Levitt, J. "Responses of Plants to Environmental Stresses." Acad. Press, New York (1972).

11. Levitt, J. *In* "Regulation of Cell Membrane Activities in Plants" (E. Marre and D. Cifferi, eds), pp. 103-119. Amsterdam: North-Holland Publishing Company (1977).
12. Liebermann, M., Craft, C. C., Audia, W. V. and Wilcox, M. S. *Plant Physiol. 23,* 307-311 (1958).
13. Loshchilova, E. and Karvaly, B. *Biochim. Biophys. Acta 470,* 492-493 (1978).
14. Lyons, J. M., Wheaton, T. A. and Pratt, H. K. *Plant Physiol. 39,* 262-268 (1964).
15. Lyons, J. M. *Ann. Rev. Plant Physiol. 24,* 445-466 (1973).
16. Marschner, H. and Richter, Ch. *Z. Pflanzenernahr. Bodenkd. 135,* 1-5 (1973).
17. Mengel, K. *Z. Pflanzenernahr. Dung. Bodenkd. 103,* 193-206 (1964).
18. Mengel, K. and Herwig, K. *Z. Pflanzenphysiol. 60,* 147-155 (1969).
19. Mengel, K. and Pfluger, R. *Plant Physiol. 49,* 16-19 (1972).
20. Minchin, A. and Simon, E. W. *J. Exp. Bot. 24,* 1231-1235 (1973).
21. Nielsen, K. F. *In* "The Plant Root and Its Environment" (E. V. Carson, ed.), pp. 293-333. Univ. Press of Virginia (1974).
22. Nobel, P. S. "Introduction to Biophysical Plant Physiology." Freeman and Company, San Francisco (1974).
23. Smolenska, G. and Kuiper, P. J. *Physiol. Plant. 41,* 29-35 (1977).
24. Toth, E. T., Vigh, L., Karvaly, B. and Farkas, T. *Physiol. Plant.* (In press) (1979).
25. Viets, F. G. *Plant Physiol. 19,* 446-486 (1944).
26. Willemot, C. *Plant Physiol. 60,* 1-4 (1977).
27. Wilson, J. M. and Crawford, R. M. M. *New Phytol. 73,* 805-820 (1974).
28. Wilson, J. M. and Rinne, R. W. *Plant Physiol. 57,* 270-273 (1976).
29. Yoshida, S. and Sakai, A. *Plant Cell Physiol. 14,* 353-359 (1973).
30. Zsoldos, F. *Z. Pflanzenernahr. Bodenkd. 119,* 169-173 (1968a).
31. Zsoldos, F. *Z. Pflanzenphysiol. 60,* 1-4 (1968b).
32. Zsoldos, F. *Acta Agr. Acad. Sci. Hung. 18,* 121-126 (1969).
33. Zsoldos, F. *Z. Pflanzenernahr. Bodenkd. 126,* 210-217 (1970).
34. Zsoldos, F. *Plant and Soil 37,* 469-478 (1972a).
35. Zsoldos, F. *Acta Biol. Szeged. 18,* 121-129 (1972b).
36. Zsoldos, F. *Proc. 10th Congr. Intern. Potash Inst., Budapest 1974,* pp. 197-204. Publ. IPI Berne, Switzerland (1975).
37. Zsoldos, F. and Karvaly, B. *Experientia 31,* 75-76 (1975).
38. Zsoldos, F. and Karvaly, B. *Physiol. Plant. 43,* 326-330 (1978a).
39. Zsoldos, F. and Karvaly, B. *Physiol. Plant. 43,* 331-336 (1978b).
40. Zsoldos, F. and Gulyas, S. *Acta Biol. Szeged.* (In press) (1979).

ION LEAKAGE IN CHILLED PLANT TISSUES

Takao Murata and Yasuo Tatsumi

Faculty of Agriculture
Shizuoka University
Oya, Shizuoka 422, Japan

There are a number of symptoms of chilling injury of fruit and vegetables which may be attributed to the denaturation of bio-membranes. Browning, pitting, scald, watery breakdown, shriveling may be classified in this category of symptoms. There may be changes in membrane permeability of phenol substances before or during the occurrence of browning. Changes in permeability of water must take place in the course of shriveling of chilling injured tissues. Since Lieberman et al (3) indicated a continuous increase in membrane permeability of potassium ion of the tissues of sweet potato root stored at chilling temperature several investigations concerning membrane permeability and chilling injury have been published (1, 2, 4, 6, 7, 9, 10, 11, 12). However, the relationship between the changes in membrane permeability and the occurrence of chilling injury of fruit and vegetables still remains to be clarified.

In this section, the effect of temperatures on ion leakage from tissue slices of chilling sensitive and insensitive plants will be reported in the form of Arrhenius plots of the rate of potassium ion leakage and changes in ion leakage during storage at chilling temperatures.

I. MATERIALS AND METHODS

Most plant materials used in this study were purchased in local markets. Cucurbits, bell pepper and egg-plant fruits were freshly picked at the table ripe stage of maturity from plants grown in soil in a greenhouse or on a farm of the Faculty of Agriculture, Shizuoka University.

To determine the effect of detergents on the occurrence of chilling injury, cucumber fruit was stored at $1^\circ C$ after dipping into aqueous solution of 0.1% Triton X-100 (HLB 13.5) for 3 hours.

Discs of 4mm diameter punched from the tissues of fruit and vegetables were cut into slices of 5 mm thickness unless otherwise mentioned. In the cases of bell pepper and tomato fruits, thickness of the discs was 2-3mm and 3-5mm, depending on the thickness of the outer wall of the fruits, respectively. Samples of discs were immersed in deionized water or in 0.4 M mannitol solution at different temperatures in the ranges of 0-30°C. After incubation for 2 hours, the content of leaked ions of potassium, sodium and magnesium in the incubation medium was measured using a Hitachi 207 type atomic absorption spectrophotometer.

Electrolyte content of the incubation medium was measured with a TOA conductivity meter at a constant temperature of 20°C during the course of incubation. After incubation for 4 hours, the medium containing the discs was boiled for 15 minutes to leach all of the ions from the discs of tissues into the medium. The rate of ion leakage was the amount of leaked ions expressed as a percentage of the total amount of ions of the tissues.

II.   RESULTS AND DISCUSSION

A.   Arrhenius Plots of Rate of Potassium Ion Leakage

1.   Arrhenius Plots of Ion Leakage from Cucurbitaceae Fruit Tissues. Fruit of cucurbits are usually chilling sensitive and they have to be stored at 7-10°C. Figure 1 shows the Arrhenius plots of potassium ion leakage from the discs of tissues of cucumber (*Cucumis sativus* L.), oriental pickling melon (*Cucumis melo* L. common MAKI - NO), pumpkin (*Cucurbita moschata* DUCH), summer squash (*Cucurbita pepo* L.) and chayote (choko; *Sechium edule* SWARTZ) for the temperature range of 0° to 30°C. In all cases, there were break points at 5-10°C that corresponded closely with the critical temperatures for chilling injury of these fruits during storage. Higher rates of potassium ion leakage were observed at the lower temperatures below the critical points.

Figure 2 shows the Arrhenius plots of rate of potassium ion leakage from the discs of cucumber fruit stored at 5°C and 10°C for 1 to 9 days. Break points were observed in Arrhenius plots from the tissues of fruit stored at 10°C that was safe from chilling injury. Break points became obscure during storage and the points became hard to distinguish after storage for 9 days at 5°C. However, clear break points in the Arrhenius plots were obtained from the fruit which had been stored at temperatures above the critical point (10°C).

Arrhenius plots of rate of potassium ion leakage from the tissues of bell pepper fruit and sweet potato root (chilling sensitive) also exhibited a similar tendency which involved typical break points.

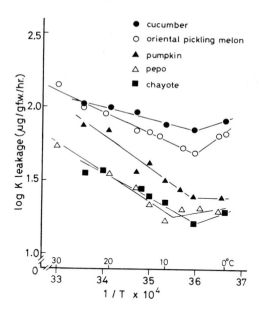

FIGURE 1. *Arrhenius plots of rate of potassium ion leakage from the discs of Cucurbitaceae fruits (* ●————● *; cucumber, O————O ; oriental pickling melon,* ▲————▲ *; pumpkin, △————△; summer squash (pepo),* ■————■ *; chayote).*

2. *Arrhenius Plots of Ion Leakage from Chilling Insensitive Plant Tissues.* Figure 3 shows the Arrhenius plots of rate of potassium ion leakage from the discs of potato tuber tissues *(Solanum tubersum L.)* which is chilling insensitive. A break point was not observed in the range between 0° to 25°C, so that the Arrhenius plots showed a linear line. However, in the ases of onion bulb *(Allium cepa L.)* and carrot root *(Daucus carrota L. sativa DC)* which are also chilling insensitive, the shapes of the lines in the Arrhenius plots were different from those observed with potato tubers. For onion bulb and carrot root, breakpoints were observed which seemed not to correspond with the temperatures of any kind of physiological characters of these plant tissues.

3. *Effect of Detergent on Arrhenius Plots.* If denaturation of bio-membranes at low temperatures is the primary cause of chilling injury, it is possible that treatment with some kinds of detergents may enhance the occurrence of chilling injury of fruit and vegetables at low temperatures. Arrhenius plots of the rate of leakage of potassium ion from the discs of cucumber fruit which were pretreated with 0.1%

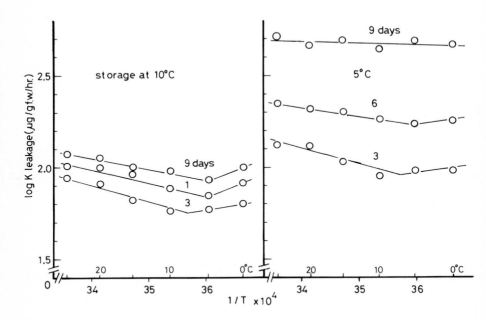

FIGURE 2. Arrhenius plots of rate of potassium ion leakage from the discs of cucumber fruit (cv. Natsuakihushinari) stored at 5°C or 10°C for 1-9 days.

Triton X-100 are shown in Figure 4. Occurrence of chilling injury of cucumber fruit during storage at low temperature was increased slightly by the treatment with the detergent, but there was no significant effect of detergent on the shapes of Arrhenius plots shown in Figure 4; breakpoints remained in the Arrhenius plots of rate of leakage of potassium ion from the discs of treated tissues during storage at 1°C.

It is reasonable to consider that the phase transition (and/or phase separation) in the membranes of tissues of chilling sensitive plants occurs at a critical temperature. However, further clarification is required because break points are also found to occur in chilling insensitive species.

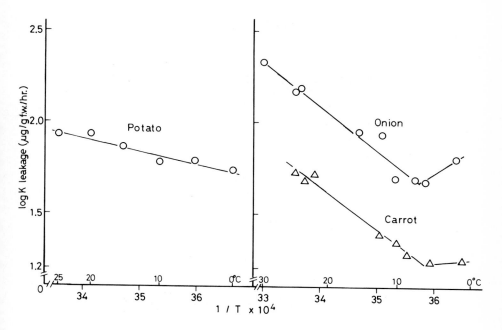

FIGURE 3. *Arrhenius plots of rate of potassium ion leakage from chilling insensitive plant tissues (O———O ; potato tuber, O———O ; onion bulb, Δ———Δ ; carrot root).*

B. Changes in the Rate of Ion Leakage

1. Changes in Ion Leakage from Chilling Sensitive Plant Tissues during Storage. The rate of ion leakage into deionized water by discs of cucumber fruit did not change during storage at $10°C$ and $20°C$ (i.e. above the critical temperature). However, the rate of ion leakage from the discs of fruit stored at $0°C$ and $5°C$ started to increase abruptly on the seventh day of storage and reached approximately twice the rate of those after 12 days storage at $10°C$ or $20°C$ (Fig. 5 left). Rate of ion leakage into aqueous mannitol from the discs of fruit stored at $5°C$ or $20°C$ showed basically the same tendency as that into deionized water (Fig. 5 right). Also, a similar increased rate of ion leakage was observed in chilled tissues of snap bean pod which is a chilling sensitive plant.

The rate of leakage of potassium, sodium and magnesium ions into water from discs of sweet potato root stored at $20°C$ showed a relatively constant level throughout the storage period. However, the rate at $5°C$ started to rise suddenly after storage for 2 weeks (Fig. 6).

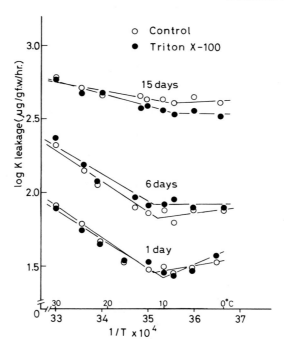

FIGURE 4. Arrhenius plots of rate of potassium ion leakage from discs of cucumber fruit (cv. Horai) stored at 1°C for 15 days with and without treatment with 0.1% Triton X-100.

The rate of leakage of potassium, sodium and magnesium ions from sweet potato root stored at 5°C was about 5 times that from roots stored at 20°C after 6 weeks.

Changes in the rate of ion leakage from the discs of bell pepper (*Capsicum annum* L.) and tomato (*Lycopersicon esculentum* Mill) fruits during storage at 2°C and 12.5° (i.e. below and above the critical temperature) exhibited different curves compared with those of cucumber and sweet potato (Fig. 7 left). A dramatic rise in the rate of ion leakage from the discs of bell pepper fruit was not observed during storage at 2°C or 12.5°C. Changes in the ion leakage from the tissues of egg plant fruit exhibited a similar tendency to that of bell pepper. These results suggest that some chilling sensitive plant tissues show little change in membrane permeability during storage at chilling temperatures. In the case of tomato fruit (mature green-breaker stages), the rate of ion leakage from the discs of fruit stored at 2°C and 12.5°C continued to increase during storage. The same tendency was observed in the changes of the leakage from banana fruit tissues (half green-green tip stages).

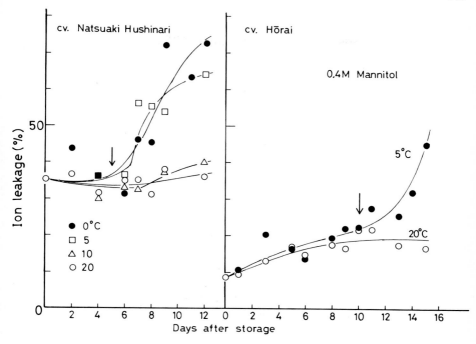

FIGURE 5.   Changes in the rate of ion leakage into deionized water (Left; cv. Natsuakihushinari) and 0.4 M mannitol solution (Right ; cv. Horai) from discs of cucumber fruit stored at 0-20°C. (% of leaked ion to total ion after incubation for 4 hours).

    2.   Changes in Ion Leakage from Chilling Insensitive Plant Tissues during Storage. Changes in the rate of ion leakage from discs of potato tubers stored at 2°C and 12.5°C are shown in Figure 7-right. Rates were very low compared with those of other plant tissues. There was no sudden rise of permeability during storage at 2°C and 12.5°C. No significant difference was observed in the rates of leakage from the discs of potato tuber stored at 2°C and 12.5°C at the end of storage.

    There are large fluctuations of chilling sensitivity of seeds in pod of Leguminosae such as soybean, pea and snap bean. In the case of pod tissues of pea, which is a chilling insensitive plant, no increase of ion leakage was observed during storage at 2°C. However, a continuous increase of ion leakage was observed in the tissues of pods of young

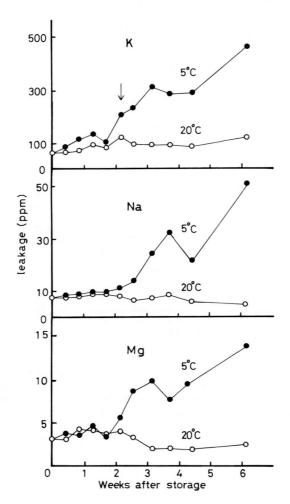

FIGURE 6. Changes in the rate of leakage of potassium, sodium and magnesium ions from discs of sweet potato root stored at 5°C or 20°C.

soybean (chilling insensitive) and of snap bean (chilling sensitive) during storage at 2°C.

Increased rates of ion leakage of chilled tissues have been found in the tissues of fruit and vegetables (2, 3, 10) and leaves (1, 4, 7, 9, 11, 12). Regarding leaf tissues, it is known that relative humidity during growth of plants affected the rate of ion leakage from the tissues (11, 12). Wright (11) has indicated that chilling alone, or water deficit

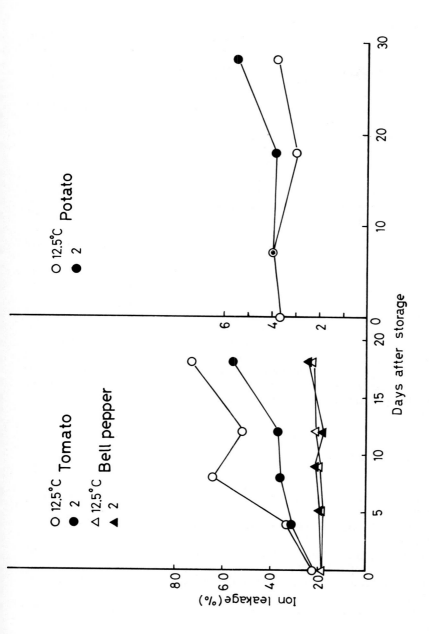

FIGURE 7. Changes in the rate of ion leakage from discs of bell pepper and tomato fruits (Left) and potato tuber (Right) stored at $2^{0}C$ and $12.5^{0}C$.

alone, did not lead to increased leakage of leaves of seedlings of *Phaseolus vulgaris* L. Simon (8) has discussed the effect of chilling treatment at different humidities on membrane permeability of leaf tissues. Morris and Platenius (5) have shown that the severity of chilling injury of cucumber fruit was controlled by raising the relative humidity during storage. It is well known that low humidity enhances the occurrence of visual symptoms of chilling injury of citrus fruit and bell pepper fruit. However, little information is available concerning the effect of water deficits on the changes in electrolyte leakage of chilled tissues of fruit and vegetables.

Our results obtained for the changes in electrolyte leakage of tissues of cucumber fruit, oriental pickling melon fruit, snap bean pod and sweet potato root coincide with the results of previous reports. However, no increased rate of ion leakage is observed in chilled tissues of bell pepper and egg plant fruits which are chilling sensitive plants. In the cases of tomato and banana fruits, a high rate of ion leakage from the tissues is observed even at temperatures above the critical point during storage, probably due to their ripening.

Nevertheless, all of the published investigations on electrolyte leakage have suggested increased rates of leakage from chilling sensitive plant tissues occur at low temperatures. Our data suggest the existence of chilling sensitive plant tissues, such as bell pepper and egg plant fruits, which exhibit no increase of electrolyte leakage at temperatures below the critical point. Therefore, it is reasonable to infer that high rates of electrolyte loss at chilling temperatures is not necessarily a general property of chilling sensitive plant tissues. It might be considered that different types of mechanisms regarding electrolyte loss from tissues may exist in the chilling injury of fruit and vegetables.

III. REFERENCES

1. Creencia, R. P., and Bramlage, W. J. *Plant Physiol. 47*, 389-392 (1971).
2. Lewis, T. L., and Workman, M. *Aust. J. Biol. Sci. 17*, 147-152 (1964).
3. Lieberman, M., Craft, C. C., Audia, W. V. and Wilcox, M. S. *Plant Physiol. 33*, 307-311 (1958).
4. Minchin, A., and Simon, E. W. *Jour. Expt. Bot. 24*, 1231-1235 (1973).
5. Morris, L. L. and Platenius, H. *Proc. Amer. Soc. Hort. Sci. 36*, 609--613 (1938).
6. Nobel, P. S. *Planta 115*, 369-372 (1974).
7. Patterson, B. D., Murata, T., and Graham, D. *Aust. J. Plant Physiol. 3*, 435-442 (1976).
8. Simon, E. W. *New Phytol. 73*, 377-420 (1974).
9. Tanczos, O. G. *Physiol. Planta. 41*, 289-292 (1977).

10. Tatsumi, Y. and Murata, T. *J. Japan. Soc. Hort. Sci. 47,* 105–110 (1978).
11. Wright, M. *Planta. 120,* 63–69 (1974).
12. Wright, M., and Simon, E. W. *Jour. Expt. Bot. 24,* 400–411 (1973).

EFFECTS OF CHILLING ON MEMBRANE POTENTIALS OF MAIZE AND OAT LEAF CELLS[1]

Paul H. Jennings

Department of Plant and Soil Sciences

and

Terry A. Tattar

Department of Plant Pathology
University of Massachusetts
Amherst, Massachusetts

I. INTRODUCTION

The introduction to this symposium reviews a large body of information encompassing the various responses of plants, organs, tissues and biological mechanisms affected by low-temperature chilling injury. We will not repeat or attempt to elaborate on that excellent overview but will point out an approach which we consider to have potential in providing an additional technique in studying chilling injury in plants.

Various approaches have been used in studying low-temperature effects on plants, ranging from observations of visible injury, such as wilting, chlorosis and necrosis, to techniques which measure electrolyte leakage from various plant parts in response to chilling, and currently to sophisticated studies employing spin-labeling of the membrane (9, 13, 19). An approach which has not received as much attention in low-temperature studies is effects on membrane potential.

Maize was chosen for study partly as a result of other current low-temperature studies being conducted with it but also because it is generally recognized as an important chilling-sensitive crop plant. Oats, as a non-chilling sensitive grass, was chosen for comparison in these

[1]*Paper No. 2287, Massachusetts Agricultural Experiment Station, University of Massachusetts at Amherst. This research supported from Experiment Station Projects No. Hatch 364 and Hatch 426.*

studies. Maize exhibits visible low-temperature injury in the form of chlorotic and necrotic lesions which form bands across the leaf (15). This banding pattern, sometimes referred to as Faris bands (3), is apparently a response of leaf cells at a critical stage in development to low-temperature stress (16). The result is failure of various cells and cellular components to develop resulting in the very characteristic chlorotic band which eventually becomes necrotic. Other dysfunctions in maize have been observed including increased electrolyte leakage from stressed leaf tissue as well as various metabolic changes (1). The duration of stress in these latter examples determined the extent and reversibility of the response. With these results and others with different plant material it seemed that studies of the effects of chilling injury on membrane potential might provide additional useful information in attempts to elucidate mechanisms of low-temperature injury.

An electrical potential (E) exists between the inside of the plant cell and the outside medium. This potential is maintained primarily across the plasma membrane and is usually termed a membrane potential (Em) (6). Considerable evidence supports the theory that much of the Em in plants is maintained at the expense of metabolic energy supplied by respiration and/or photosynthesis (6, 12). It has also been demonstrated that the electrical energy gradient in the Em is used to perform transport work across the plasma membrane (12, 17). More recently changes in Em have been related to uptake of amino acids and carbohydrates (2, 11).

Little information is available on the effects of chilling on Em. Some investigators, however, have used low temperature (less than 10°C) incubation to inhibit metabolism and to demonstrate its depolarizing effect on Em, much the same way as potassium cyanide and sodium azide have been used to depolarize Em chemically (4, 5, 8). We report on the effects of chilling on the Em of maize and oats as well as the interactive effects of various metabolic inhibitors and naturally occurring organic compounds.

II. MATERIALS AND METHODS

A. Tissue Preparation

The plant material used in these studies consisted of commercial varieties of maize *(Zea mays* cv Seneca Chief) obtained from the Joseph Harris Seed Company and oats *(Avena sativa* cv Garry) obtained from Agway (a farmers' cooperative). Seed were soaked overnight in water and germinated in vermiculite in a growth chamber under the following conditions: 15/9 hr day/night light cycle; light intensity of 110 Wm^{-2} provided by cool white fluorescent and incandescent bulbs; $24^{\circ}/24^{\circ}$C day/night temperature cycle. Plants were watered with a complete Hoagland solution (7) modified to contain 3X iron. In experiments set up to determine low growth temperature effects on poststress membrane potential a growth temperature cycle of $24^{\circ}/8^{\circ}$C was used. Plant tissue for membrane potential measurements was obtained using the first leaf when the plants were between 12 and 16 days old. Tissue was prepared by removing the leaf from the plant with a razor blade and in the case of maize cutting the leaf in half longitudinally along the mid-vein. The lower epidermis was then removed using a pair of electron microscopy forceps. Sections of stripped leaf material (approx 2 x 8 mm) were then floated stripped surface down in petri dishes containing 1 mM potassium phosphate buffer pH 6.5 with 1 mM calcium chloride. This same buffer was used to perfuse the tissue in the perfusion chamber described below. The leaf sections were incubated 20-24 hr under the growth chamber conditions described above.

B.   Measurement of Em

Tissue sections were perfused with buffer in a Lucite chamber that was attached to a horizontally mounted microscope (2). Glass capillary microelectrodes, tip diam less than 1 μm, were prepared from fiber-filled glass capillaries with an electrode puller (Industrial Science). Microelectrodes were filled with 3 M potassium chloride and had a tip resistance of 10-20 Meg ohms. The Em was measured between an electrode filled with 3 M potassium chloride in 2% agar, in the bathing solution, and the microelectrode in the cell. Both electrodes were connected to an amplifier (WPI Instruments, Model 725) and the Em recorded on a strip chart recorder.

C.   Effects of Light, Inhibitors, Glycine and Sucrose

All initial experiments were conducted with buffer at $21^{\circ}C$ and stable Em measurements were achieved for 5 minutes in the dark before any treatments were imposed. In light experiments tissue was illuminated with a light intensity of 670 Wm^{-2} for 5 minutes and then the light was turned off for 5 minutes. In inhibitor experiments, the standard 1 mM potassium phosphate buffer was made 1 mM with respect to either potassium cyanide or sodium azide, and the tissue exposed to inhibitor by switching from perfusion with buffer alone to buffer plus inhibitor without interruption of liquid flow. In some experiments with potassium cyanide the tissue was exposed to light after maximum depolarization had occurred. Tissue was perfused with sucrose (50 mM) or glycine (50 mM) in buffer in a similar manner for 5 minutes and was followed by 5 minutes of buffer alone.

D.   Effects of Cold Buffer

Buffer at $0\text{-}2^{\circ}C$ was passed through the perfusion chamber containing the maize or oat leaf sections in order to measure direct effects of chilling temperature on Em. The temperature in the chamber was continuously monitored with a thermoprobe (YSI Tele-thermometer). Effects of light vs dark on Em and effects of 50 mM glycine or 50 mM sucrose were determined in cold buffer (8°C) as previously described for 21°C buffer.

III. RESULTS

A. Effects of Light

No differences were detected in the Em of control or chilled maize or oats, grown at either $24^{o}/24^{o}$C or $24^{o}/8^{o}$C, when Em measurements were made with 21°C buffer (Table I). There were also no differences in response of Em to light with either control or chilled leaf sections of maize with 21°C buffer (data not shown). Therefore, all subsequent experiments were conducted with plant material grown at $24^{o}/24^{o}$C, except where noted. Incubation with 8°C buffer in the dark caused a 30% depolarization of Em in both maize and oat leaf cells, compared with 21°C buffer (Fig. 1). The Em of oat leaf cells at 8°C in the dark, however, would often slowly hyperpolarize to levels similar to Em at 21°C in the dark (data not shown). When oat and maize tissues were illuminated in 8°C buffer, oat leaf cells again hyperpolarized to an Em comparable to cells of that species in 21°C buffer, but the Em of maize leaf cells exposed to light at 8°C depolarized and did not recover even to their initial Em in the dark.

TABLE I. *Effects of Inhibitors and Growth Temperatures on Em of Corn and Oat Leaf Cells*[a]

	Em in mV			
	Maize		Oats	
Treatment	$24^{o}/24^{o}$	$24^{o}/8^{o}$	$24^{o}/24^{o}$	$24^{o}/8^{o}$
Control	120 ± 2	125 ± 7	147 ± 4	142 ± 6
CN	68 ± 13	58 ± 13	-	-
N_3	57 ± 9	63 ± 6	69 ± 6	73 ± 6

[a]*Em measurements were made with buffer at 21°C in the dark. Plants were grown with a 15/9 hr day/night light cycle.*

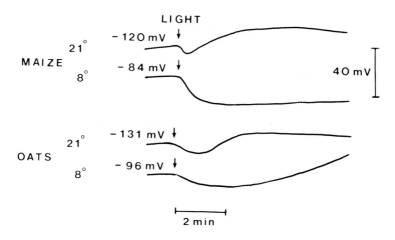

FIGURE 1. Effect of light and temperature on leaf cell Em of maize and oats.

B.   Effects of Metabolic Inhibitors

Addition of 1 mM potassium cyanide or 1 mM sodium azide at 21°C caused an immediate collapse of Em to approximately 50% of the initial levels in both maize and oats (Table 1). A similar Em response followed addition of sodium azide in oats. In KCN experiments where maize was illuminated after maximum Em depolarization had been achieved, the Em of both chilled and control sections recovered to approximately the initial dark levels in less than 10 minutes (data not shown). No differences in Em were observed between the control and the chilled maize leaves in response to either of these inhibitors when Em was determined using 21°C buffer. Similarly, no difference in Em response was detected between control and chilled oats. The Em of maize leaf cells depolarized to a slightly greater extent when the tissue was perfused with 8°C buffer containing 1 mM potassium cyanide in the dark (data not shown).

C.   Effects of Glycine

After the addition of 50 mM glycine in buffer at 21°C the Em of both species immediately depolarized but began to recover after 2-3 minutes and recovered more quickly after the glycine was removed (Fig. 2). No major differences were detected between the Em response of maize and of oats to the addition of 50 mM glycine at 21°C. However, when maize

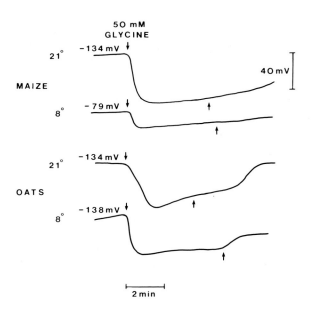

FIGURE 2. Effect of 50 mM glycine on maize and oat leaf cell Em at 21° and $8^{o}C$.

leaf sections were perfused with 50 mM glycine in 8°C buffer, only a minor depolarization occurred followed by a slight recovery, while in oat leaf cells depolarization and recovery were similar to results with 21°C buffer. The Em of oat leaf cells was often found to be unstable in the dark and, consequently, many 50 mM glycine experiments were also conducted totally in light. The results of these experiments, however, were comparable with those in the dark although the Em tended to be higher. Similar experiments in the light were also conducted with maize at 8°C, but the results were comparable to those in the dark.

D.   Effects of Sucrose

Addition of 50 mM sucrose in buffer at 21°C caused an immediate depolarization of the Em of both species, followed by a slow recovery after 2–3 minutes (Fig. 3). A rapid recovery occurred when the sucrose was removed, and ultimately resulted in a hyperpolarization. A similar pattern was observed in oats at 8°C but the amount of initial Em depolarization was reduced approximately 50% compared to the 21°C treatment. If experiments at 8°C were conducted in light, however, initial Em depolarization in oats was comparable to the dark experiments

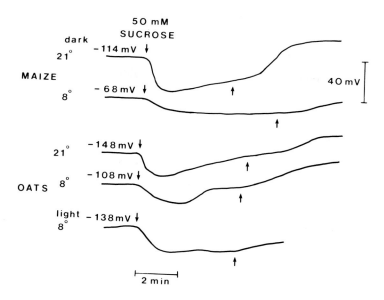

FIGURE 3. Effect of 50 mM sucrose on maize and oat leaf cell Em at $21^{\circ}C$ and $8^{\circ}C$.

at 21°C. Light had no apparent effect on results with 50 mM sucrose in maize at either 8°C or 21°C. In this species at 8°C, the initial Em depolarization with sucrose was substantially reduced and very little recovery could be detected even after removal of sucrose.

IV. DISCUSSION

Since one component of Em appears to be at least partially dependent on a metabolic energy supply, it is not surprising that cells of tissue exposed to low temperatures would depolarize as a result of decreased respiration and consequently oxidative phosphorylation. Our results have shown that this temperature effect on Em depolarization is readily and rapidly reversible with both maize and oats. At 21°C there was consistently no difference in Em response between tissues grown at $24^{\circ}/24^{\circ}$C and $24^{\circ}/8^{\circ}$ C of both species. Thus, when Em was measured during low-temperature treatment of leaf tissue, a depolarization was observed in both oats and maize in the dark. However, when the tissue was exposed to light, oat leaf cells were able to re-establish an Em at 8°C comparable to similar tissue at 21°C in the light whereas maize was apparently unable to utilize light energy to recover its Em at 8°C. This chilling effect with maize is not surprising since Taylor *et al.* (18)

found that several key enzyme systems in maize involved with photoconversion of light to chemical energy were severely inhibited during chilling. In addition, Novacky and Karr (10) found that suppression of Em responses to light was a common pathological alteration in some plant diseases.

Amino acid uptake was correlated with degree of Em depolarization in oat coleoptiles (2) and in oat leaves (14), while uptake of several sugars, including sucrose, was correlated with Em depolarization in *Lemna gibba* (11). The change in Em in both oats and maize following addition of glycine or of sucrose was, therefore, most likely caused by uptake of these materials across the plasma membrane. The major difference between the effects of glycine and sucrose on Em of maize at $21^{\circ}C$ and at $8^{\circ}C$ probably indicates a strong inhibition of uptake of these materials at $8^{\circ}C$. The minor difference between Em response at $21^{\circ}C$ and $8^{\circ}C$ in oats suggests that uptake of glycine and sucrose were only slightly inhibited at $8^{\circ}C$.

Novacky *et al.* (11) found a strong correlation between Em and intracellular ATP concentration in *L. gibba* cells. These researchers also found that hyperpolarization of Em, following sucrose addition, was related to increased ATP production from metabolism of this sugar. Leaf cells of oats at $8^{\circ}C$ were able to recover Em in sucrose and to hyperpolarize Em following sucrose removal, while those of maize could not, even though cells of maize were able to recover Em in sucrose and to hyperpolarize Em after sucrose removal in similar experiments at $21^{\circ}C$. Chilling sensitivity in maize, therefore, may be due to both suppressed uptake and the inability to rapidly metabolize compounds that are transported into the cell. Using Em we have been able to show a difference in response between representatives of chilling and non-chilling sensitive plant species. The initial response in the respiratory energy generating system to low temperature was similar in the two grass species used in this study. However, the energy generating system associated with photosynthesis is apparently very sensitive to chilling in maize and virtually unaffected in oats, at least in terms of maintenance of Em.

We anticipate that further studies of the photosynthetic system could be profitably pursued using various metabolic probes in conjunction with Em measurements to determine the site(s) affected by low temperature exposure. In addition, genetic differences might also be employed to help elucidate chilling injury mechanisms, again using Em as a monitor of plant response. The use of Em might also provide a new and useful approach to screening of genetic material for chilling resistance in a plant breeding program. To further elaborate on the mechanism of chilling sensitivity, Em could be determined over a range of temperatures which spans the chilling sensitive range with comparisons made between chilling and non-chilling sensitive plant material. Other approaches will undoubtedly develop to use Em in chilling injury studies.

V. REFERENCES

1. Creencia, R. P. and Bramlage, W. J. *Plant Physiol. 47*, 389-392 (1971).
2. Etherton, B. and Rubinstein, B. *Plant Physiol. 61*, 933-937 (1978).
3. Faris, J. A. *Phytopath. 16*, 885-891 (1927).
4. Gradmann, D. *Planta 93*, 323-353 (1970).
5. Gradmann, D. and Bentrup, F. W. *Naturwisenshaften 57*, 46-47 (1970).
6. Higinbotham, N. *Plant Physiol. 24*, 25-46 (1973).
7. Hoagland, D. R. and Arnon, D. I. *Calif. Agric. Expt. Sta. Cir. 347*, pp. 1-32 (1950).
8. Ichino, K., Katou, K., and Okamoto, H. *Plant and Cell Physiol. 14*, 127-137 (1973).
9. Lyons, J. M. *Plant Physiol. 24*, 445-466 (1973).
10. Novacky, A., and Karr, A. L. *In* "Regulation of Cell Membrane Activities in Plants." (E. Marre and O. Ciferri, eds.), pp. 137-144, Elsevier/North Holland, Amsterdam (1977).
11. Novacky, A., Ullrich-Eberius, C. I., and Luttge, U. *Planta 138* 263- 270 (1978).
12. Poole, R. J. *Plant Physiol. 29*, 437-460 (1978).
13. Raison, J. K., Lyons, J. M., Mehlhorn, R. J., and Keith, A. D. *J. Biol. Chem. 246* 4036-4040 (1971).
14. Rubinstein, B. and Tattar, T. A. *Plant Physiol. 61* (Suppl.) 107 (1978).
15. Sellschop, J. P. F. and Salmon, S. C. *J. Agric. Res. 37*, 315-338 (1928).
16. Slack, C. R., Roughan, P. G., and Bassett, H. C. M. *Planta 118*, 57-77 (1974).
17. Slayman, C. L. *In* "Membrane Transport in Plants." (U. Zimmermann and J. Dainty, eds.), pp. 107-119, Springer-Verlag (1974).
18. Taylor, A. O., Slack, C. R., and McPherson, H. G. *Plant Physiol. 54*, 696-701 (1974).
19. Wade, N. L, Breidenbach, R. W., Lyons, J. M., and Keith, A. D. *Plant Physiol. 54*, 320-323 (1974).

TEMPERATURE SENSITIVITY OF ION-STIMULATED ATPases ASSOCIATED WITH SOME PLANT MEMBRANES

Edward J. McMurchie

Plant Physiology Unit
CSIRO Division of Food Research
and School of Biological Sciences
Macquarie University
North Ryde, N.S.W., Australia

I. INTRODUCTION

The hypothesis that changes in the molecular ordering of membrane lipids are the initial events in chilling injury (7, 13) has come primarily from results obtained using isolated plant mitochondria (8, 13). These studies have shown that membrane-associated respiratory enzymes of mitochondria isolated from chilling-sensitive plants undergo a change in apparent activation energy (E_a) at some critical temperature. This change in E_a is not observed in mitochondria isolated from chilling-resistant plants (8). The change in E_a for membrane-associated enzymes correlates with a change in membrane lipid order or fluidity as determined by spin label (13, 14) or fluorescent probe techniques (12). However, these techniques have not defined the precise nature of the molecular change in the membrane lipids which is responsible for the change in enzyme E_a. Although a change in molecular ordering of membrane lipid is observed at the critical temperature for chilling, its relationship to such processes as lipid phase transitions and/or lipid phase separations, which have been implicated as initial events in the phenomenon of chilling injury (7, 13) is not clear.

In addition to differences in the thermal behaviour of mitochondrial membranes from chilling-sensitive and chilling-resistant plants, it is possible that other cellular membranes undergo changes of a similar nature. In this regard changes in E_a have been reported for some membrane-associated photosynthetic activities of chloroplasts isolated from chilling-sensitive plants (17). Differences in the thermal behaviour of plasma membranes from chilling-sensitive and chilling-resistant plants have not been reported, although changes in the ion permeability of some plant tissues subject to chilling temperatures, have been observed (2, 19). Such changes in ion permeability may reflect changes in the temperature sensitivity of ion-stimulated ATPases associated with the plasma membrane, as these ATPases have been implicated in the ion transport process (3, 4, 18). Thus temperature-induced changes in the plasma membrane and associated enzymes may also play a role in chilling sensitivity.

In this study the temperature sensitivity of ion-stimulated ATPase was examined using a post-mitochondrial membrane preparation isolated from tissues of chilling-sensitive tomato and cucumber and from chilling-resistant cauliflower. Changes were observed in the E_a of ion-stimulated ATPase which may indicate that the plasma membrane undergoes changes which are similar to those observed for mitochondrial membranes. As a means of correlating these changes in enzyme E_a with the possible presence of phase transitions in the membrane, membranes and membrane lipid extracts were examined by differential scanning calorimetry.

II. MATERIALS AND METHODS

A. Membrane Isolation

Green tomato fruit (*Lycopersicon esculentum*,var.   Red   China), cauliflower florets (*Brassica oleraceae*) and cucumber fruit (*Cucumis sativus*) were obtained from commercial sources. For tomato and cucumber, only the pericarp tissue was used. Tissues were homogenized in a commercial juice extractor using an ice-cold medium of 300 mM sucrose, 0.5% (w/v) polyvinylpyrolidone, 1 mg/ml bovine serum albumin, 3 mM EGTA, 25 mM Tris, 2 mM Hepes, 4 mM dithiothreitol, all adjusted to pH 7.4 with HCl. A solution:tissue ratio of 2:1 (v/w) was used. The brei was maintained at pH 7.2 with NaOH during the extraction. The brei was then strained through 4 layers of cheesecloth and centrifuged at 13,000 g for 15 min. The supernatant was then poured through 4 layers of cheesecloth and centrifuged at 135,000 g for 30 min. The pellet was washed twice in TM buffer (25 mM Tris·HCl, 1.5 mM $MgSO_4$, pH 7.5) by resuspension and centrifugation at 135,000 g for 30 min.   The final pellet was resuspended in TM buffer at a concentration of 3 to 6 mg protein per ml. This material was used immediately for ATPase assays.

B. Assays

1.   ATPase activity was measured in a 1.0 ml final volume containing 25 mM Tris·HCl, 1.5 mM $MgSO_4$, pH 8.0 (Mg^{2+} ATPase) or the same buffer containing 40 mM KCl adjusted to pH 8.0 (K^+ stimulated Mg^{2+} ATPase). The reaction contained 300 to 600 µg membrane protein and was started by the addition of Tris·ATP at a final

concentration of 3.0 mM. The reaction time for any set of assays was chosen so that no more than 20% of the substrate was converted to inorganic phosphate during the course of the assay. The reaction was terminated by the addition of 50 µl of 70% (v/v) perchloric acid and inorganic phosphate was determined using the procedure of Rosenthal and Matheson (15). The temperature sensitivity of Mg^{2+} ATPase and K^+ stimulated Mg^{2+} ATPase was determined using a shaking thermogradient aluminum block capable of maintaining 14 separate assay temperatures at any accuracy of $\pm 0.2°C$ over the experimental period. Both enzyme activities were determined simultaneously using duplicate assays. Assays contained identical amounts of membrane protein from the same membrane preparation. Final rates of ATPase activity have been corrected for possible effects of temperature on the blank value and the rate of non-enzymic ATP hydrolysis.

2. Membrane protein was determined by the method of Lowry et al., (6).

C. Lipid Extraction

Membrane preparations suspended in TM buffer were extracted overnight with 21 vol. chloroform:methanol (2:1) containing 1 mg butylated hydroxytoluene as antioxidant. The extract was then partitioned against 0.2 vol. of 0.73% (w/v) NaCl and the aqueous layer removed. The lipid extract was dried, taken up in chloroform and stored at $-20°C$.

D. Electron Microscopy

Membrane preparations in TM buffer were centrifuged (135,000 g, 30 min) and resuspended in 50 mM potassium phosphate buffer, pH 7.5 containing 4% (v/v) glutaraldehyde. After 30 min the membranes were centrifuged as above and sectioned specimens were prepared by standard procedures.

E. Differential Scanning Calorimetry

Differential scanning calorimetry (DSC) was performed using a Perkin-Elmer DSC-2 calorimeter. Membrane preparations in TM buffer were centrifuged at 135,000 g for 30 min and the pellet was loaded into a 20 µl capacity gold pan and then sealed. Membrane lipid samples in chloroform were dried and then suspended in 25 mM Tris·HCl, 10 mM EDTA, pH 8.0 by vortexing followed by a 2 min sonication period. The lipid suspension was centrifuged and loaded into the pan as described for the membrane samples. Dry weights of the samples were determined at the completion of the experiment. Scans were performed at a rate of 10°C/min with an instrument operating sensitivity of 0.2 mcal/sec. Transition enthalpies were determined from the areas of the endothermic or exothermic peaks using Indium as a standard.

III.   RESULTS

A.   *Ion-Stimulated ATPase Activity*

The procedure used for membrane isolation yielded membrane vesicles from all three tissues which were similar in appearance to those shown for cauliflower (Figure 1). These membrane preparations contained Mg^{2+} dependent ATPase activity which could be further stimulated by monovalent cations. ATPase activity could be almost completely inhibited by Ag^{+} ions and protected against such inhibition by dithiothreitol. These properties are all consistent with the known properties of plasma membrane-associated ATPases (3). As well as containing plasma membranes, the membrane preparation would be expected to contain membrane fragments derived from the tonoplast and endoplasmic reticulum. As an estimate of the level of mitochondrial ATPase, the total ATPase activity of the membrane preparation was inhibited by no more than 15% in the presence of 50 µg/ml oligomycin (results not shown).

Arrhenius plots of Mg^{2+}ATPase and K^{+}stimulated Mg^{2+}ATPase activity from tomato are shown in Figure 2. Both plots exhibited a discontinuity at about 20°C and they remained non-linear below this temperature. However, no clear assignment of a critical temperature at about 10°C, which would be in agreement with the critical temperature observed for mitochondrial respiratory activity of tomato (8), could be made from either plot.

Arrhenius plots of Mg^{2+}ATPase and K^{+}stimulated Mg^{2+}ATPase activity from cucumber each exhibited a discontinuity at about 10°C (Figure 3). This temperature is in agreement with the temperature at which changes in membrane-associated respiratory activity are observed in cucumber mitochondria (8). The discontinuity in the Arrhenius plot was less pronounced for the K^{+}stimulated ATPase which may indicate that this enzyme is less sensitive to low temperatures when assayed in the presence of K^{+}.

It has previously been shown that for mitochondria isolated from chilling-resistant plants, membrane-associated respiratory activities do not undergo abrupt changes in E_a in the temperature range from 0° to 25°C (8). However ion-stimulated ATPase activity associated with membranes isolated from the chilling-resistant cauliflower did not display linear Arrhenius kinetics, i.e., a constant E_a when assayed over the range 3° to 28°C (Figure 4). The Arrhenius plot of Mg^{2+}ATPase was triphasic with discontinuities occurring at about 8° and 15°C. In

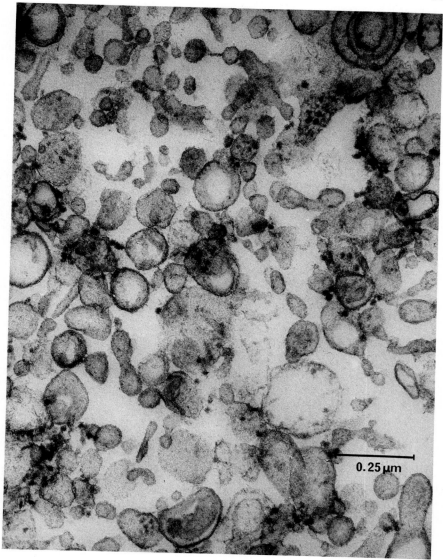

FIGURE 1. Electron micrograph of a section of a pelleted membrane fraction isolated from cauliflower buds. (Electronmicrograph, courtesy of Dr. J. Bain, CSIRO Division of Food Research).

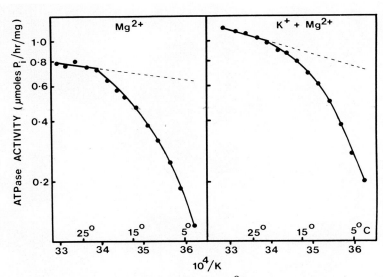

FIGURE 2. Arrhenius plots of Mg^{2+}ATPase and K^+ stimulated Mg^{2+}ATPase activities associated with membranes isolated from green tomato fruit.

contrast, only one discontinuity at about $15^{\circ}C$ was observed for the K^+stimulated Mg^{2+}ATPase activity. At all temperatures, the sensitivity of the Mg^{2+}ATPase to stimulation by K^+ was far greater than that observed for either the tomato or the cucumber ATPases. In addition, the sensitivity of the cauliflower enzyme to temperatures below $8^{\circ}C$ was greatly decreased when the enzyme was assayed in the presence of K^+ (Figure 4) or Rb^+ (results not shown).

The data for cauliflower and tomato ATPases was replotted as the K^+dependent ATPase activity which represents the difference in rate between the K^+stimulated Mg^{2+}ATPase and the Mg^{2+}ATPase. Arrhenius plots of the K^+dependent ATPase activity for tomato and cauliflower membranes are shown in Figure 5. Apart from the 10-fold difference in the specific activity of the respective ATPases, the Arrhenius plots are remarkably similar. Both plots exhibited a discontinuity at a similar temperature and the values of the Arrhenius activation energy when compared either above or below the discontinuity were almost identical.

B. Differential Scanning Calorimetry

The thermal behaviour of membranes and membrane lipids from tomato, cucumber and cauliflower was investigated using differential

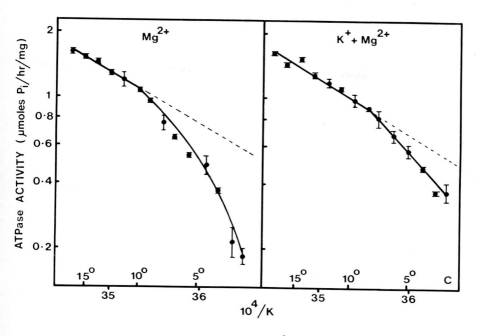

FIGURE 3. Arrhenius plots of Mg^{2+} ATPase and K^+ stimulated Mg^{2+} ATPase activities associated with membranes isolated from cucumber. Error bars represent the limits obtained using duplicate ATPase assays.

scanning calorimetry (DSC). This technique has been used to detect thermotropic lipid phase transitions in a variety of biological membranes (11).

Lipids isolated from tomato membrane preparations exhibited a reversible phase transition as shown in Figure 6. The temperature limits of the transition differed depending on the scanning mode. However, the enthalpies of the transitions obtained using either mode were approximately the same (Table I). No transition was detected in the total membrane lipids of cauliflower over the temperature range 52° to $-5^{\circ}C$, (DSC scan not shown). For membranes isolated from cauliflower and cucumber a small exothermic transition was evident by DSC in both membrane preparations (Figure 7). These transitions were similar both in regard to their temperature limits and their transition enthalpy (Table I). No transition was detected in the membrane preparations from tomato when examined under similar conditions, (DSC scan not shown).

The thermal behaviour of membranes from chilling-sensitive tissues could be altered by heating the membrane sample at $60^{\circ}C$ for 3 min as shown in Figure 8. For cucumber membrane preparations, heating the

E. J. McMurchie

TABLE I.   DSC Analysis of Cauliflower, Cucumber and Tomato
            Membrane Preparations.

Membrane preparation	Transition range (^oC)			ΔH (cal/gm lipid)
Cauliflower	9.0^o to	2.0^o	(exotherm)	0.04
"	9.1^o to	18.8^o	(endotherm)	0.04
" (denatured)		N.D.		--
" (total lipids)		N.D.		--
Cucumber	10.5^o to	0.4^o	(exotherm)	0.04
" (denatured)	21.3^o to	0°	(exotherm)	0.47
Tomato		N.D.		--
" (denatured)	27.9^o to	0°	(exotherm)	0.66
" (total lipids)	12.7^o to	-6.1^o	(exotherm)	0.15
" (total lipids)	3.2^o to	22.4^o	(endotherm)	0.13

N.D. = Transition not detected.

sample resulted in an approximate doubling of the temperature range of
the transition and a ten-fold increase in the enthalpy of the transition
(Table I). Whereas no transition was observed for tomato membranes
before heating, a transition of similar enthalpy to that of the heated
cucumber membranes, was detected after heating (Table I). For the
tomato, the range of the transition in the heated membranes exceeded
that observed for the corresponding membrane lipids and the
temperature limits of the transition were comparatively higher. In
contrast to the membranes from cucumber and tomato, cauliflower
membranes did not exhibit a transition after heating (Figure 8). Indeed
the small transition evident before denaturation (Figure 7), was no longer
detectable.

IV. DISCUSSION

The temperature observed for the change in the E_a for Mg^{2+} ATPase
and K^+ stimulated Mg^{2+} ATPase activities isolated from cucumber are in
close agreement with the critical temperature for chilling injury in this
tissue. Such agreement is not observed for the tomato when the ATPase

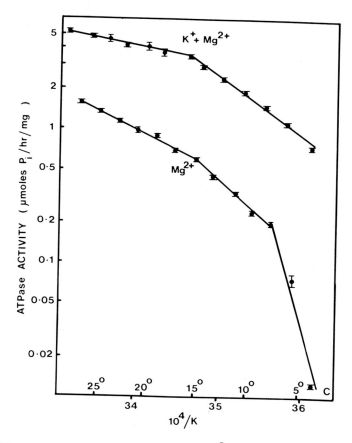

$_{Mg^{2+}}$FIGURE 4. Arrhenius plots of Mg^{2+} ATPase and K^+ stimulated Mg^{2+} ATPase activities associated with membranes isolated from cauliflower buds. Error bars are as described in Figure 3.

activities are presented as Arrhenius plots. However, when the data is presented in the form shown in Figure 5 (i.e., as the difference in the activity of the Mg^2 ATPase due to the presence or absence of K^+), changes in the tomato ATPase are observed at temperatures which approximate the critical temperature for chilling. It is likely that the effect of temperature on the ion stimulation of the ATPase may be an important physiological factor to consider particularly in regard to those ion transport processes such as K^+ uptake, which are probably mediated by membrane-associated ATPases (3).

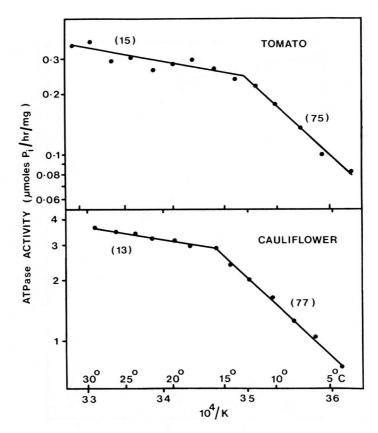

FIGURE 5. Arrhenius plots of the K^+ dependent ATPase activity associated with membranes isolated from tomato fruit and cauliflower buds. The K^+ dependent ATPase activity is the difference between the Mg^{2+} ATPase activities measured in the presence and absence of 40 mM KCl. Numbers in brackets are the Arrhenius activation energy (kJ/mole) for each linear segment of the Arrhenius plot.

Unlike the situation observed for mitochondrial respiratory activity in chilling–sensitive and chilling–resistant plants (13), the changes in the temperature sensitivity of ion–stimulated ATPases are not confined solely to the chilling–sensitive plants, but are observed in the chilling–resistant cauliflower. For the cauliflower Arrhenius plots of Mg^{2+} ATPase and K^+ stimulated Mg^{2+} ATPase are non linear, and the Arrhenius plot of the K^+ dependent ATPase activity (Figure 5) is remarkably similar to that obtained for the tomato. The K^+ dependent ATPase may be reflecting that component of the enzyme activity which

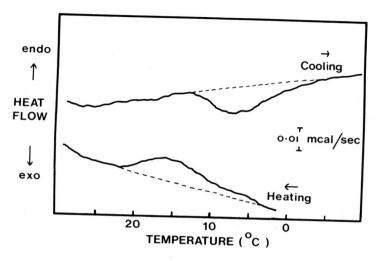

FIGURE 6.   DSC scans of a total lipid extract from tomato membrane preparations.   Endo and exo refer to the endothermic or exothermic peaks obtained from heating or cooling scans respectively.

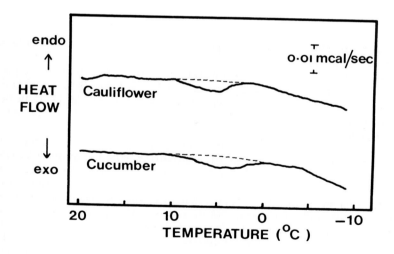

FIGURE 7.   DSC cooling scans obtained for membrane preparation from cauliflower and cucumber.

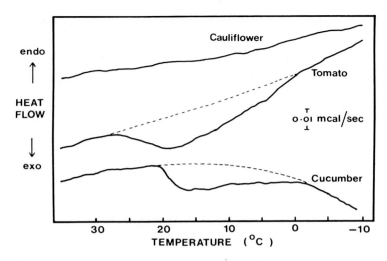

FIGURE 8. DSC cooling scans obtained for heat-denatured membranes from cauliflower, tomato and cucumber. Membranes were heat-denatured at 60°C for 3 min.

is involved in the active transport of K^+ ions across the plasma membrane. If this is so, changes in temperature may affect ion transport in some chilling-sensitive and chilling-resistant plants in a similar manner. Indeed the observation that Rb^+ uptake into root tissue undergoes an increase in E_a at temperatures below about 12°C in both corn (chilling-sensitive) and barley (chilling-resistant), (1) is consistent with this proposal.

The thermal response of membrane preparations from tomato and cucumber were different to those of cauliflower, when examined by DSC, but only under certain conditions. This may relate to differences in the chilling sensitivity between these two groups of plants. Phase transitions were only observed in the extracted membrane lipids of the chilling-sensitive tomato and cucumber, and these transitions occurred within the chilling temperature range for these tissues. However, the values obtained for the enthalpy of the transitions indicated that less than 10% of the membrane lipids were actually involved in the transition. Furthermore, the transitions were greatly enhanced in the native membranes when the samples were heated at 60°C for 3 min. This suggests that although some of the membrane lipids of chilling-sensitive plants are capable of undergoing a phase transition, some form of association between the lipids and the intrinsic membrane proteins may prevent much of the lipid from undergoing a phase transition. Lipid-protein interactions which could account for the above situation have been observed in reconstitution studies (5). In these studies it has been demonstrated that lipids can become sequestered around membrane proteins to form an immobilized, disordered boundary layer, which might not undergo a phase transition. Protein denaturation might abolish this interaction, and thus allow the bulk of the lipids to undergo a phase transition.

In comparison to the relatively greater transitions observed after heating, the transitions which occurred in unheated cauliflower and cucumber membranes had an extremely low enthalpy value and involved no more than 1% of the membrane lipids. As these transitions were observed in the membranes of both types of plants, their possible role in the chilling reponse is uncertain.

It is clear from the presence of exothermic/endothermic transitions that some of the membrane lipids of chilling-sensitive plants are capable of forming a gel or crystalline phase within the range of temperatures at which chilling injury is observed. The proportion of gel-phase lipid which exists at the critical chilling temperature, may be small in proportion to the amount of lipid in the fluid or liquid-crystalline phase. However, the presence of some gel-phase lipid in the membrane may be sufficient to cause a lateral separation of the two lipid phases within the membrane, and lead to changes in membrane function. The presence of gel-phase lipid at physiological temperatures has been observed in microsomal membranes of bean cotyledons by X-ray diffraction studies (9). In these membranes the amount of gel-phase lipid was influenced by the presence of both protein and neutral lipid (10). Whereas gel-phase lipid was not detected in the isolated phospholipids of bean cotyledon microsomal membranes at temperatures greater than $5^{\circ}C$, the addition of extracted neutral lipid was able to induce the formation of gel-phase lipid in the isolated phospholipids up to temperatures of $40^{\circ}C$ (10). In the present study, as both neutral lipids and phospholipids were present together in the membrane lipid extracts examined by DSC, some form of lipid-lipid interaction, similar to that described above, may contribute to the phase transition (and gel-phase lipid), which extends into the chilling temperature range. Furthermore, the absence of phase transitions in membranes from chilling-resistant plants may be the result of a changed neutral lipid-phospholipid interaction, rather than to a change peculiar to the phospholipids.

No difference in the thermal behaviour of the untreated membranes from the two groups of plants was observed by DSC. Indeed the small transition which involved less than 1% of the lipid was not observed with membranes prepared from tomato. As transitions were only clearly seen after heating of the membranes or in the extracted membrane lipids, it follows that the change in the E_a of membrane-associated ion-stimulated ATPase which was observed in untreated membranes from both groups of plants, is probably not directly related to the presence of phase transitions in the bulk membrane lipids. In this regard a number of other membrane-associated enzymes have been shown to undergo changes in E_a which are independent of changes occurring in the surrounding membrane lipids (16).

V. REFERENCES

1. Carey, R. W. and Berry, J. A. *Plant Physiol. 61,* 858-860 (1978).
2. Drew, M. C. and Biddulph, O. *Plant Physiol. 48,* 426-432 (1971).
3. Hodges, T. K. *In* "Transport in Plants II, Part A, Cells." (U. Luttge and M. G. Pitman, eds.), Vol. 2, pp. 260-283 Springer-Verlag, Berlin (1978).
4. Hodges, T. K., Leonard, R. T., Bracker, C. E., and Keenan, T. W. *Proc. Nat. Acad. Sci. USA 69,* 3307-3311 (1972).
5. Jost, P. C., Griffith, O. H., Capaldi, R. A., and Vanderkooi, G. *Biochim, Biophys. Acta. 311,* 141-152 (1973).
6. Lowry, O. H., Rosebrough, N. J., Farr, A. L., and Randall, R. J. *J. Biol. Chem. 193,* 265-275 (1951).
7. Lyons, J. M. *Ann. Rev. Plant Physiol. 24,* 445-466 (1973).
8. Lyons, J. M., and Raison, J. K. *Plant Physiol. 45,* 386-389 (1970).
9. McKersie, B. D., and Thompson, J. E. *Plant Physiol. 59,* 803-807 (1977).
10. McKersie, B. D., and Thompson, J. E. *Biochim. Biophys. Acta. 550,* 48-58 (1979).
11. Melchoir, D. L., and Steim, J. M. *Ann. Rev. Biophys. Bioeng. 5,* 205-238 (1976).
12. Pike, C., Raison, J. K., and Berry, J. This volume (1979).
13. Raison, J. K. *In* "Mechanisms of Regulation of Plant Growth." (R. L. Bieleski, A. R. Ferguson, and M. Cresswell, eds.) *Bull. 12,* pp. 487-497. The Royal Society of New Zealand (1974).
14. Raison, J. K., and Chapman, E. A. *Aust. J. Plant Physiol. 3,* 291-299 (1976).
15. Rosenthal, S. L., and Matheson, A. *Biochim. Biophys. Acta 318,* 252-261 (1973).
16. Sandermann, H. *Biochim. Biophys. Acta 515,* 209-237 (1978).
17. Shneyour, A., Raison, J. K., and Smillie, R. M. *Biochim. Biophys. Acta 292,* 152-161 (1973).
18. Sze, H., and Hodges, T. K. *Plant Physiol. 59,* 641-646 (1977).
19. Wheaton, T. A., and Morris, L. L. *Proc. Am. Soc. Hort. Sci. 91,* 529-533 (1967).

MEMRANE LIPID TRANSITIONS: THEIR CORRELATION WITH THE CLIMATIC DISTRIBUTION OF PLANTS

John K. Raison, Elza A. Chapman and Lesley C. Wright

Plant Physiology Unit
CSIRO Division of Food Research
and School of Biological Sciences
Macquarie University
North Ryde, N.S.W. 2113, Australia

S. W. L. Jacobs

Royal Botanic Gardens
Sydney, N.S.W. 2000, Australia

I. INTRODUCTION

For sensitive plants a very good correlation has been established between physiological disorders which occur at chilling temperatures and changes in both the structure and function of mitochondrial and chloroplast membranes (4, 8). This has led to the hypothesis that a temperature-induced alteration in the molecular ordering of membrane lipids is the primary event in chilling injury (4, 8). The metabolic disorders, such as the imbalance between respiration and glycolysis (13), which are observed when sensitive plants are exposed to chilling temperatures are considered secondary events following the primary change in membrane lipid structure. The magnitude of these secondary events depends on both the age of the plant, the tissue, the degree of chilling and the time of exposure (4).

The critical temperature for the primary change in membrane lipid structure has been referred to as T_s, because, by analogy with pure phospholipids, it was thought to represent the temperature for the completion of the liquid-crystalline to gel transition (9). T_f was considered the temperature for the initiation of this transition. However from data using the change in fluorescence of parinaric acid (Figure 7 and Pike *et al.* this volume), T_s is more likely the temperature of the initiation of this transition, if a melting transition is involved at all. T_s

has in most instances been determined as the point at which there is a change in the temperature coefficient of motion of a spin label intercalated into either membrane lipids or into vesicles made from membrane polar lipids. T_s can also be detected as a change in the Arrhenius activation energy (Ea) of some membrane-associated enzymes such as succinate oxidase of mitochondria (5, 6) and photoreduction of NADP by chloroplasts (12). All chilling-sensitive plants examined by these methods show the changes in membrane structure and function in the temperature range of about 9° to $15^\circ C$. For chilling-resistant plants no such changes are observed in the temperature range 0° to $25^\circ C$ but a structural change occurs at or below $0^\circ C$ (10). This correlation between the temperature response of membrane lipids and the ability of plants to withstand chilling has been observed mainly with cereal and vegetable crops. It is possible that selective breeding of these crops might have favoured development of a membrane with specific physical properties favourable to a particular environment. To determine whether the same correlation exists in native plant species, where natural selection would dominate, values for T_s and T_f have been obtained for the leaf polar lipids of plants from different climatic regions. The results show that in general T_s is relatively high (9° to $17^\circ C$) for plants from tropical regions and low for winter growing plants from temperate regions. Thus a good correlation exists between T_s and the climate of the plant's habitat.

II. METHODS

Leaves were removed from growing plants, frozen immediately in liquid nitrogen and kept frozen until extracted with chloroform:methanol as described (7). The polar lipids were separated and isolated by chromatography on acid-washed florisil. The polar lipids were dispersed in 0.1 M tris-acetate buffer, pH 7.2 containing 5 mM EDTA at a concentration of 15 mg lipids/ml. The temperatures of T_s and T_f were usually determined by intercalating a spin label into the lipid dispersion and measuring the rotational correlation time τ_0 as previously described (3). The spin labels used were the methyl esters of 12- and 16-nitroxide stearate (12NSMe and 16NSMe), 5- and 9-nitroxide stearic acid (5NS and 9NS), 11-nitroxide heneicosanane (11N21) and 5-nitroxide decane (5N10).

III. RESULTS

Before describing the results of the survey of native plants it is important to emphasize the rationale underlying the use of Arrhenius plots of enzyme activity and the spin-labelling methods for determining T_s.

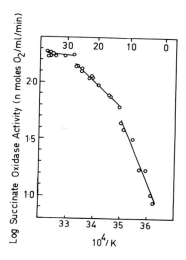

FIGURE 1. Changes in the succinate oxidase activity of maize root mitochondria as a function of temperature. The mitochondria were isolated from 6 day-old roots by the method described (9). Measurements were made on four preparations of mitochondria each at six different temperatures. The composite plot was made by normalizing the rates of the preparations at $25^\circ C$. The straight lines were fitted by regression analysis (9). The Ea above T_f ($27^\circ C$) is 1 kcal/mole, above T_s ($12^\circ C$) 10 kcal/mole and below T_s 25 kcal/mole.

For example, Figure 1 shows the change in rate of succinate-oxidase activity of mitochondria from maize-root tissue as a function of temperature. Since there are abrupt vertical displacements in rates at about $27^\circ C$ and $12^\circ C$ the data is best described by three, straight lines, which delineate three different Ea values for this reaction. As shown in Figure 2, changes in the temperature coefficient of motion occur at two temperatures; at $27^\circ C$ detected by 12NSMe and at $12^\circ C$ detected by 16NSME. The temperatures obtained for T_f and T_s using the plots of enzyme activity and fitting straight lines always coincide with those obtained from spin-label motion even when there is no indication of non-intersecting discontinuities in the oxidative plots. This "straight-line" approach for representing the change in the rate of an enzyme reaction with temperature thus provides a comparative measure of the temperatures at which structural and functional changes occur and allows the formulation of a working hypothesis to describe chilling injury in molecular terms. While it might be statistically more correct to fit polynomial curves to the points in Figures 1 and 2 as Bagnall and Wolfe (1) did with their results, it does not provide information of assistance in explaining chilling injury at the molecular level. Data relating the temperature response of the secondary events which follow from the primary event might be more accurately described in terms of curves and

TABLE I.   The Temperatures T_s and T_f of Leaf Polar Lipids for Native Plants Sampled in their Natural Habitat and for Crop Plants Grown under Glasshouse Conditions.

Region	Name	T_s (^{o}C)	T_f (^{o}C)
TEMPERATE (Winter grown)	Hordeum vulgare	-1	29
	Triticum aestivum	0	30
	Atriplex nummularia	1	28
	Atriplex vesicaria	1	28
	Beta vulgaris	0	30
	Rhagodia spinescens	0	27
	Suaeda australis	1	32
	Pisum sativum	0	26
	Trifolium repens	2	30
	Solanum tuberosum	0	30
	Helianthus tuberosus (dormant tubers)	-5	9
TEMPERATE (Summer grown)	Triodia irritans	1	30
	Zostera capricorni (A)	15	28
	Zostera capricorni (B)	9	23
	Zostera muelleri	1	30
	Phaseolus vulgaris	14	0
	Helianthus tuberosus (mature plant)	3	27
SUB-TROPICAL	Eragrostis phillipica	1	
	Hemarthria uncinnata	6	30
	Themeda australis	11	30
	Zea mays	12	30
	Vigna radiata cv. Mungo	15	28
	Vigna radiata cv. Berken	10	23
	Cucumis sativus [+]	11	25
	Passiflora edulis [+]	2	26
	Lycopersicon esculentum[*]	12	28
	Lycopersicon esculentum	7	23
TROPICAL	Saccharum officinarum	10	28
	Themeda arguens	16	-
	Strongylodon sp. [+]	17	32
	Passiflora flavicarpa [+]	10	27

[+]
[*] From (7).
cv. Grosse lisse (10007D, Yates).

gradual increases in the temperature coefficient with decreasing temperature as suggested by Bagnall and Wolfe (1). The rates of reactions associated with the secondary events are however time dependent and it is likely that the kinetics of these processes are altered to a greater or lesser extent with time of exposure to chilling. The kinetics can become even more complicated since some processes increase their relative rates at low temperature as a result of the chilling injury (5). Considering these factors, the inability to fit straight lines to Arrhenius-type plots of physiological events and the absence of distinct vertical discontinuities cannot be used as "evidence" (1) to refute the existence of a temperature-induced change in the molecular order of membrane lipids in chilling-sensitive plants. Indeed sharp changes occur in the curvature of the graphs expressing growth rates as a function of temperature and these are evident at temperatures which approximate T_s for the plants examined by Bagnall and Wolfe (1). Furthermore this temperature is "critical" for the plant, because, as noted by these authors, morning glory grown at $28^{\circ}C$ and transferred to below $9.8^{\circ}C$ died whereas plants transferred to $10.8^{\circ}C$ showed appreciable gain in dry weight.

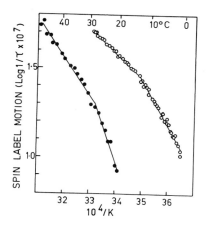

FIGURE 2. The effect of temperature on the motion of spin labels intercalated into the membranes of mitochondria from maize root tissue. The spin label 16NSMe was used in the temperature range 0° to $30^{\circ}C$ (O) and 12NSMe in the range 20° to $50^{\circ}C$ (●). T_o was calculated as described (3) from first derivative spectra and straight lines were fitted to the data by regression analysis as previously described (9). The points of intersection of the straight lines using 16NSMe give T_s ($12^{\circ}C$) and using 12NSMe, T_f ($27^{\circ}C$).

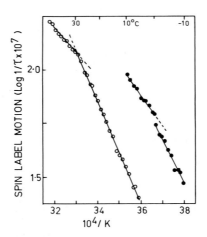

FIGURE 3. Effect of temperature on the motion of spin labels intercalated into membrane lipids from wheat mitochondria. The labels used were 16NSMe (●) between $-11^{o}C$ and $10^{o}C$ and 12NSMe (O) between 10° and $40^{o}C$ indicating a T_s of $0^{o}C$ and a T_f of 30° respectively.

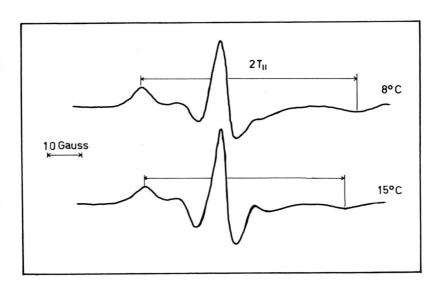

FIGURE 4. First derivative esr absorption spectra for 5NS intercalated into the membranes of wheat mitochondria. The separation of the outer extrema is shown as $2T_{11}$ for spectra at $8^{o}C$ and at $15^{o}C$. [Data from Raison, et al., (10)].

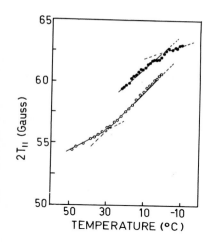

FIGURE 5. Plot of $2T_{11}$ as a function of temperature for $^{1}5NS$ in wheat mitochondrial membranes. The sample was in 50% v/v ethylene glycol for data in the temperature range -10° to $20^{O}C$ (●) and in the aqueous medium (O) for the range 1^{O} to $50^{O}C$. The lines drawn were fitted by regression analysis and indicate changes (T_{s}) at $0^{O}C$ and (T_{f}) at $30^{O}C$.

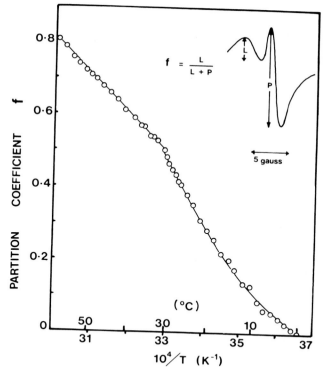

FIGURE 6. The change in the partition coefficient (f) of 5N10 as a function of temperature for a suspension of wheat mitochondrial membranes. The inset shows the line heights used to calculate the partition coefficient, f. T_{f} is the point of departure from the linear relationship between f and $1/T$ at $30^{O}C$. [Data from Raison, et al (10)].

A recent review has been critical of methods used in spin labelling (11). For the data shown in Table I the polar lipids were analyzed by the spin-label method and T_f and T_s were detected as the temperatures at which the temperature coefficient of spin-label motion changed. This is shown in Figure 2 with mitochondrial membranes of maize. The correlation time for the labels used in this study were between 8×10^{-9} sec and 1×10^{-7} sec, an order of magnitude faster than the 5 nitroxide stearate label used by Schreier *et al.* (11) as an example of the problems encountered in the calculation of motion parameters. Using these more mobile labels, correlation times are less likely to be artificially increased as the lipids became more ordered at low temperatures. However, in some plots relating spin-label motion to temperature, it is difficult to determine a clear change in slope, especially in membranes or in lipids from those plants where there is a 30C degree temperature range between the two transitions as shown in Figure 3 for membranes from wheat. In samples such as these it is possible to measure the hyperfine splitting of a spin label which is undergoing anisotropic motion. The measurement is made as shown in Figure 4 and is the separation, in gauss, between the outer extrema of first derivative spectra. As shown in Figure 5 this also shows a change at about 0° and 30°C and does not depend on Arrhenius-type plots. The determination of T_s and T_f does however depend on fitting straight lines to the data. Another parameter which can be used to detect changes in ordering is the partition coefficient. This measures the relative proportions of spin label in the lipid and aqueous phases in membrane preparations or lipid suspensions. It is determined from the relative heights of the peaks in the high-field line of a first derivative spectrum as shown in Figure 6. Although the partition coefficient should be calculated from the relative concentration of label in the two phases this would involve determining line width as well as height. In most instances the line width for the peak corresponding to label in lipid is not resolved and there is probably less error in relating line heights, especially in a temperature range above the liquid to gel transition (11). Using this technique, a change in the partition coefficient is observed with lipids from wheat mitochondria at 30°C (Fig. 6) corresponding with T_f as observed by the other methods. When T_s is near 0°C it cannot be detected by this method because substances used to prevent ice formation interfere with the partitioning.

Since all three methods indicate the same temperatures for T_f and T_s these temperatures were mostly determined by plotting the logarithm of spin-label motion against the reciprocal of absolute temperature and by estimating the points for changes in slopes of straight lines.

Furthermore, based on a technique using the fluorescence of parinaric acid, changes in the molecular ordering of polar lipids from some plants have been detected at temperatures coincident with those shown by the esr methods. An example is shown in Figure 7 for lipids of *Phaseolus vulgaris* where fluorescence intensity increases below about 15°C which is approximately the temperature of T_s (14°C) determined by esr spectroscopy (Table I).

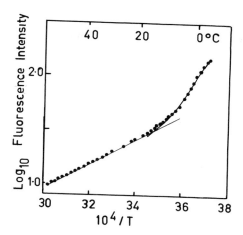

FIGURE 7. The change in the logarithm of relative fluorescence intensity of parinaric acid in polar lipids from bean as a function of temperature. (From data of Pike, et al, this volume). The increase in the coefficient of intensity below 15°C is indicative of the formation of gel phase lipids.

Using the methods outlined we determined T_f and T_s for membrane polar lipids of eighteen species of plants native to Australia and collected from various climatic regions. The results for some of these are shown in Table I together with similar data from a number of cereal and vegetable crops. The plants have been grouped according to the climatic conditions of the habitat in which they were collected. "Tropical" means a continuous warm climate with no frosts, "sub-tropical", slightly cooler than tropical with no frosts and "temperate", cool with frosts. It is apparent that for the winter growing plants from the temperate regions T_s is low and the transition range between T_s and T_f large, (approximately 30C degrees). For summer growing plants from this region T_s is sometimes higher than zero as shown by Helianthus tuber-osus. This species also provides an interesting example in that T_s for the mature plant, which grows in summer, is about 8C degrees higher than that of the dormant tuber in winter. Not only is T_s lowered but the transition range is reduced. The lowering of T_s for lipids of the tuber is progressive, starting at about 3°C at maturity, decreasing to -5°C in mid-dormancy and returning to 3°C at time of sprouting in spring (2). For the plants from the sub-tropical and tropical climates T_s is relatively high and the transition range small (24 to 13 C degrees).

The seagrasses Zostera spp. provide an interesting example of the correlation between the temperature of T_s and the temperature of the plants' habitat. These plants grow in shallow sandy regions along the coast of Australia from latitude 10° (Cape York Peninsula) to latitude 44° (southern Tasmania). *Z. capricorni* shows phenotypic, clinal variations along the coast of New South Wales with a marked change at about latitude 34° (central coast region of New South Wales). Sample A, (T_s of $15^{\circ}C$), was from this area where oceanographic records indicate a mean winter sea temperature of $13^{\circ}C$. Sample B, (T_s of $9^{\circ}C$), was from the southern limit for this species (latitude 36°) and the lower T_s reflects the lower sea temperature of this region. The closely related species *Z. muelleri* (T_s of $-1^{\circ}C$) grows at a latitude of 44° in the south Tasman Sea.

Plants with a T_s in the region of $10^{\circ}C$ are not found growing in temperate climates. Plants with a T_s near $0^{\circ}C$ could however grow in tropical climates and *Helianthus tuberosus*. (T_s of $3^{\circ}C$) is a good example as this plant is found growing successfully in tropical regions.

This correlation between T_s and the temperature of the plant's habitat indicates that the temperature response of the membrane lipids is a major factor in limiting the distribution of plant species. It also indicates that this property would be important in determining the lower temperature limit for the storage and transport of tropical fruits and vegetables.

IV. REFERENCES

1. Bagnall, D. J., and Wolfe, J. A. *J. Exp. Bot. 29,* 1231-1242 (1978).
2. Chapman, E. A., Wright, L. C., and Raison, J. K. *Plant Physiol. 63,* 363-365 (1979).
3. Henry, S. A. and Keith, A. D. *Chem. Phys. Lipids 7,* 245 (1971).
4. Lyons, J. M. *Ann. Rev. Plant Physiol. 24,*445-466 (1973).
5. Lyons, J. M. and Raison, J. K. *Plant Physiol. 45,* 386-389 (1970).
6. Lyons, J. M., Raison, J. K., and Kumamoto, J. *In* "Methods in Enzymology" (S. Fleischer, L. Packer and R. Estabrook, eds.), Vol. 32, pp. 258-262 (1974).
7. Patterson, B. D., Kenrick, J. R., and Raison, J. K. *Phytochemistry 17,* 1089-1092 (1978).
8. Raison, J. K. *In* "Mechanisms of Regulation of Plant Growth" Bulletin 12, The Royal Society of New Zealand, Wellington (R. Bieleski, A. Ferguson, and M. Cresswell, eds.), pp. 487-497 (1974).
9. Raison, J. K., and Chapman, E. A. *Aust. J. Plant Physiol. 3,* 291-299 (1976).
10. Raison, J. K., Chapman, E. A., and White, P. Y. *Plant Physiol. 59,* 623-627 (1977).
11. Schreier, S., Polnaszek, C. F., and Smith, I. C. P. *Biochim.   Biophys. Acta 515,* 375-436 (1978).
12. Shneyour, A., Raison, J. K., and Smillie, R. M. *Biochim.   Biophys. Acta 292,* 152-161 (1973).
13. Watada, A. E., and Morris, L. L. *Proc. Amer. Soc. Hort. Sci. 89,* 368-380 (1966).

THE USEFUL CHLOROPLAST: A NEW APPROACH FOR INVESTIGATING CHILLING STRESS IN PLANTS

Robert M. Smillie

Plant Physiology Unit
CSIRO Division of Food Research,
and School of Biological Sciences
P.O. Box 52, North Ryde
Sydney, N.S.W. 2113, Australia

I.   INTRODUCTION

The research described in the paper deals with effects of chilling temperatures on the chlorophyll-containing inner membranes (thylakoids) of the chloroplast. The intention is not so much to learn more about how temperature may affect the function and biogenesis of chloroplast membranes, as to use the chloroplast membrane system as an experimental tool for investigating the diverse responses of plant species to chilling stresses, their growth and survival at chilling temperatures, and the genetic adaptations involved. The chloroplast thylakoid membranes are especially suitable for this experimental approach as more is known about their composition, structure, function and assembly than any other plant membrane. Furthermore, these membranes contain high concentrations of colored lipids and proteins that may be used as intrinsic membrane probes for monitoring temperature-induced changes.

In this study, knowledge of the properties of chloroplast membranes has been applied to investigating inhibition of growth at chilling temperatures, susceptibility to chilling injury, and possible ways of genetically modifying chilling sensitivity and resistance.

II.   CHLOROPLAST ACTIVITY AND DEVELOPMENT AT DIFFERENT TEMPERATURES IN CHILLING-SENSITIVE AND RESISTANT PLANTS

Growth and development of plants originating in lowland tropical climates is inhibited at chilling temperatures.   The inhibition is reversible although prolonged exposure to low temperatures can lead to irreversible cellular damage in these chilling-sensitive plants (Section III). The different responses of chilling-sensitive and resistant plants to chilling temperatures have been investigated by comparing chloroplast photoreductive activity (in isolated chloroplasts) and chlorophyll synthesis as a function of temperature in the two groups of plants (1, 11, 12, 19, 20). Two results from these studies will be highlighted in this paper. Both concern two basic tenets put forth by Lyons and Raison (6, 7, 16) to explain the phenomenon of chilling sensitivity, namely that

1) the inhibition of growth and metabolism at chilling temperatures in chilling-sensitive plants results from a temperature-induced membrane change below some 'critical' temperature (usually in the proximity of $10^{\circ}C$) and,

2) chilling-resistant plants do not show a temperature-induced membrane change around this temperature.

Their explanation is based mainly on an observed change in slope around $10^{\circ}C$, in the case of chilling-sensitive but not chilling-resistant plants, in Arrhenius plots of membrane-bound activities of mitochondria and chloroplasts and of the motion of spin labels partitioned into mitochondrial and chloroplast membranes. However, our studies with several chilling-resistant plants have not produced the predicted linear relationship with temperature. For example, Arrhenius plots of photoreductive activity of isolated barley *(Hordeum vulgare)* chloroplasts, using either photosystem 1 or photosystem 2 electron acceptors, showed an increase in slope below $9^{\circ}C$ (11). Temperature coefficients for the rate of chlorophyll synthesis in barley and the rate of elongation of the primary leaf increased below $10^{\circ}C$ (19).

The experiments with chilling-sensitive plants were also revealing. In cucumber *(Cucumis sativus),* for instance, the Q_{10} for both the rate of synthesis of chlorophyll in the cotyledons and the rate of hypocotyl extension in dark-grown seedlings increased markedly around $22^{\circ}C$ (19, 20). The decrease in these rates with decreasing temperature below $22^{\circ}C$ was such that both growth processes were barely detectable once temperatures of 10° to $12^{\circ}C$ had been reached. One is left with the question: Is it necessary to postulate the existence of membrane changes below $10-12^{\circ}C$ (whether due to lipid or protein) to explain inhibition of growth at chilling temperatures, when growth and development have almost ceased before these temperatures are reached?

III. CHILLING INJURY

A continuing problem for studies of chilling injury in plants has been the lack of suitable methods for measuring susceptibility to chilling stresses. Reliance on visible symptoms of injury is generally unsatisfactory, since these tend to be late manifestations of cellular damage caused by chilling. Functional properties of the chloroplast thylakoid membranes are known to change during chilling (5, 9) and it was decided to see if these changes could be used to detect and measure cellular injury well before irreversible damage and visible symptoms of injury to tissues become apparent. Before assays for cellular chilling injury are described, it is first necessary to examine the changes in chloroplast function taking place in chilling-sensitive plants at low temperatures.

A.   Changes in Chloroplast Function at Chilling Temperatures

In chilling-sensitive plants, a number of chloroplast functions deteriorate at chilling temperatures; the particular functions affected depend upon experimental conditions and the species of plant (Table I).
Photoreductive activity declines in leaves of bean, cucumber and tomato at temperatures near $0°C$ in either dark or light. The lesion lies on the oxygen-evolving side of photosystem 2 (8). In illuminated chilled leaves the region of the electron transfer pathway between the two photosystems is also affected with the rapid loss of photophosphorylative and proton uptake capacities. At very high light intensities chilling conditions also interfere with chlorophyll turnover in cucumber and bleaching of the chlorophyll occurs. The situation is different in leaves of the tropical golden passionfruit *(Passiflora edulis forma flavicarpa)* in which photosystem 1 and photosystem 2 activities decline concomitantly. Photobleaching at chilling temperatures is not observed.

TABLE I.   *The Effect of Chilling on Chloroplast Function*

Reaction	Conditions	Plants	References
Decline in capacity for oxygen evolution	Light or dark	Bean	(9) (5)
		Cucumber	(8) (2)
Loss of photophosphorylation and proton uptake	Light	Tomato	(4) (22)
Decline in activity of photosystems 1 and 2	Light or dark	Passionfruit	(unpublished results)
Photobleaching of chlorophyll	High light intensity	Cucumber	(25)

B.   Assays for Chilling Injury

A feature common to the chloroplast membranes of all chilling-sensitive plants species investigated so far is the progressive loss of capacity for oxygen evolution at chilling temperatures (Table I). The decline in this capacity can be followed by measuring Hill reaction activity in isolated chloroplasts. Spectrophotometric or fluorometric measurements can be used to monitor changes in photosynthetic electron transfer activity in intact leaves.

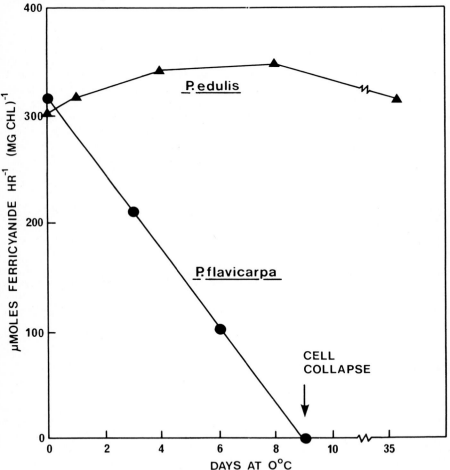

FIGURE 1.   Photoreductive activity of chloroplasts isolated from chilled leaves of the purple (P. edulis) and the yellow (P. flavicarpa) passionfruit. Leaves were detached and stored in darkness at $0^{o}C$ for the times indicated. Chloroplasts were isolated and photoreductive activity [μ moles ferricyanide photoreduced/hr^{-1}/(mg chlorophyll)$^{-1}$] measured as described by Critchley et al (1).

1. Isolated Chloroplasts. An assay for chilling injury based on the decline in photoreductive activity in isolated chloroplasts has been described by Smillie and Nott (22) and used to compare susceptibility to chilling injury in cultivars of the domestic tomato *(Lycopersicon esculentum)* and high and low altitudinal forms of a wild tomato *(L. hirsutum).* Tomato leaves were chilled at $9^{\circ}C$ and after various periods of chilling, the chloroplasts were isolated and oxygen evolving capacity ascertained by the photoreduction of ferricyanide in the presence of gramicidin D, an uncoupler of photophosphorylation.   presence of gramicidin D, an uncoupler of photophosphorylation.

Figure 1 shows this assay applied to leaves of the chilling-sensitive golden or yellow passionfruit *(P. edulis forma flavicarpa)* and leaves of the purple passionfruit *(P. edulis),* which is chilling resistant (13). There was little change in activity of chloroplasts isolated from leaves of the purple passionfruit, even after 36 days at $0^{\circ}C$. In contrast, there was a progressive decline in activity for leaves of the yellow passionfruit kept at $0^{\circ}C$. First symptoms of massive chilling injury (loss of turgor in the leaves; changes in chloroplast membrane structure revealed by electron microscopy) coincided with loss of photoreductive activity, i.e. after 8 to 9 days at $0^{\circ}C$. Photoreductive activity did not decline in leaves of either passionfruit held at $10^{\circ}C$ for 9 days. Thus these measurements clearly distinguish between the species susceptible to chilling injury at $0^{\circ}C$ and the non-susceptible species. There appeared to be little or no lag in the onset of the chloroplast membrane changes which lead to loss of photoreductive activity in the susceptible species at $0^{\circ}C$.

2. Intact Tissue. Changes in chlorophyll fluorescence, the light-dependent absorbance change at 518 nm and the photooxidation and reduction of cytochrome f have been used to follow development of chilling injury in intact green tissues. An example using cytochrome changes is given in Section IV, C. The chlorophyll fluorescence method will be described here since it is probably the most versatile of the methods developed and can be adapted to screening a relatively large number of samples.

Figure 2 shows changes in chlorophyll fluorescence in peanut *(Arachis hypogaea)* and pea *(Pisum sativum)* leaves chilled at $0^{\circ}C$. The fluorescence measurements were also made while the leaves were at $0^{\circ}C$; there was no re-warming of the leaves before measurement. Upon illumination with red light an initial level of fluorescence (F_o) was established which did not change appreciably as chilling injury developed. In instances of more severe damage to the chloroplast membrane, as in the case of heat-induced damage, F_o increased and has been used to follow the development of heat-induced changes in chloroplast membranes and to compare the heat sensitivities of plant species (17, 21, 23, 24).

There was a further rise in chlorophyll fluorescence (variabl chlorophyll fluorescence) to a new level (F_{max}) as a component of the photosystem 2 complex, Q, becomes reduced.   Q, which quenches

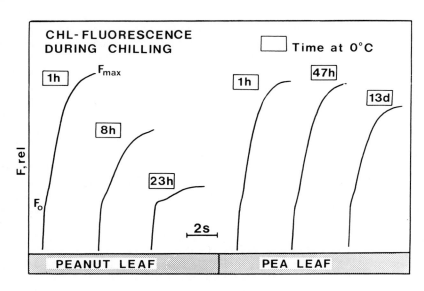

FIGURE 2. *Chlorophyll fluorescence rise at $0°C$ in dark-adapted peanut and pea leaves held at $0°C$ for various periods of time. F_o indicates the initial fluorescence level, F_{max} the maximum reached. h = hours; d = days.*

chlorophyll fluorescence when in the oxidized state, is reduced as the result of water splitting and oxidized by components of the electron transfer pathway located between photosystem 2 and photosystem 1. Thus factors which inhibit on the reducing side of photosystem 2 (e.g. chilling injury) would be expected to decrease the rise in fluorescence above F_o, while factors which inhibit on the oxidizing side of photosystem 2 (e.g. DCMU) might be expected to increase it. Figure 2 shows that both the rate of the variable fluorescence rise and the extent of this rise (F_{max} $-F_o$) had decreased substantially in peanut leaf after one day at $0°C$. Only small changes occurred in fluorescence in leaves of the chilling-resistant pea, even after 13 days at $0°C$.

In routine measurements only the rate of the rise in variable chlorophyll fluorescence was recorded. Up to 96 samples of leaves or leaf segments were positioned on an aluminium block which was then set in ice contained in a darkened, insulated box. Using a portable fluorescence sensor probe (18), it was possible to record the fluorescence rise in the 96 different samples of chilled leaf tissue in 10 to 12 minutes. Fig. 3 compares the time course for the change in rate of the fluorescence rise in peanut and pea leaves during storage at $0°C$.

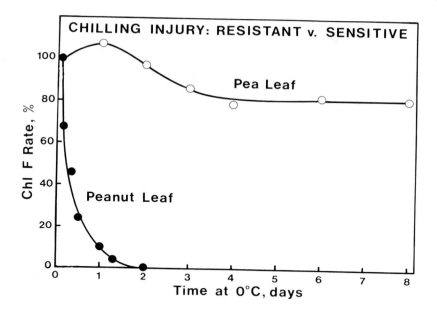

FIGURE 3. *The rate of the rise in variable fluorescence at $0°C$ in dark-adapted leaves of pea and peanut stored at $0°C$ for various periods of time. The leaves were detached, positioned on an aluminium block, and kept in darkness for 15 min. The block was embedded in ice, except for the upper surface which was covered with insulation material. The first measurement of the rate of fluorescence rise was made about 1 hour later. Subsequent rates measured are expressed as a percentage of this first measured rate. Each point is the average value for 8 different leaf samples.*

The chlorophyll fluorescence method can also be used to compare the effects of chilling on closely related species, including instances where both species are susceptible to chilling injury. Figure 4 shows a comparison between a tropical guava *(Psidium guajava)* and a temperate species of *Psidium (P. cattleyanum)*. In Fig. 5 *Capsicum annum* is compared with *C. rocoto,* a species also from tropical latitudes but found at higher altitudes. Both are susceptible species but *C. rocoto,* on the basis of the chlorophyll fluorescence measurements, is less susceptible than *C. annum.* Similar results were obtained in a comparison of the paw paw *(Carica papaya)* with the mountain paw paw *(C. pubescens).*

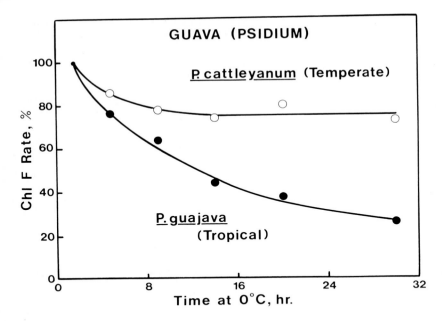

FIGURE 4. Development of chilling injury in leaves of two species of Psidium (chlorophyll fluorescence method). Conditions are as described in Figure 3.

The use of chlorophyll fluorescence changes to monitor the progressive damage to chloroplast membranes caused by chilling temperatures has made it possible to embark on a wide range of studies of chilling injury which were hitherto not feasible. The ranking of plants for susceptibility to chilling injury is being extended to a wide range of species. The kinetics of chilling injury, its reversibility and the effects of transient re-warming are being studied. By making measurements at different temperatures, it should be possible to determine the Q_{10} for chilling injury in different plants. The effects of environmental and other factors (light, age, hormones, diurnal and seasonal changes, etc.) on the development of chilling injury can also be investigated. Detailed analysis of the chlorophyll fluorescence changes should yield new information of mechanisms involved in chilling injury. The simplicity and rapidity of measuring the change in induced fluorescence makes it suitable for studies of acclimation and adaptation to chilling temperatures, and perhaps coupled with infra-red photography, as a screening method for the selection of chilling tolerant plants in plant breeding and mutation programs.

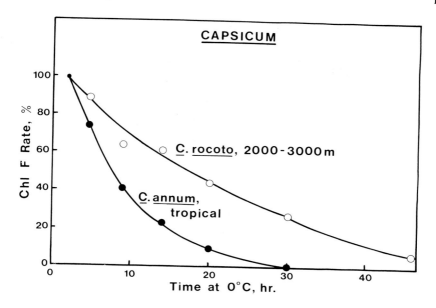

FIGURE 5. Development of chilling injury in leaves of two species of Capsicum (chlorophyll fluorescence method). Conditions are as described in Figure 3.

The chlorophyll fluorescence method also works well with the green peel of fruit, so that measurements of chilling injury can be carried out using green fruit as well as leaves.

IV. GENETIC MODIFICATION OF CHILLING SENSITIVITY

For many chilling-sensitive plants of commercial importance, there is much interest in increasing their resistance to chilling temperatures, in order to extend the growing season and range of climates in which crops can be raised economically and to make low temperature storage of the fruit a feasible proposition. Genetically-controlled responses of plants to chilling temperatures may be altered by changing either the genetic information or the expression of this information. Little is known at present about regulation of gene expression for chilling sensitivity by biochemical and environmental factors. Of obvious interest is the role of temperature itself as a potential regulator of gene expression and it is known that some chilling-sensitive plants can be 'hardened' to withstand moderately low chilling temperatures.

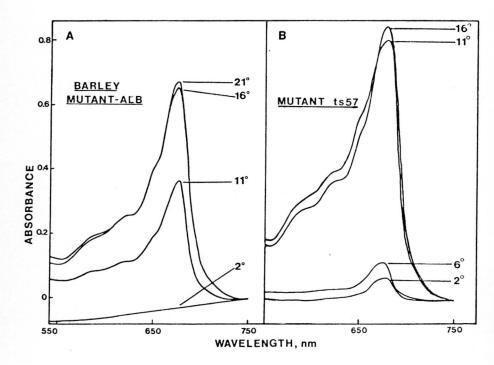

FIGURE 6. Absorption spectra of leaves of barley mutants grown at different temperatures. Each spectrum was recorded for the region of the primary leaf 3 to 4 cm below the leaf tip. A. A temperature-dependent albino mutant originally produced at the Danish Atomic Energy Commission's Research Establishment at Risø. B. The mutant viridis-zj^{ts57}, from the stock of mutants at the Carlsberg Research Center.

Genetic information determining chilling sensitivity or resistance may be altered by mutation or by the introduction of new genes. Three recent studies in this area are outlined below.

A.   Barley Mutants Chilling-Sensitive for Chloroplast Development

Can selected growth or development processes in chilling resistant plants be rendered chilling-sensitive by mutation of nuclear genes? This question was explored by surveying a number of photosynthetic mutants of barley from among stock held at the Carlsberg Research Center, Copenhagen, to see if any were temperature-dependent for chloroplast development. Seedlings of mutant plants were raised in glass cylinders immersed in water baths (24) held at a range of temperatures starting at $2°C$. Several mutants were identified in which chloroplast development was totally or severely inhibited at chilling temperatures although rate of growth of the young seedlings was not affected when compared with the wild type. Figure 6 shows absorption spectra of the primary leaf of two of these mutants grown at different temperatures. In both mutants chloroplast development begins to be inhibited below 10 to $12°C$. Chlorophyll synthesis was undetectable in the albino mutant (Fig. 6A) at severe chilling temperatures, while it was greatly reduced in the mutant $viridis$-zj^{ts57} (Fig. 6B). This mutant is also heat-sensitive for chloroplast development above $27°C$ (24). The location of the block in chloroplast development in these mutants is unknown. It occurs at a very early stage in the plastid biogenesis and may possibly result from low-temperature inhibition of the synthesis of chloroplast DNA or its transcription during plastid replication.

B.   Interspecific Sexual Hybridization: Tomato Species

While mutation can provide interesting material for studies of chilling sensitivity, it does not at present appear to be a practical way of increasing the chilling tolerance of sensitive species. A more promising approach is the introduction of new genes for chilling resistance into a domestic species by utilizing the genetic variability that often exists in wild forms of the same species, or in closely related species with which it can by hybridized sexually. Yet another approach lies in the introduction of new genes for chilling resistance from more distantly related species by somatic hybridization (Section IV, C).

Dr. B. D. Patterson in our laboratory is investigating the feasibility of increasing the chilling resistance of the domestic tomato (L. esculentum) by hybridization with high altitudinal forms of the wild tomato (L. hirsutum) (14, 15). The first step in this investigation was to use a number of methods to establish whether some forms of L. hirsutum were indeed more chilling resistant than L. esculentum and whether the chilling resistance was correlated with altitude. Figure 7 compares susceptibility to chilling injury at $0°C$, using the chlorophyll fluorescence

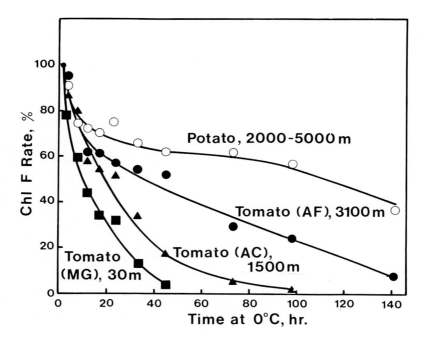

FIGURE 7. Development of chilling injury (chlorophyll fluorescence method) in leaves of potato and three altitudinal forms of wild tomato, L. hirsutum. The tomato plants are derived from seed collected at 30, 1500 and 3100 meters; the code letters MG, AC and AF, respectively, refer to the localities where the seed was collected [see Patterson et al., (14)]. Assay conditions as in Figure 3.

method (Section III, B, 2), of three forms of L. hirsutum raised from seed originally collected from indigenous plants growing at 30, 1500 and 3100 meters, in Ecuador or Peru. The low altitude form collected near sea-level was very susceptible and was comparable in this respect to domestic cultivars of L. esculentum. The form from 1500 meters was less susceptible and the one from 3100 meters considerably less susceptible. These results agreed with estimates of relative chilling sensitivity using other methods (14). The potato (Solanum tuberosum), the wild forms of which are indigenous to the Sierras of Peru and Bolivia at altitudes above 2000 meters (3), was more resistant again. The second stage of the investigation, to cross L. esculentum with the most chilling-resistant forms of L. hirsutum, is in progress. Hybrids of L. esculentum and L. hirsutum (AF, 3100 meters) and first and second

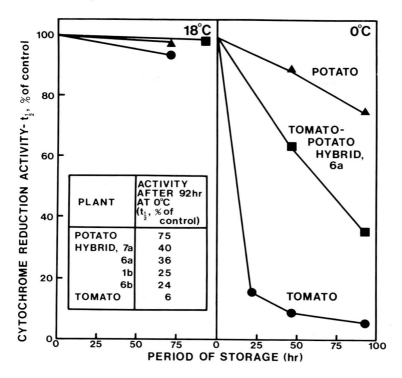

The figure contains the following table:

PLANT	ACTIVITY AFTER 92hr AT 0°C ($t_{\frac{1}{2}}$, % of control)
POTATO	75
HYBRID, 7a	40
6a	36
1b	25
6b	24
TOMATO	6

FIGURE 8. Susceptibility of tomato-potato hybrids to chilling injury as measured by changes in reduction of cytochrome f. Leaves were detached and stored in darkness at 0°C or 18°C. Changes in photosynthetic electron transfer activity in individual leaves were assayed by rapidly warming the leaves to 23°C and recording the rate of dark reduction of cytochrome f (cytochrome-554) following a 30-second period of illumination with red light. The absorbance increase at 554 nm coinciding with cytochrome reduction was monitored using an Aminco DW-2a spectrophotometer operated in the dual wavelength mode (reference wavelength 541 nm). The values for the rate of cytochrome reduction are the $T_{\frac{1}{2}}$ for the reduction in seconds and are expressed as a percentage of the initial rate obtained with freshly detached leaves. Changes in rate of cytochrome reduction were small in leaves stored at 18°C. ▲, potato; ■, the tomato-potato hybrid 6a; ●, tomato. The four somatic tomato-potato hybrids are identified as 1b, 6a, 6b, and 7b; these designations refer to the protocols of the callus transfers and shoot regenerations used [Melchers et al., (10)].

backcrosses of selected hybrids to *L. esculentum* have been produced
and are currently being tested.

C.   *Intergeneric Somatic Hybridization: The Tomato-Potato Hybrid*

Melchers *et al.* (10) have recently produced several somatic hybrid
plants of tomato and potato regenerated from fused tomato and potato
protoplasts. Fortuitously, the new plants are hybrids of a chilling-
sensitive plant and a chilling-resistant one and it was of considerable
interest therefore to determine their chilling resistance. Susceptibility
to chilling injury, as indicated by changes in the dark reduction of
cytochrome f in leaves stored at $0^{\circ}C$ was measured in four of the
somatic tomato-potato hybrids (Fig. 8). All four showed enhanced
resistance compared with tomato, but were less resistant than potato. If
genetic exchange between the tomato and potato chromosomes of the
hybrid can be induced, either by homoeologous pairing followed by
crossing-over or by reciprocal translocations, and subsequent separation
of the two genomes obtained, these hybrids should be useful for
transferring genes for chilling resistance into the domestic tomato.
Since the chlorophyll fluorescence assay for chilling injury is relatively
simple and can be applied to small leaf samples, it should be of value in
the identification of chilling-resistant recombinants in the tomato
genome.

V.   CONCLUSIONS

A change in slope in Arrhenius plots of the photoreductive activity
of isolated chloroplasts at chilling temperatures does not necessarily
predict chilling sensitivity. Chloroplasts isolated from both chilling-
sensitive and resistant plants showed non-linear Arrhenius plots between
0 and $20^{\circ}C$.

The rate of chloroplast development in cucumber *(Cucumis sativus)* cotyledon and growth of cucumber hypocotyl decrease rapidly with
falling temperatures below $22^{\circ}C$ and have almost ceased before the
chilling-sensitive temperature range is reached. The high Q_{10} between
12° and $22^{\circ}C$ is sufficient to account for the apparent inhibition of
growth and development in cucumber by chilling temperatures.

The development of cellular chilling injury in leaves and green fruit
can be conveniently followed by measurements of chlorophyll
fluorescence or oxidation of cytochrome f. These methods were used to
rank plants for susceptibility to chilling injury and in studies aimed at
increasing chilling tolerance in tomato *(Lycopersicon esculentum)* via
interspecific hybridization with high altitude forms of a wild tomato
(L. hirsutum) and intergeneric somatic hybridization with potato *(Solanum tuberosum)*.

ACKNOWLEDGMENTS

I should like to thank Drs. B. D. Patterson, W. G. Nolan, C. Critchley and Miss R. Nott who have contributed to various aspects of this work. The studies described in Sections IV, A and IV, C are joint research projects with the Carlsberg Research Center, Copenhagen, and the cooperation and support of Professor D. von Wettstein is gratefully acknowledged.

VI. REFERENCES

1. Critchley, C., Smillie, R. M., and Patterson, B. D. *Aust. J. Plant Physiol.* 5, 443-448 (1978).
2. Garber, M. P. *Plant Physiol.* 59, 981-985 (1977).
3. Hawkes, J. G. *In* 'The Potato Crop'. (Ed. P. M. Harris.) pp. 1-14 Chapman & Hall, London (1978).
4. Kaniuga, Z., Sochanowicz, B., Zabek, J., and Krzystniak, K. *Planta* 140, 121-128 (1978).
5. Kislyuk, I. M., and Vas'kovskii, M. D. *Sov. Plant Physiol.* 19, 813-818 (1972).
6. Lyons, J. M. *Annu. Rev. Plant Physiol.* 24, 445-466 (1973).
7. Lyons, J. M., and Raison, J. K. *Plant Physiol.* 45, 386-389 (1970).
8. Margulies, M. M. *Biochim. Biophys. Acta* 267, 96-103 (1972).
9. Margulies, M. M. and Jagendorf, A. T. *Arch. Biochem. Biophys.* 90, 176-183 (1960).
10. Melchers, G., Sacristan, M. D., and Holder, A. A. *Carlsberg Res. Commun.* 43, 203-218 (1978).
11. Nolan, W. G., and Smillie, R. M. *Biochim. Biophys. Acta* 440, 461-475 (1976).
12. Nolan, W. G., and Smillie, R. M. *Plant Physiol.* 59, 1141-1145 (1977).
13. Patterson, B. D., Murata, T., and Graham, D. *Aust. J. Plant Physiol.* 3, 435-442 (1976).
14. Patterson, B. D., Paull, R., and Smillie, R. M. *Aust. J. Plant Physiol.* 5, 609-617 (1978).
15. Patterson, B. D., Paull, R., and Graham, D. This volume (1979).
16. Raison, J. K., and Chapman, E. A. *Aust. J. Plant Physiol.* 3, 291-299 (1976).
17. Schreiber, U., and Berry, J. A. *Planta* 136, 233-238 (1977).
18. Schreiber, U., Groberman, L., and Vidaver, W. *Rev. Sci. Instrum.* 46, 538 542 (1975).
19. Smillie, R. M. *In* 'Genetics and Biogenesis of Chloroplasts and Mitochondria'. (Eds Th. Bucher *et al.*) pp. 103-110. Elsevier:Amsterdam (1976).

20. Smillie, R. M. "Temperature Regulation of Chloroplast Development and Photosynthesis. Genetic, Biochemical and Applied Aspects. Proc. XIV Int. Congr. Genetics, Moscow 1978, In Press (1979).
21. Smillie, R. M. *Aust. J. Plant Physiol. 6*, 121-133 (1979).
22. Smillie, R. M., and Nott, R. *Plant Physiol. 63*, 796-801 (1979).
23. Smillie, R. M., and Nott, R. *Aust. J. Plant Physiol. 6*, 135-141 (1979).
24. Smillie, R. M., Henningsen, K. W., Bain, J. W., Critchley, C., Fester, T., and von Wettstein, D. *Carlsberg Res. Commun. 43*, 351 364 (1978).
25. van Hasselt, P. R. *Acta Bot. Neerl. 21*, 539-548 (1972).

LOW TEMPERATURE RESPONSE
OF CHLOROPLAST THYLAKOIDS

M. P. Garber

Weyerhaeuser Company
Southern Forestry Research Center
Hot Springs, Arkansas

Low temperature is considered by many to be the single most limiting factor in the geographical distribution of plants. Physiological dysfunctions are elicited primarily by two segments of the low temperature spectrum: (1) chilling temperatures (0° to approximately $12^\circ C$), and (2) subfreezing temperatures ($< 0^\circ C$).

The physiological dysfunctions at chilling temperatures are collectively referred to as "chilling injury" and plants which exhibit dysfunctions are referred to as "chilling-sensitive" plants, which includes many economically important crops. Although most of the investigations of chilling injury have involved chilling in the dark, there is substantial evidence suggesting that chilling in the light is more damaging to the photosynthetic apparatus of chilling-sensitive plants (3, 9, 21). Since under field conditions chilling temperatures are likely to occur in the presence of light, information pertaining to chilling in the light may provide a more accurate explanation of reduced plant growth under chilling conditions. Pertinent questions on light and chilling injury that are addressed in this paper include: (1) what are the similarities and differences of chilling in the presence and absence of light? (2) are the requirements for recovery from light and chilling the same as for dark and chilling? and (3) is the reduced growth of field-grown chilling-sensitive plants more accurately predicted by light and chilling injury and the requirement of reactivation?

The low temperature response of plants will be discussed in reference to thylakoids, the chloroplast membrane system. Thylakoids have been utilized in the study of low temperature injury due to their ease of isolation and many well defined functions. Also, photosynthesis is affected by low temperatures and thylakoids play an integral role in the photosynthetic process. However, the ultimate utility of the isolated thylakoid system for the study of low temperature injury may depend on how well thylakoids parallel whole plant responses. For instance, is the mechanism of inactivation of isolated thylakoids exposed to chilling temperatures similar to thylakoids treated *in situ?*

203

I. EFFECT OF CHILLING TEMPERATURES ON CHLOROPLAST THYLAKOIDS

A. Chilling in the Dark

Most studies have observed chilling injury of thylakoids in the absence of light and have generally utilized one of two methods of treating thylakoids: (1) leaf tissue is exposed to chilling temperatures and thylakoids subsequently isolated and membrane functions measured above the chilling range -- *in situ* treatment; and (2) thylakoids are first isolated from leaf tissue and then exposed to chilling temperatures with the measurement of membrane functions at or above the chilling range -- *in vitro* treatment.

1. Thylakoid Response to in situ Chilling. Thylakoids from leaves of chilling-sensitive plants chilled in the dark exhibit low Hill reaction activity but good photosynthetic phosphorylation, as measured by PMS-dependent ATP synthesis (12), and the reduction in PSII activity correlated well with the reduction in photosynthesis (9), suggestive that PSII is the most sensitive part of the thylakoid. Additional evidence for PSII sensitivity is the low fluorescence yields, even at high light intensity, in the presence of DCMU and low donor-dependent DCMU-sensitive reduction of $NADP^+$ (13). The reduction in PSII activity was associated with a decrease in loosely bound manganese and restoration of PSII activity was associated with an increase in bound manganese (7, 13). Spinach, a chilling-resistant plant, did not lose Hill activity after chilling in the dark (9).

In the *in situ* studies, thylakoid functions, such as PMS-dependent ATP synthesis, were measured at $16\text{-}20^\circ C$. Therefore, the absence of a lesion in a membrane function does not imply that the function was unaffected by chilling temperatures but rather, any alterations that occurred at low temperatures may be readily and completely reversed at above chilling temperatures. Conversely the appearance of a lesion suggests that the alteration at chilling temperatures is not readily reversed. Therefore, might a lesion such as decreased PSII activity be a rate-limiting step in the recovery of chloroplasts from chilling injury?

2. Thylakoid Response to in vitro Chilling. The *in vitro* results have generally been expressed as an Arrhenius plot, with a change in slope of the Arrhenius plot correlating with a change in membrane fluidity (8). A change in the slope of the Arrhenius plot (increased activation energy) for the photoreduction of $NADP^+$ from water by chloroplasts from the chilling-sensitive bean and tomato occurred at $12^\circ C$ (18). A similar change was not observed for chloroplasts from chilling-resistant plants. Temperature-induced phase changes were observed by esr spectroscopy in thylakoids from chilling-sensitive tomato and corn, but not in thylakoids from chilling-resistant plants (17).

The two temperature-induced changes (increased activation energy of electron transport and membrane phase change) occurred at approximately the same temperature ($12^{o}C$), which corresponds to the critical chilling temperature of the plant. The results for chloroplast thylakoids are consistent with the hypothesis for mitochondria that the primary effect of chilling temperatures is a membrane phase change (11).

The existence of a phase change and increased Arrhenius activation energy has generally been considered synonymous with chilling-sensitive plants while chilling-resistant plants characteristically lack such alterations. There are, however, several studies which question the concept that a membrane phase transition is the primary mechanism of chilling injury in chloroplast thylakoids. Although a phase transition was not detected in tomato leaf thylakoids, there were alterations in tomato thylakoid functions at chilling temperatures (7, 18), suggestive that the lipid phase transition is not a prerequisite to membrane dysfunction. The Ea for the photoreduction of DCIP and ferricyanide for chilling-resistant and chilling-sensitive plants was constant in the chilling range. However, in the presence of an uncoupler of photophosphorylation thylakoids of both groups of plants exhibited an increase Ea in the chilling range. The increased Ea is considered to be a manifestation of a phase change of lipids but the results of Nolan and Smillie (14, 15) suggest that it is not possible to distinguish chilling-sensitive and chilling-resistant thylakoids on the basis of changes in Ea or phase transitions in the lipids. Additional evidence for a lack of correlation between the phase transition and chilling injury comes from a comparison of *in vitro* (18) and *in situ* studies (9, 13). The *in vitro* study showed a constant Ea for PSII activity of chilling-sensitive plants over the range of 3° to $25^{o}C$. However, PSII activity was the most sensitive thylakoid function under *in situ* treatment and its inhibition paralleled the loss of photosynthesis. The loss of PSII activity was not a secondary effect of chilling injury or the result of altered metabolism since a chloroplast preparation from a mixture of fresh and cold stored leaves had a specific activity about half that of the sum of the two prepared separately (12). Consequently the absence of a change in Ea for PSII activity did not correctly predict the response of PSII to *in situ* chilling.

B.   Chilling in the Light

Chilling in the light is more damaging to the photosynthetic apparatus of chilling-sensitive plants than is chilling in the dark (3, 9, 21). Initial studies emphasized enzymatic dysfunctions (21, 22, 23) as opposed to membrane controlled processes.

Inactivation of thylakoids differs in the presence and absence of light. The reduction in photosynthesis under $0.5^{o}C$ and light correlated better with the loss of cyclic and non-cyclic phosphorylation than with the slower loss of Hill activity (9). However, the authors found that the loss of photosynthesis due to chilling in the dark correlated better with the loss of Hill activity than with PSI phosphorylation. Unfortunately, PSII phosphorylation was not measured. Photosystem I activity was not affected by chilling in the dark (3, 9, 12) but was the most sensitive thylakoid function to chilling in light (Figure 1). The loss of cyclic phosphorylation preceded the loss of cyclic electron transport and proton uptake. The available evidence suggests that PSII is the thylakoid function most sensitive to chilling in the dark whereas PSI phosphorylation is the thylakoid function most sensitive to chilling in the light. There is, however, need for a more defined comparative approach to determine the photochemical function of thylakoids most sensitive to chilling in the presence and absence of light. Additional evidence for a difference in the inactivation of thylakoids due to dark and light chilling is that the ATPase activity of CF_1 is inhibited at chilling temperatures in the presence of light but not in the dark (3).

Available information on the mechanism of light and chilling injury suggests that the inactivation of thylakoids is probably a photo-oxidative process since the rate of inactivation of proton uptake (Figure 2), cyclic electron flow (Figure 2), Ca^{++}-ATPase activity, and PMS-phosphorylation is decreased in the presence of N_2 and increased in the presence of O_2. The loss of Ca^{++}-ATPase activity at $4^\circ C$ in light is due

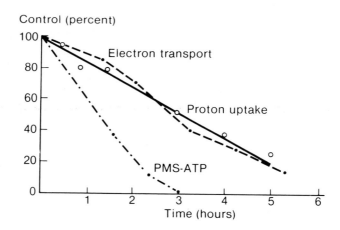

FIGURE 1. *Cucumber thylakoids pretreated in vitro at $4^\circ C$ in light. The rate of inactivation of PMS-phosphorylation exceeded that of cyclic electron transport and proton uptake, which were inactivated at the same rate. The control activities were: PMS-phosphorylation, 370μ moles Pi esterified/mg Chl/hr; proton uptake, 0.85 meq H^+ accum/mg Chl; cyclic electron transport (DAD/Ascorbate $\rightarrow$-methyl viologen), 1182 μ moles O_2 consumed/mg Chl/hr.*

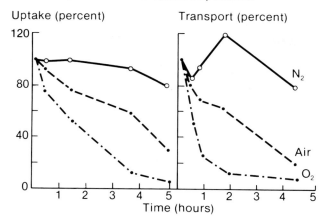

FIGURE 2. Isolated cucumber thylakoids exposed to $4^{o}C$ in light in the presence of nitrogen (N_2), air, or oxygen (O_2). The thylakoid suspensions were bubbled with N_2, air, or O_2. The inhibition of proton uptake (A) and cyclic electron transport (B) is reduced in the presence of N_2 and accelerated in the presence of O_2, suggestive that the inhibition is a photo-oxidative process.

to inactivation of the CF_1 enzyme as compared to a simple release of CF_1 particles from the membrane. Electron transport, or the lack thereof, has been implicated in the development of light and chilling injury. vanHasselt (25) suggests that photo-oxidation of chlorophyll is due to the generation of the harmful chlorophyll triplet state when energy from absorbed light quanta is not deflected by electron transport. It was further suggested that the lag phase in chlorophyll degradation is the time required for inhibition of electron transport. It is unlikely that loss of electron transport is the only prerequisite to chlorophyll degradation since electron transport is inhibited much sooner than the initiation of chlorophyll degradation (3). However, that non-cyclic electron transport plays a key role in preventing photo-oxidative damage is demonstrated by the treatment of thylakoids with NH_2OH (Figure 3). The NH_2OH treatment inhibits PSII activity yet it accelerated the loss of proton uptake, a photosystem I function.

Detailed investigations into the mechanism of light and chilling injury of thylakoids of chilling-sensitive plants would be facilitated if in vitro treatment could be substituted for in situ treatment. The validity of using in vitro results to make inferences about in situ phenomena has been established for the study of freezing injury (6). Garber (3) exposed cucumber thylakoids to $4^{o}C$ and light, in situ and in vitro, and several membrane functions or components including, PMS-dependent phosphorylation, proton uptake, osmotic response to sucrose, Ca^{++}-ATPase activity and chlorophyll content were monitored. The sequence of activities and components lost during inactivation was the same with in vitro and in situ treated thylakoids. Also, the

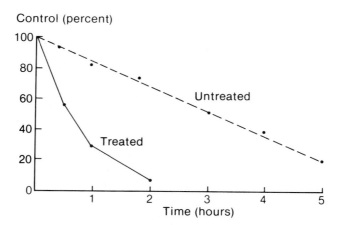

FIGURE 3.   Cucumber thylakoids treated with NH_2OH prior to in vitro exposure to $4°C$ in light.   The NH_2OH treatment specifically inhibits PSII but not PSI. Apparently non-cyclic electron transport can decrease the rate of inactivation of proton uptake, a PSI function.

mechanism of in vitro and in situ inactivation appeared to be the same.   The addition of DCCD to partly inhibited thylakoids did not stimulate proton uptake despite a reduction in the Ca^{++}-ATPase activity, suggesting that the CF_1 enzyme is inactivated and not simply released from the membrane. In addition, all thylakoids were inactivated to varying degrees as opposed to the existence of completely inactivated or normal thylakoids. The results suggest there is validity in using in vitro results to make inferences about in situ phenomena for cucumber thylakoids exposed to $4°C$ in light. In summary, isolated thylakoids of chilling-sensitive plants can be used for the study of chilling injury. Also, the manifestations of chilling injury of thylakoids are dependent on the presence or absence of light.

C.   Recovery from Chilling Injury

The reduced growth of chilling-sensitive plants due to chilling injury will be the net result of the extent of inactivation and degree of recovery.   The previous section indicated that the inactivation of thylakoids is more rapid in the presence of light and that the mechanism of inactivation differs from chilling in the dark. The possibility exists that the requirements for reactivation differ, and if so, may provide a basis for a more accurate explanation of reduced growth of chilling-sensitive plants under chilling conditions.

The longer cucumber discs are exposed to $4°C$ in light the slower is the rate of recovery at $20°C$ (Figure 4) which is in agreement with the recovery of bean leaves from chilling in the dark (12).   The rate of recovery of proton uptake was faster in the dark than in the light,

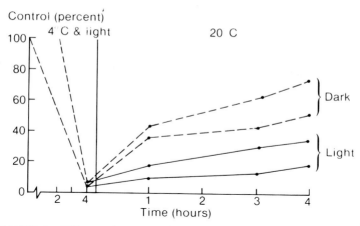

FIGURE 4. Effect of the duration of exposure of cucumber discs to $4^{o}C$ and light on the subsequent reactivation of proton uptake. Reactivation was carried out in the presence or absence of light.

suggestive that light is inhibitory to the recovery of cucumber thylakoids from light and chilling injury. However, the potential for recovery exists even in the presence of light. If cucumber thylakoids are allowed to recover at $20^{o}C$ in light and subsequently transferred to $20^{o}C$ in dark, the rate of recovery approaches that of the dark contol (Figure 5). In contrast to recovery from chilling in the light, the recovery from chilling in the dark proceeds faster in the presence of light (7, 12).

The enhanced thylakoid inactivation due to light and chilling conditions and the subsequent requirements for recovery suggests that much of the chilling-injury associated with field-grown chilling-sensitive plants can be more accurately explained in terms of exposure to chilling temperatures in the presence of light. The inhibited photosynthetic apparatus would recover slowly during the daylight hours (even if temperatures rise above the chilling range) with greatest recovery during the next dark period, provided temperatures are sufficiently high. The results provide an explanation for the lack of recovery from chilling injury even though the day temperatures are above chilling. A warm dark period would be required for optimum recovery from chilling injury.

D. Alterations in Thylakoids during Cold Acclimation

Chilling-sensitive plants exhibit a variety of morphological symptoms when exposed to chilling temperatures, yet a common underlying mechanism of chilling injury has been proposed -- a temperature induced physical change in the lipids of membranes. If the chilling-sensitivity of plants is due to a common event, then might the chilling-resistance of plants, such as spinach, also be the result of a

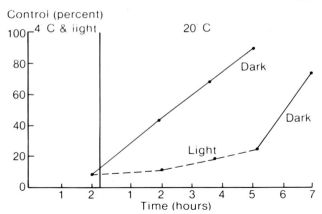

FIGURE 5.   In vivo reactivation of proton uptake of cucumber
thylakoids, following in vivo exposure to 4°C and light, as affected by the
presence or absence of light. Reactivation occurred in the dark only, or
in the light and then the dark.

common trait? A possibility with even greater ramifications is that the
chilling-resistance and freezing-resistance (acclimated tissue) of
spinach be the result of a common property of the cell. For instance, is
the chilling and freezing resistance of the spinach photosynthetic
apparatus due to a "low-temperature-resistant" membrane?

Spinach seedlings are known to undergo cold acclimation in response
to chilling temperatures, facilitating their survival at subfreezing
temperatures.   The possibility of alteration in thylakoids during
acclimation resulting in a more freeze-resistant membrane has been an
issue of some controversy (4, 5, 19). Steponkus, et al., (20) raised an
interesting question in regard to alterations in cellular organelles, "if the
plasmalemma is the site of freezing injury and if during cold acclimation
protection of this membrane is achieved, is it not reasonable to assume
that other membrane systems acclimate, lest they become the primary
site of freezing injury?"

Alterations in the hydrophobic region of thylakoids following
acclimation have been demonstrated with the freeze-fracture technique
(2). There are 850 ± 60 particles/μ/m^2 on the inner fracture face of
acclimated thylakoids compared to 1650 ± 110 particles for non-
acclimated thylakoids. Also, the inner fracture face of non-acclimated
thylakoids have two particle size groups, ± 100 A and ± 165 A in diameter
while acclimated thylakoids have one size group, ± 140 A.   If the
ultrastructural changes are causally related to the cold acclimation
process and consequently increased freezing resistance of the
membrane, then the freeze protection of thylakoids by sucrose should be
different for thylakoids isolated from acclimated and non-acclimated
spinach leaves. In fact, acclimated thylakoids require a 25 mM sucrose
solution for maximum protection of proton uptake whereas non-

acclimated thylakoids require 50 mM sucrose. The difference in sucrose protection of proton uptake for acclimated and non-acclimated thylakoids suggests that the freezing resistance of acclimated thylakoids is at least in part due to a more freeze-resistant membrane.

A comparison of *in vitro* and *in situ* chilling of spinach thylakoids, in the light, with *in vitro* and *in situ* treated cucumber thylakoids suggested that the resistance of the spinach photosynthetic apparatus to chilling in the light is not solely a function of the thylakoid membrane (Table I). The chilling resistance of spinach thylakoids may be due to the interaction of the thylakoids and components of the stroma since "intact" chloroplasts resuspended in an osmoticum were inactivated more slowly than thylakoids washed free of stromal components. However, the osmoticum (sucrose) was not effective in protecting washed thylakoids. In summary, the freezing and chilling resistance of spinach chloroplasts is apparently a function of a "low-temperature-resistant" membrane but the low temperature resistance is only manifested in the presence of selective stromal components. The freezing resistance, but not the chilling resistance, of spinach thylakoids is expressed in the presence of sucrose. The prevention by sucrose of freezing injury but not chilling injury may implicate a differential stress for the freezing and chilling processes.

II. EFFECT OF SUBFREEZING TEMPERATURES ON SPINACH THYLAKOIDS

The differential response of plants to low temperatures has been clearly defined in studies which demonstrate that thylakoids of cucumber are inactivated by chilling temperatures (3) whereas spinach thylakoids are inactivated by freezing temperatures but not chilling temperatures (1,2). A common site of injury apparently exists for both chilling and freezing. The cellular membranes have been implicated as the primary site of chilling (10, 11) and freezing (5, 24) injury. However, no comparison has been made of membrane lesions resulting from chilling and freezing. The nature of chilling injury in cucumber thylakoids has been discussed above and a discussion of freezing injury in spinach thylakoids follows. It should be mentioned that a primary impetus for previous studies on freezing injury in spinach thylakoids was the realization that identification of specific membrane alterations causal to the cold acclimation process requires elucidation of the primary site of freezing injury.

Garber and Steponkus (1) subjected isolated thylakoids to a freeze-thaw cycle with the rate comparable to natural conditions. Two response regions to sucrose were evident suggesting the occurrence of at least two lesions in proton uptake. Negatively stained thylakoids had a reduced number of CF_1 particles when frozen in the absence or presence of low concentrations of sucrose, suggestive that CF_1 is released (1). Post-thaw determination of proton uptake and Ca^{++}-ATPase activity on the same samples confirmed that the reduced proton was due to CF_1 release (20). The proton uptake of thylakoids frozen in 15 to 100 mM sucrose and reconstituted with CF_1 was only 50% of the control, suggesting a second lesion. Thylakoids frozen in the presence of $100 \mu M$ plastocyanin form-ed a proton gradient equal to 33% of the unfrozen controls suggestive that plastocyanin is released, a second lesion during freezing. The inability of CF_1 to stimulate the proton gradient of thylakoids frozen in the absence or presence of 2 or 5 mM sucrose, and the inability of DCCD to maximally stimulate proton uptake below 15 mM suggested that a third lesion contributed to decreased proton uptake. If thylakoids were frozen in 0 to 20 mM sucrose the majority of the thylakoids were osmotically unresponsive whereas, the majority were osmotically responsive if frozen in sucrose concentrations greater than 20 mM. The proportion of osmotically active vesicles paralleled the responsiveness of the vesicles to DCCD, suggestive that the third lesion in proton uptake is the loss of membrane semipermeability. If the three identified lesions, loss of semipermeability, loss of CF_1, and loss of plastocyanin are the cause of reduced proton uptake then amelioration of the lesions should result in a proton uptake comparable to the unfrozen control. In fact, if thylakoids are frozen in the presence of $100 \mu M$ plastocyanin, sufficient to prevent a net loss, and 100 mM sucrose, sufficient to prevent the loss of semipermeability, and after thawing are reconstituted with CF_1, they exhibit a proton uptake equal to the unfrozen control. The identification of specific membrane lesions, particularly the release of a specific protein such as CF_1, takes us one step closer to the elucidation of the primary site of freezing injury in cellular membranes. Attention should now be directed to an understanding of which membrane component(s) retain CF_1 on the membrane and are altered by the freeze-thaw cycle. Alteration of such components during cold acclimation could then be implicated as causal to the increased freezing resistance of acclimated spinach thylakoids.

   Identification of freezing lesions in spinach thylakoids permits a comparison with the mechanism of light and chilling injury in cucumber thylakoids. In both the freezing (5, 24) and chilling (3) stresses, the loss of ATP synthesis is a primary lesion. However, the loss of ATP synthesis due to freezing may be due to altered membrane permeability (26) whereas under light and chilling conditions the altered membrane semipermeability was detected subsequent to the loss of ATP synthesis (3). In addition, freezing injury causes the release of CF_1 particles (2) while chilling in the light results in inactivation of the CF_1 enzyme but not the release from the thylakoid (3). The freezing stress causes the uncoupling of ATP synthesis from electron transport whereas light and chilling inhibits ATP synthesis without stimulating electron transport. The available information indicates that the response of cucumber thylakoids to chilling injury and the response of spinach thylakoids to freezing are different. Therefore, the stress during freezing is probably different than the stress during chilling.

TABLE I.   Pretreatment of Cucumber and Spinach Thylakoid In Vitro

Cucumber and spinach thylakoids were exposed to 4 C in light (21,000 lux) or dark before measurement of membrane functions. The results are expressed as percent of the control (not exposed to chilling temperatures). The control activities for cucumber and spinach thylakoids are, respectively: 421 and 492 μmoles Pi esterified/mg Chl/hr; 0.86 and 0.92 μeq H^+ accum/mg Chl; 185 and 210 μmoles Pi liberated/mg Chl/hr and 200 μg Chl/ml.

Time of Pretreatment (hr)

Sample	Pretreatment Conditions		ATP						Proton uptake				Ca^{++}-ATPase					Chl				
	Light	Dark	1	2	3	4	5	6	2	4	6	8	4	8	12	16	24	8	16	24	32	40
Cucumber	+		90		97			99	104	101	98	101	92	86	96		106	100	98	99	96	100
		+	26	10		2			73	40	24	12	80	62		34	4	98	102	101	80	51
Spinach	+		108	104			92		96	96	97	98	106	104		106	100	103	101	99	102	100
		+	86	44	26		2		70	50	20	10	74	52		30	6	97	99	101	83	48

III. REFERENCES

1. Garber, M. P. and Steponkus, P. L. *Plant Physiol. 57,* 673-680 (1976a).
2. Garber, M. P. and Steponkus, P. L. *Plant Physiol. 57,* 681-686 (1976b).
3. Garber, M. P. *Plant Physiol. 59,* 981-985 (1977).
4. Heber, U. W. and Santarius, K. A. *Plant Physiol. 39,* 712-719 (1964).
5. Heber, U. W. *Plant Physiol. 42,* 1343-1350 (1967).
6. Heber, U. W., Tyankova, L., and Santarius, K. A. *Biochem. Biophys. Acta 291,* 23-37 (1973).
7. Kaniuga, Z., Sochanowicz, B., Zabek, J., and Krzystyniak, K. *P Planta 140,* 121-128 (1978).
8. Keith, A. D. and Melhorn, R. J. *Chem. Phys. Lipids 8,* 314-317 (1972).
9. Kislyuk, I. M., Vas'kovskii, M. D. *Sov. Plant Physiol. 191,* 688-692 (1972).
10. Lyons, J. M., and Raison, J. K. *Plant Physiol. 45,* 386-389 (1970).
11. Lyons, J. M. *Ann. Rev. Plant Physiol. 24,* 445-466 (1973).
12. Margulies, M. M., and Jagendorf, A. T. *Arch. Biochem. Biophys. 90,* 176-183 (1960).
13. Margulies, M. M. *Biochim. Biophys. Acta 267,* 96-103 (1972).
14. Nolan, W. G. and Smillie, R. M. *Biochim. Biophys. Acta 440,* 461-475 (1976).
15. Nolan, W. G. and Smillie, R. M. *Plant Physiol. 59,* 1141-1145 (1977).
16. Raison, J. K., Lyons, J. M., Mehlhorn, R. J., and Keith, A. D. *J. Biol. Chem. 246,* 4036-4040 (1971a).
17. Raison, J. K. *In* "Mechanisms of Regulation of Plant Growth" (R. L. Bieleski, A. R. Ferguson, and M. M. Cresswell, eds.), pp 487-497. The Royal Society of New Zealand, Wellington (1974).
18. Shneyour, A., Raison, J. K., and Smillie, R. M. *Biochim. Biophys. Acta 292,* 152-161 (1973).
19. Steponkus, P. L. *Plant Physiol. 47,* 175-180 (1971).
20. Steponkus, P. L., Garber, M. P., Myers, S. P., and Lineberger, R. D. *Cryobiolgoy 14,* 303-321 (1977).
21. Taylor, A. O. and Rowley, J. A. *Plant Physiol. 47,* 713-718 (1971).
22. Taylor, A. O., Jepsen, N. M., and Christeller, J. T. *Plant Physiol. 49,* 798-802 (1972).
23. Taylor, A. O., Slack, C. R., and McPherson, H. G. *Plant Physiol. 54,* 696-701 (1974).
24. Uribe, E. G. and Jagendorf, A. T. *Arch. Biochem. Biophys. 128,* 351--359 (1968).
25. Van Hasselt, Ph. R. *Acta Bot. Neerl. 21,* 539-548 (1972).

THE INFLUENCE OF CHANGES IN THE PHYSICAL PHASE OF THYLAKOID MEMBRANE LIPIDS ON PHOTOSYNTHETIC ACTIVITY[1]

David C. Fork

Department of Plant Biology
Carnegie Institution of Washington
Stanford, California

I. INTRODUCTION

The energy-conserving reactions of photosynthesis, as well as the energy-liberating reactions of respiration are found within the confines of a lipid bilayer (51). Components essential to the photosynthetic process are in many cases associated with proteins, and are embedded or attached to the thylakoid membrane. In this category would be the reaction centers of photosystems I and II along with their respective acceptors "Q" and the iron-sulfur centers. The highly lipophilic electron carrier, plastoquinone, seems to be separated vectorily across the membrane from other carriers such as cytochrome f and plastocyanin. The photosynthetically-active pigments such as chlorophylls a and b, the phycobilins and, to some extent, the carotenoids are also located in ordered pigment-protein arrays so that efficient energy capture and transfer can occur. The enzyme causing ATP formation by dissipating the H^+ gradients formed in the light across photosynthetic membranes protrudes from the surface of the membrane where it is weakly held to a protein subunit that spans the membrane and is firmly bound within the membrane (1). It would be expected that some of the components essential to photosynthesis would be affected by alterations in the physical state of thylakoid membrane lipids that can occur with changes of temperature.

[1]Carnegie Institution of Washington - Department of Plant Biology Publication No. 664.

II.   THERMOTROPIC PHASE TRANSITIONS

Membrane lipids undergo a phase transition from a random or fluid phase at high temperature to an ordered crystalline (gel or solid) phase at low temperature. The temperature (T_c) at which phase transitions begin depends upon the kind of lipid, the length of the acyl hydrocarbon chains, their saturation and the hydration of the sample. Lipids in the fluid state can undergo rotational isomerization about the C-C bonds of the fatty acid chains. In this state the lipid molecules can diffuse rapidly within the plane of the bilayer (23) but not across the membrane (45). Lipid phase transitions can occur over a very narrow or broad temperature range depending upon the purity of the system. In a model membrane system composed of 1,2 dipalmitoyl phosphatidyl choline, for example, the temperatures of onset (T_c) and completion of melting occur within several degrees (8). In a membrane system composed of two or more lipid types the changes in state can occur at two or more temperature regions characteristic of the individual lipids or be fused to form one broad temperature region that begins with the onset and ends with the completion of melting.

A.   *Fluorescent Probes in Model Membrane Systems*

Figure 1 shows a diagrammatic presentation of phase transitions in model membranes composed of two lipid species. The upper part of the figures shows the result expected using differential scanning calorimetry (DSC) and the bottom part shows the expected temperature dependence of fluorescence yield using a fluorescent probe. In A the endothermic transition occurs in two distinctly separated regions. In B the phase transition occurs continuously over a wide range of temperatures. Above T_{high} in both cases the lipids are in the liquid crystalline state and below T_{low} the lipids are in the gel or solid state. At temperatures between T_{low} and T_{high} the lipids consist of a mixture of the solid and liquid crystalline states. The lower part of the figure represents diagrammatically what would be expected for the yield of fluorescence emitted from fluorescent probes. In Fig. 1A two sigmoid curves would be seen. The beginning of the first steep fluorescence increase corresponds to the onset of the steep rise of the first endothermic transition. The peak of the endotherm and the mid-point of the steep fluorescence rise correspond to the temperature where the transition occurs at its maximum rate. The fluorescence maximum corresponds to the end of the gel to liquid crystalline transition. In the situation illustrated in Fig. 1B the lipid composition is such that only one broad endotherm is seen. The fluorescence shows a broad minimum and maximum at T_{low} and T_{high} corresponding to the beginning and end of the gel to liquid crystalline states respectively.

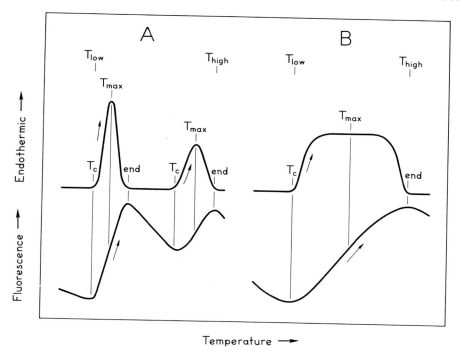

FIGURE 1.   Diagrammatic representation of phase transitions
detected by calorimetric and fluorescent probe techniques in model
membrane systems composed to two lipid species. See text for details.

B.   Chlorophyll a – an Extrinsic and Intrinsic Fluorescent Probe

Colbow (9) found chlorophyll a to be a sensitive probe of phase
transition in phospholipid vesicles. Heating liposomes prepared with
chlorophyll and dipalmitoyl phosphatidyl choline (DPPC) produced a
large sigmoidal fluorescence increase beginning at $41^{\circ}C$, the
temperature of transition of DPPC from the gel to the liquid crystalline
state. Colbow (9) and Lee (24) suggest that chlorophyll molecules in the
gel state of DPPC are so closely spaced that a concentration quenching
effect results in low fluorescence. Increasing temperatures give rise to a
decreasing quenching and a fluorescence increase since the chlorophyll
molecules are not as closely packed in the larger volume available in the
liquid crystalline state.

Another explanation of this effect (32) used the analogy to the behavior of artificial fluorescent probes such as 1-anilinonaphthalene-8-sulfonate or N -phenyl-1-naphthalene that are partitioned in model membranes between the hydrophobic and the aqueous phase and fluoresce when they are bound to the membrane (41). In the gel state fluorescence is low because the hydrophobicity is low and the number of bound probes is low. Transition to the more hydrophobic liquid crystalline state results in more bound probes and increased fluorescence. It is possible that both of these ideas may be used to some extent to explain the fluorescence behavior of chlorophyll a when it is associated with lipid membranes.

It appears that chlorophyll acts as an intrinsic fluorescent probe in natural membranes. This is fortunate since chlorophyll is localized exclusively in the thylakoid membrane and added artificial fluorescent probes have so far not worked well in photosynthetic systems (32). The use of chlorophyll as a natural fluorescent probe rests on the assumption that it behaves in the intact system as in the membrane systems described above and on the correspondence between phase transition temperatures determined by the fluorescence technique and those determined by studies on X-ray diffraction and differential thermal analyses (N. Murata, personal communication). A further correspondence was also noted between phase transition temperatures as determined by fluorescence measurements and those obtained using the spin label 5-SAL that was added to thylakoid membrane fragments prepared from *Anacystis nidulans* grown at 28 and $38^{\circ}C$ (36).

III. THE EFFECT OF PHASE TRANSITIONS ON PHOTOSYNTHETIC ACTIVITY

The studies described here were aimed at identifying temperature induced changes in the physical phase of the thylakoid membrane lipids by examining a wide variety of plants that are capable of carrying out photosynthesis at temperatures ranging from $75^{\circ}C$ to near freezing and studying the effect these phase transitions have on photosynthetic function.

A.   Higher Plants

1.   *Phase Transitions Detected by Chlorophyll Fluorescence.* A number of studies have been directed at identifying the phase transitions of membrane lipids in higher plants using chlorophyll a as an intrinsic fluorescent probe. With the exception of the thermo and xerophilic desert annual *Tidestromia oblongifolia* that had a transition near $5^{\circ}C$ (33), no indications of phase changes were seen in the region from $0^{\circ}C$ to about $22^{\circ}C$ for spinach (field grown) or for lettuce

and tomato grown at 15 and 25°C (31). Chloroplasts extracted from spinach and lettuce leaves and resuspended in 50% ethylene glycol and 5 mM Mg^{++} showed broad fluorescence maxima centered around -31°C. Interestingly, when the Mg^{++} was omitted and 5 mM Na^+ substituted in the medium these maxima shifted to higher temperatures [to -20°C for spinach and to -11°C for lettuce (34)]. This lowering of the freezing point by divalent cations is opposite to its effect in model systems (53) where addition of Mg^{++} normally raises the temperature for the lipid phase transition presumably because Mg^{++} stabilizes the gel phase by removing the electrostatic repulsion between polar head groups. Calcium ions induced a phase separation of phosphatidylserine and phosphatidylcholine membranes into a solid phase of phosphatidylserine bridged by calcium chelation and a fluid phase of phosphatidylcholine molecules (39).

Since lipids of thylakoid membranes from chloroplasts of higher plants and algae (with the exception of some blue-green and red algae) have high contents of unsaturated lipids such as linolenic acid (16), it is reasonable to expect that the co-operative phase transitions in these plants would occur at temperatures below freezing. Shipley *et al.*, (49) used purified lipids from a higher plant and found phase transitions at -30°C for monogalactosyldiglyceride (MGDG) and at -50°C for digalactosyldiglyceride (DGDG).

Studies described above that used the fluorescence of chlorophyll *a* of leaves and chloroplasts and those done on purified lipids from higher plants suggest that the major phase transitions of the bulk lipids occur *in vivo* in most plant thylakoid membranes at very low temperatures.

2.   *Electron Transfer Reactions.* The Hill reaction measured in lettuce chloroplasts grown at 25°C showed no breaks in the Arrhenius plot of the initial rate of reduction of 2,6-dichlorophenol-indophenol (DCIP) as a function of temperature over the range from about 8 to 33°C (35).

Inoue (20) found breaks in Arrhenius plots of DCIP photoreduction occurring at -9°C in spinach fragments (without adding Mg^{++}) that were obtained from plants grown at low and high temperatures. Another break was seen at 10°C but only in fragments prepared from spinach grown in chilling temperatures. The 10°C break was not considered to be produced by a membrane phase transition since it was still seen even after the membrane was disrupted by digitonin treatment. Detergent treatment is known (13) to abolish effects that require intact membranes such as the 515 nm carotenoid shift (discussed later).

The temperature dependence of electron transport in lettuce chloroplasts was also followed by measuring the rate at which oxidized P700 can be reduced in the dark upon excitation of photosystem II with a 3 μsec flash of high intensity red actinic light that resulted in the production of one electron (32). This experiment avoids the need of adding exogenous electron carriers and, unlike the Hill reaction with DCIP, involves all the electron transport carriers such as "Q", the primary acceptor of photosystem II, secondary acceptors, R and plastoquinone, cytochrome f and plastocyanin that function between photosystems I and II. Over the range from 9 to 23°C, the Arrhenius plot also showed no breaks or inflection points. The different portions of the electron transport chain involved in the Hill reaction with DCIP and the electron transport *in vivo* seem to be reflected in the different activation energies found - 10 kcal/mole for the DCIP Hill reaction and 28 kcal/mole for the *in vivo* P700 reduction. Reduction of DCIP may be rate limited at the step involving the dark oxidation of the lipophilic electron transport carrier plastoquinone while the *in vivo* system, as explained above, involves all the electron carriers and may have a different rate-limiting step.

Measurements were made by Cox (10) of the rates of dark reduction of cytochrome f and P700 that were previously oxidized in the light. Breaks were seen in the Arrhenius plots near -35°C for spinach chloroplasts suspended in a medium containing ethylene glycol and Mg^{++}. This temperature corresponds to the phase transition seen at -31°C in spinach chloroplasts as described earlier.

The photoreduction of NADP also requires the complete electron transport sequence and the cooperation of both photosystems. Shneyour *et al.*, (50) using chloroplasts prepared from chilling sensitive plants found a change in slope near 12°C in the Arrhenius plot of the activity of ferredoxin-NADP. By contrast, Riov and Brown (44) using chloroplasts prepared from hardened and non-hardened (chill sensitive) wheat and found no breaks in the Arrhenius plot for NADP reductase activity.

3.   *Cation Induced Membrane Effects.* Addition of divalent cations to chloroplasts in a low salt medium causes increased yield of chlorophyll a fluorescence apparently as a result of configurational changes that alter the efficiency of excitation energy transfer between the two pigment systems of photosynthesis (18, 29, 31, 35). In both lettuce and spinach chloroplasts the rate of this cation-induced fluorescence increase was strongly temperature dependent. The Arrhenius plot for spinach and lettuce (grown at 15 or 25°C) gave no evidence for discontinuities in the temperature range from about 5 to 35°C (36).

4.   *Pigment and Ion Permeability Changes in Thylakoid Membranes.* The absorption of light by the reaction centers of photosynthesis gives rise to negative charges on the outside and positive charges on the inside of the thylakoid membrane. The electric field that is generated by these charge separations can be of such magnitude that the absorption maxima of pigments that are embedded in the membrane such as carotenoids and chlorophylls are shifted to longer wavelengths in the light (55). In the dark the maxima return to their original positions. These absorbance changes are large and can easily be seen in all photosynthetic organisms so far examined (including the photosynthetic bacteria but excluding the blue-green algae).

Changes of this type are most commonly measured at the 515 nm positive maximum in the green algae and higher plants and seem to be caused mainly by β-carotene. The kinetics of the 515 nm change (or "shift") are complex. The initial rise in the light occurs within 20 nsec (56). But continuous illumination leads to a steady state positive 515 nm change that appears to be produced by the formation of a transient diffusion potential across the membrane as a result of the differential permeability of the thylakoid membrane to certain ions (21). Protons accumulate inside the thylakoid space in the light and are released to the outside medium in the dark (37, 47, 48).

In intact leaves the decay of the 515 nm absorption change is fast, about 100 msec (33), probably as a result of the dissipation of the light-induced H^+ gradient that is tightly coupled to phosphorylation. The decay of the 515 nm change in isolated chloroplasts can be much slower [10 sec, Wakamatsu et al., (54)] since phosphorylation would be expected to be less well coupled in chloroplasts as compared to intact leaves. The same trend in decay times is seen in chloroplasts given flash illumination. The decay time of the 515 nm change under non-phosphorylating conditions is around 200 msec but is accelerated up to about 5 times to 40 msec upon addition of ADP, inorganic phosphate and Mg^{++} to promote phosphorylation (56).

The temperature dependence of the dark decay of the 515 nm absorbance change in leaves of the chilling resistant plants spinach and lettuce (grown at 15 and 25°C) showed no discontinuities in the Arrhenius plots in the temperature range from near freezing to about 28°C (33). But leaves of Tidestromia did, however, show a break at 5°C in the Arrhenius plot for the decay of the 515 nm change. This discontinuity was seen both on decreasing and increasing the temperature of the leaf. Interestingly, the decay of the 515 nm change was monophasic when the membrane lipids were in the liquid crystalline state. But below about 5°C, corresponding to the beginning of the phase separation state as revealed by fluorescence measurements, the decay of the 515 nm change became biphasic with the appearance of another, more rapid phase.

Leaves from chilling sensitive plants tomato and bean were also used (33). The Arrhenius curve for the rate of the dark decay of the 515 nm change obtained using tomato grown at 15°C had a break near 12°C. It was seen, particularly in the curves for bean and tomato grown at 25°C, that at the break points the lines not only changed slope but also exhibited an abrupt displacement to new positions on the graph.

So far no clear indications of phase transitions have been seen above 0°C using tomato or bean in measurements of the fluorescence of chlorophyll a. However, Raison (43) has suggested on the basis of studies of epr with spin probes that phase transitions do occur around 10°C in chilling sensitive plants.

Since it is known that ions and small molecules diffuse more rapidly across model membranes that are in the phase separation state (3, 4, 19, 38, 42), the abrupt increase of the decay rate of the 515 nm change below the phase transition temperature suggests that the thylakoid membrane becomes leaky to ions in the phase-separation state. If phosphorylation is coupled to ion movements across membranes as suggested by the Mitchell (26, 27) hypothesis then then it would appear that phosphorylation would be greatly impaired at low temperatures. Ono and Murata (40) have recently shown that phosphorylation in *Anacystis* declines at a temperature corresponding to the onset of the phase transition from the liquid crystalline to the phase separation states. This observation suggests that either the ion gradient across the thylakoid membrane and/or the activity of CF_1, the coupling factor for photophosphorylation, is affected by this phase transition.

The fluorescent indicator 9-amino acridine (9AA) can be used to follow the rate of dark decay of the pH gradient produced previously across the thylakoid membrane in the light (47). When added to a chloroplast suspension this lipid-soluble amine can diffuse across the membrane to establish equilibrium between the inside and outside of the thylakoid space. Protons that accumulate in the light within the thylakoid space combine with the free amine to form the non-fluorescent protonated form that cannot diffuse across the membrane. Thus chloroplasts in the presence of 9AA show a fluorescence decrease upon illumination. Fluorescence increases again in the dark when the protonated amine disassociates and the protons accumulated in the light and bound to 9AA are released and diffuse out of the thylakoid space. This proton release in the dark can be followed as a fluorescence increase at 500 nm. The initial rates of fluorescence increase were taken as a measure of the rate of proton efflux from chloroplast preparations obtained from chilling sensitive and chilling resistant plants.

Measurements starting at 10°C and increasing the temperature in the desert shrub *Tidestromia* gave an Arrhenius plot having a straight line up to about 27°C when a steep increase was seen (2). The break at 27°C seems not to be related to a phase transition but rather to an irreversible inactivation effect since the points above 27°C followed a continuously-upward sloping curve. Decreasing the temperature from points above 27°C did not produce the discontinuity at 27°C again. Instead, the points followed a straight line that represented faster decay times than seen before. If the temperature was raised just to the 27°C point and then lowered immediately in the *Tidestromia* chloroplast fragments then the two lines followed almost the same slope. If the starting temperature was around 8°C and was increased only up to about 20°C and then lowered again, the points of the Arrhenius plot followed along the same line. In this case changes in slope of the line appeared near 9.5 and 5°C. The 5°C point was observed in several other experiments with *Tidestromia* and seems to be a reflection of a phase transition as revealed by the fluorescence measurements described above. Similar measurements were also made using bean, another chilling sensitive plant. Here an anomoly was seen in the Arrhenius plot near 5°C where the line jumped to a new position on the graph. Above about 18°C the points for the rates of proton efflux began to rise steeply. This temperature would seem to correspond to the 27°C point described above for *Tidestromia*.

Measurements of 9AA fluorescence made upon cooling chloroplast fragments of a chilling resistant plant (spinach) from 17° to near 3°C showed no breaks in the Arrhenius plot. Also, no discontinuities were seen upon increasing the temperature again; the points fell on the same line as obtained before. Decreasing the temperature from temperatures above 20°C gave points following a straight line that represented faster decay times than those observed initially. This result suggests an irreversible inactivation of spinach chloroplasts like that described above for *Tidestromia*.

These results using 9AA, and those described above for the 515 nm change, suggest that below the phase transition temperature the membrane becomes leaky to ions and that clear differences can be seen between chilling sensitive and chilling resistant plants when measuring these effects.

B. Blue-Green Algae

1. Phase Transitions Detected by Chlorophyll Fluorescence.
Probably the most extreme case of a chilling sensitive plant occurs in the obligate thermophilic blue-green alga *Synechococcus lividus*. This alga has been reported (6, 7, 22) to photosynthesize up to 75°C, the highest temperatures so far recorded for this process. These algae are obligate thermophiles (25). They will not survive in the light at room temperature but can be kept in the dark at this temperature for some weeks (R. Castenholz, personal communication).

Phase transitions were measured (12) for a number of different "strains" of this alga that were isolated from several different temperature niches in neutral and alkaline hot springs. The curves for the temperature dependence of chlorophyll *a* fluorescence showed maxima (or shoulders) from 40 to 42°C on decreasing and increasing the temperature respectively in cells that were grown at temperatures ranging from 55 to 65°C (Fig. 2D). Cells growing at 55°C were adapted to grow at 38°C. The temperature dependence of fluorescence in these cells showed breaks near 22 and 25°C upon decreasing and increasing the temperature respectively (15).

Figure 2A shows examples of the use of chlorophyll *a* fluorescence to detect phase transitions in intact cells of *Anacystis* grown at 38°C. In this case a shoulder near 24°C and a broad maximum around 16°C were seen upon cooling the cells. Reheating produced a broad maximum near 21°C.

2. Phase Transitions in Phospho- and Glycolipids from Algae.
Vesicles prepared by suspending in buffer the phospholipids extracted from *Synechococcus* grown at 55°C showed phase transitions at 39 and 38°C upon increasing and decreasing the temperature respectively. Vesicles prepared from the glycolipid fraction had maxima at 40 and 39°C upon increasing and decreasing the temperature respectively.

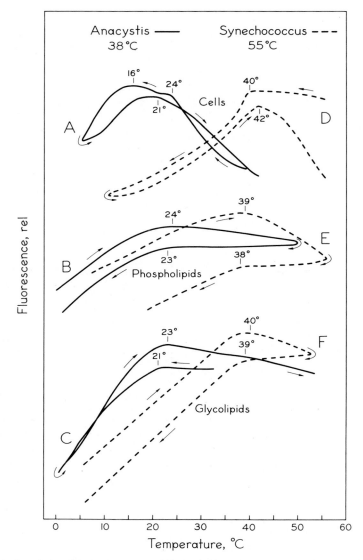

FIGURE 2. Temperature dependence of chlorophyll a fluorescence
in cells of the blue-green algae Anacystis nidulans (grown at 38°C) (31),
and Synechococcus lividus (grown at 55°C (15) and in vesicles made from
lipids extracted from both types of algae.

Partially purified phospho- and glycolipids (mono and
digalactosyldiglyceride and sulfoquinovosyl diglyceride from Anacystis
that contained a trace of chlorophyll a were suspended in 10 mM
phosphate buffer (pH 7.6), shaken with glass beads and sonicated for
several minutes to form vesicles. Vesicles were made in a similar way

These phase transition temperatures in the model systems are somewhat lower for both lipid classes compared to the intact cells.

Figure 2B and C shows that chlorophyll *a*-containing vesicles prepared using phospholipid and glycolipids extracted from *Anacystis* grown at 38°C showed phase transitions in the temperature range from 21 to 24°C similar to the temperatures seen for whole cells.

The vesicles used in these studies were prepared from the phospho- and glycolipid fractions obtained from the whole cells. Presumably, the lipids originated mainly from the thylakoids since in blue-green algae there is an extensive thylakoid membrane system dispersed throughout the cell of which the cytoplasmic membranes make up only a small fraction.

All the phase transitions illustrated in Figure 2 as observed by fluorescence measurements are reversible. These transitions in the vesicles (as well as intact cells) show a hysteresis effect like that described by Trauble and Overath (52) for biological and model membrane systems.

The fluorescence curves shown in Figure 2 are like example B of Figure 1 where the fluorescence maximum represents the beginning of the phase transition from the liquid-crystalline to the phase separation state. No minima were seen in the curves shown in Fig. 2 (except perhaps for curves given in D). This suggests that the transition from the phase separation to the solid states had not occurred within the temperature range used. Fluorescence versus temperature curves of the type shown in A of Fig. 2 are not common. So far only the alga *Cyanidium caldarium* has been seen to have this type of curve (11, 32).

3.   *The Effect of Growth Temperature Changes on Lipid Composition.* The fatty acid composition of *Synechococcus* is like that of the related blue-green alga *Anacystis nidulans* in having only saturated and monounsaturated fatty acids. Polyunsaturated fatty acids

from the more purified lipids extracted from Synechococcus but in this case chlorophyll a extracted from sunflower leaves was added to give a chlorophyll: lipid ratio of 1:400.

Whole cells were treated with DCMU to inhibit photosynthetic electron transport reactions. The fluorescence of cells was excited with 430 nm light and measured at 684 nm. For the lipid vesicles the excitation wavelength was 433 nm and the fluorescence measured at 672 nm. The arrows indicate the direction of temperature change. The experiments with lipid vesicles were done in collaboration with G. van Ginkel.

such as linoleic and linolenic are absent from both of these thermophilic algae. *Synechococcus* contains only four major lipids (15) as do other blue-green algae: Mono- and digalactosyldiglyceride (MGDG, DGDG), sulfoquinovosyldiglyceride (SQDG) and phosphatidylglycerol (PG) (17, 46).

When *Synechococcus* growing at 55°C was adapted to grow at 38°C there was an increase of the more fluid fatty acids (unsaturated and short chains) in all the lipid types (15). In the total lipids the ratio of unsaturated/saturated fatty acids went from 0.31 to 1.31 and the ratio of C_{16} to C_{18} fatty acids increased about 2 times. The negatively-charged lipids SQDG and PG obtained from cells grown at the lower temperature had more unsaturated fatty acids 16:1 and 18:1 and less saturated fatty acid 18:0 while 16:0 did not change. In the uncharged lipids MGDG and DGDG only the unsaturated fatty acid 16:1 increased when the growth temperature was lowered and 18:1 remained the same. However, both the saturated fatty acids 16:0 and 18:0 decreased in MGDG and DGDG at the lower growth temperature.

4. *The Effect of Growth Temperature Changes on Pigment Composition.* Adaptation of *Synechococcus* to grow at 38°C instead of 55°C not only produced distinct alterations in the fatty acid composition of the lipids but also produced marked alterations in their pigment composition (14). The difference spectrum between the 38 and 55°C grown cells showed that growth at the lower temperature produced more of a pigment that appeared to be a carotenoid since peaks and a shoulder could be seen at 480, 452 and 430 nm respectively. The 38°C cells contained less phycocyanin as well as was shown by the decrease at 618 nm. Differences in the region of chlorophyll absorption could also be seen between the two cultures. It appears, in addition to the lipid changes described above, that alteration of pigment composition (particularly of carotenoids) is an important part of the process of adaptation to decreased growth temperatures.

5. *Electron Transfer Reactions.* In order to see if phase transitions in thylakoid membranes of *Synechococcus* can be correlated with changes of photosynthetic functioning, measurements were made on the temperature dependence of electron transport between the two photosystems as well as on the temperature dependence of alterations in the state of pigments in the photosynthetic membrane that lead to changes in the distribution of quanta between the two pigment systems of photosynthesis (the so-called pigment state 1 - state 2 shift).

The temperature dependence of electron transfer can be followed readily in *Synechococcus* by measuring the rates of a transient reduction of the f type cytochrome that functions as a carrier between the two light reactions of photosynthesis. The Arrhenius plots showed discontinuities near 43 and 26°C for cells grown at 55°C and at 37 and 27°C for cells grown at 38°C (15). When the cells were warmed again the breaks were also seen again near these same temperatures (with a hysteresis effect). The discontinuity observed at 43°C in the curve for electron transport corresponds to the appearance of a phase transition between the liquid crystalline and the phase separation states seen near this temperature in the fluorescence versus temperature curve for cells, phospho- and glycolipids (Fig. 2 D, E, F). So far no clear discontinuity has been seen near 37°C in the fluorescence curve; although an indistinct shoulder sometimes appeared near this temperature. Even though it has not yet been clearly demonstrated, it is assumed that a phase transition occurs near 37°C and is responsible for the 37°C break seen in the plot of electron transport in the cells grown at 38°C.

The characteristic points seen near 26°C for both the cells grown at 55 and 38°C was not reflected in a clear way in the fluorescence versus temperature curve; although sometimes a shoulder was seen at this temperature. The origins of these characteristic points at the lower temperatures are still being investigated.

6.   *Pigment State Changes.* Illumination of algal cells with wavelengths of light that cause unbalanced excitation of the two photosystems of photosynthesis is overcome to some extent by a kind of self-regulating system that allows transfer of light quanta from the photosystem that receives too many quanta to the photosystem that receives too few (5, 28, 30). This mechanism presumably involves changes in orientation and/or in distance between photosynthetic pigments embedded in the thylakoid membrane so that energy transfer probabilities are altered. Illumination of the blue-green alga *Synecho-coccus* with blue light that excites largely only photosystem I gives rise to a condition (termed state 1) whereby some of the excess blue quanta are transferred to photosystem II thus exciting both photosystems more evenly and causing a more efficient photosynthesis. If these cells are suddenly switched to green light that is absorbed preferentially by photosystem II then a transient high level of chlorophyll *a* fluorescence is seen since the cells in state 1 cause quanta to be transferred to or remain in system II and to be emitted as fluorescence. At 53°C this transient high fluorescence decreases with a half time about 0.1 sec for cells grown at 55°C as the algae pass from state 1 to state 2. In state 2 the excess quanta arriving in photosystem II are transferred to photosystem I.

We found that this state 1 to state 2 transition was highly temperature sensitive. Measurement of the rates of change of this membrane state as a function of the reciprocal of the absolute temperature for *Synechococcus* grown at 55 and 38°C revealed clear discontinuities at 44 and 37°C respectively (15). The line obtained for the 55°C cells had an abrupt change of slope and jumped to a new position at a temperature that corresponded to the beginning of the transition from the liquid crystalline to the phase separation states. For cells grown at 38°C a possible second break was seen near 25°C; but this observation rests on only one data point. The breaks in the Arrhenius plots were a reflection of a reversible event since they were again seen near these same temperatures when the cells were heated.

IV. REFERENCES

1.   Avron, M. *Ann. Rev. Biochem. 46*, 143-155 (1977).
2.   Avron, M., and Fork, D. C. *Carnegie Inst. Year Book 76*, 231-235 (1977).
3.   Blok, M. C., van der Neut-Kok, E. C. M., van Deenen, L. L. M., and de Gier, J. *Biochim. Biophys. Acta 406*, 187-196 (1975).
4.   Blok, M. C., van Deenen, L. L. M., and de Gier, J. *Biochim. Biophys. Acta 443*, 1-12 (1976).
5.   Bonaventura, C., and Myers, J. *Biochim. Biophys. Acta 189*, 366-383 (1969).
6.   Brock, T. D. *Nature 214*, 882-885 (1967).
7.   Castenholz, R. W. *In* "Taxonomy and Biology of Blue Green Algae" (T. V. Desikachary, ed.), pp 406-418. Univ. of Madras, Madras (1972).
8.   Chapman, D., Urbina, J. and Keough, K. M. *J. Biol. Chem. 249*, 2512-2521 (1974).
9.   Colbow, K. *Biochim. Biophys. Acta. 318*, 4-9 (1973).
10. Cox, R. P. *Eur. J. Biochem. 55*, 625-631 (1975).
11. Fork, D. C. and Murata, N. *Special Issue of Pl. & Cell Physiol.* 427-436 (1977).
12. Fork, D. C., and Murata, N. *Carnegie Inst. Year Book 76*, 222-226 (1977a).
13. Fork, D. C., Amesz, J., and Anderson, J. M. *Brookhaven Symp. Biol. 19*, 81-94 (1967).
14. Fork, D. C., Murata, N. and Sato, N. *Carnegie Inst. Year Book 77*, 283-289 (1978).
15. Fork, D. C., Murata, N., and Sato, N. *Pl. Physiol. 63*, 524-530 (1979).
16. Hitchcock, C., and Nichols, B. W. "Plant Lipid Biochemistry." Academic Press, London, New York (1971).
17. Holton, R. W., Blecker, H. H and Onore, M. *Phytochem. 3*, 595-602 (1964).
18. Homann, P. *Pl. Physiol. 44*, 932-936 (1969).
19. Inoue, K. *Biochim. Biophys. Acta 339*, 390-402 (1974).
20. Inoue, H. *Pl. and Cell Physiol. 19*, 355-363 (1978).
21. Jackson, J. B., and Crofts, A. R. *FEBS Lett. 4*, 185-189 (1969).
22. Kempner, E. E. *Science, 142*, 1318-1319 (1963).
23. Kornberg, R. F., and McConnell, H. M. *Proc. Nat. Acad. Sci. USA 68*, 2564-2568 (1971).
24. Lee, A. G. *Biochem. 14*, 4397-4402.
25. Meeks, J. C., and Castenholz, R. W. *J. Therm. Biol. 3*, 11-18 (1978).
26. Mitchell, P. *Nature 191*, 144-148 (1961).
27. Mitchell, P. *Biol. Rev. 41*, 445-502 (1966).
28. Murata, N. *Biochim. Biophys. Acta 189*, 171-181 (1969a).
29. Murata, N. *Biochim. Biophys. Acta 189*, 171-181 (1969a).
30. Murata, N. *Biochim. Biophys. Acta 205*, 379-389 (1970).
31. Murata, N. *Biochim. Biophys. Acta 245*, 365-372 (1971).

32. Murata, N., and Fork, D. C., *Pl. Physiol. 56*, 791–796 (1975).
33. Murata, N., and Fork, D. C. *Biochim. Biophys. Acta 461*, 365–378 (1977).
34. Murata, N., and Fork, D. C. *Pl. and Cell Physiol. 18*, 1265–1271 (1977a).
35. Murata, N., Tashiro, H., and Takamiya, A. *Biochim. Biophys. Acta 197*, 250–256 (1970).
36. Murata, N., Troughton, J. H., and Fork, D. C. *Pl. Physiol. 56*, 508–517 (1975).
37. Neumann, J., and Jagendorf, A. T. *Arch. Biochem. Biophys. 107*, 109–119 (1964).
38. Nicholls, P., and Miller, N. *Biochim. Biophys. Acta 356*, 184–198 (1964).
39. Ohnishi, S., and Itoh, T. *Biochem. 13*, 881–887 (1974).
40. Ono, T., and Murata, N. *Biochim. Biophys. Acta 545*, 69–76 (1979).
41. Overath, P., and Trauble, H. *Biochem. 12*, 2625–2634 (1973).
42. Papahadjopoulos, D., Jacobson, K., Nir, S., and Isac, T. *Biochim. Biophys. Acta 311*, 330–348 (1973).
43. Raison, J. K. *J. Bioenerg. 4*, 285–309 (1973).
44. Riov, J., and Brown, G. N. *Cryobiol. 15*, 80–86 (1978).
45. Rothman, J. E., and Lenard, J. *Science 195*, 743–755 (1977).
46. Sato, N., Murata, N., Miura, Y., and Ueta, N. *Biochim. Biophys. Acta, 572*, 19–28 (1979).
47. Schuldiner, S., Rottenberg, H., and Avron, M. *Eur. J. Biochem. 25*, 64–70 (1972).
48. Schuldiner, S., Rottenberg, H., and Avron, M. *Eur. J. Biochem. 39*, 455–462 (1973).
49. Shipley, G. G., Green, J. P., and Nichols, B. W. *Biochim. Biophys. Acta 311*, 531–544 (1973).
50. Shneyour, A., Raison, J. K., and Smillie, R. *Biochim. Biophys. Acta 292*, 152–161 (1973).
51. Singer, S. J., and Nicholson, G. L. *Science 175*, 720–731 (1972).
52. Trauble, H., and Overath, P. *Biochim. Biophys. Acta 307*, 491–512 (1973).
53. van Dijck, P. W. M., Ververgaert, P. H. J. Th., Verkleij, A. J., van Deenen, L. L. M., and de Gier, J. *Biochim. Biophys. Acta 406*, 465–478 (1975).
54. Wakamatsu, K., Ikehara, N., and Nishimura, M. *Pl. and Cell Physiol. 15*, 601–610 (1974).
55. Witt, H. *Quart. Rev. Biophys. 4*, 365–477 (1971).
56. Wolff, Ch., Buchwald, H., Ruppel, H., Witt, K., and Witt, H. T. *Z. Naturforschg. 24b*, 1038–1041 (1969).

FREEZE-THAW INDUCED LESIONS
IN THE PLASMA MEMBRANE[1]

Peter L. Steponkus

Department of Agronomy
Cornell University
Ithaca, New York

Steven C. Wiest

Department of Horticulture and Forestry
Cook College -Rutgers University
New Brunswick, New Jersey

I.   INTRODUCTION

As membrane damage is commonly inferred to be the primary cause of freezing injury (see Chapter 1), characterization of membrane lesions, especially to the plasma membrane is essential to a complete understanding of the mechanism of freezing injury. A major impediment to such a characterization is the inability to isolate plant plasma membranes in both sufficiently pure form and in a state which can be shown to be identical to their state *in vivo*. An alternative is to study the plasma membrane *in situ*. We have adopted such an approach by studying the effects of a freeze-thaw cycle on isolated protoplasts. Such studies with isolated protoplasts provide an opportunity to study the repercussions of freezing on the plasma membrane under relatively standard and simplified conditions where the complexities imposed by tissue organization or cell walls are eliminated. These factors and other considerations [see Wiest and Steponkus (48)] have precluded a direct comparison of freezing of plant cells vs. mammalian cells or microorganisms. Mazur (23) has indicated that these have been two distinct fields in the area of cryobiology and little effort has been made to ". . . incorporate the results into more effective concepts of freezing injury."

[1]*Department of Agronomy Series Paper No. 1283.*

Integration of the observations in the separate areas of plant and mammalian cryobiology is of considerable importance. Mammalian cryobiology is well advanced in the delineation of physico-chemical events occurring during freezing but is deficient in specific information regarding the resultant cellular membrane lesions. In contrast, plant cryobiology is advanced in the latter area but deficient in the former. Additionally, plant cells have the capacity to increase in cold hardiness. Isolated protoplasts provide a system in which the repercussions of a freeze-thaw cycle can be directly compared with those of mammalian cells exposed to similar environments. If commanlities can be demonstrated, integration of the observations in the two areas will be enhanced considerably, and significant benefits could accrue to both fields.

A. The Freezing Process and Repercussions on the Cellular Environment

The physico-chemical events which occur during the freezing of a cell suspension and the repercussions of the freezing process on the cellular environment have been discussed in the first chapter of this volume. Since freezing results in a multitude of stresses, consideration of the freezing process as a sequential series of potentially lethal stress barriers appears especially appropriate. An initial stress barrier which must be overcome is the disequilibrium between the chemical potential of water in the cell and that of the surrounding medium following the initial extracellular nucleation event. This may be achieved either by cellular dehydration and continued extracellular ice formation or by intracellular ice formation.

Direct observations of cereal (rye) protoplasts exposed to varied freeze-thaw regimes using an electronically-programmable cryomicroscope capable of precise control of cooling rates and temperatures (16) indicate that the probability of intracellular ice formation increases if the cooling rate exceeds $3.0^{\circ}C/min$ (Fig. 1). However, the influence of cooling rate on the probability of intracellular ice formation is influenced considerably by the minimum temperature to which the cell suspension is cooled (5). If cooled to temperatures between -2° and $-5^{\circ}C$, the probability of intracellular ice formation is extremely low--regardless of the cooling rate (in the range of 2° to $120^{\circ}C/min$). If cooled to temperatures between -5° and $-20^{\circ}C$, the occurrence of intracellular ice formation is strongly influenced by the cooling rate (in the range of 1° to $80^{\circ}C/min$.). The probability is close to zero at rates less than $3^{\circ}C/min$. and increases as the cooling rate increases. At cooling rates greater than $20^{\circ}C/min$., the probability is greater than 95 percent. If cells are cooled to temperatures lower than $-20^{\circ}C$, the probability of intracellular ice formation is greater than 95 percent--regardless of the cooling rates imposed (in the range of 4° to $80^{\circ}C/min$.). A preliminary integration and analysis of these results indicates that the probability of intracellular ice formation is high if the extent of supercooling of the intracellular solution is greater than 10 degrees but is low if the extent is less than 3 degrees (15).

If cells are able to achieve equilibration by cellular dehydration, several consequences of dehydration are encountered. Each may be considered as a potentially lethal stress barrier. Lovelock (18) alluded to this possibility in considering the effect of freezing on red blood cells. Mazur (26) has similarly proposed that between the time the cells face the first extracellular ice and the time that they are returned to post-thaw conditions, they meet a sequence of events--any one of which is potentially lethal. He further suggests that the concentration of solutes (electrolytes) is the first such event and is followed by cell shrinkage. While it is true that extracellular solute concentration must occur before cellular dehydration and shrinkage, it might be more appropriate to consider that volume reduction is the first potentially lethal stress barrier encountered. On the basis of relative changes in intensity, reductions in cell volume increase in intensity at relatively warm sub-zero temperatures--before increased solute concentrations achieve appreciably high levels. This is because cellular volume decreases as an inverse function of temperature while solute concentration (osmolality) increases linearly with decreases in temperature (Fig. 2). In such a

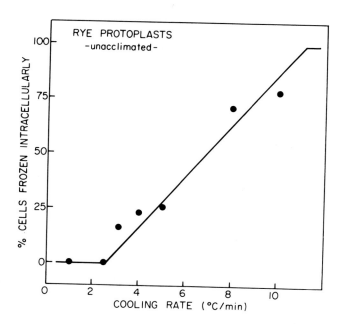

FIGURE 1. The probability of intracellular ice formation in isolated rye protoplasts as a function of cooling rate.

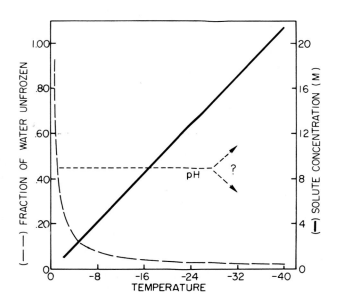

FIGURE 2. A schematic representation of changes in cell volume, solute concentration and pH during freezing to demonstrate the sequential nature of potential stress barriers.

scheme, changes in pH may be envisioned to occur abruptly--at some temperature where buffering salts precipitate. Whether reductions in cell volume reach injurious levels before concentration of electrolytes becomes injurious remains to be resolved, although Meryman (28, 30, 31) provides evidence to support this concept.

If reductions in cell volume are injurious, they may be manifested in the contracted state or manifestation of a lesion may be deferred until cell expansion occurs during thawing. Hence, the expansion process may also be considered as a potential stress barrier which must be overcome by the cell on its return to the osmotic concentrations of the thawed solution. Mazur (26) has proposed that cells are injured by different stresses depending upon the freezing conditions: specifically, injury in the absence of cryoprotectants is due to causes different than those which result in injury in the presence of cryoprotectants. We would further suggest that if a sequential series of stress barriers is encountered during a freeze-thaw cycle, it is entirely appropriate to envision different cellular lesions as a result of the different stresses. Thus, while injury may be caused by any one of several stress barriers it may be manifested by any one of several potentially lethal strains and resultant lesions. Experimental demonstration of multiple, but

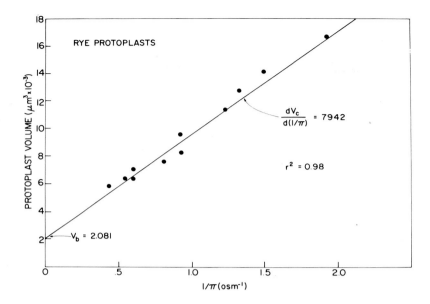

FIGURE 3. Boyle van't Hoff plot of isolated rye protoplasts after 10-min. contraction in salt (equiosmolar $CaCl_2$ + NaCl).

temporally separable, lesions in chloroplast thylakoid vesicles which preclude manifestation of light-induced proton uptake has been reported (8, 44). A similar demonstration of multiple lesions in the plasma membrane remains to be presented.

With the above conceptual framework, we would like to specifically address the stresses associated with cellular volumetric changes in protoplasts during a freeze-thaw cycle and the resultant lesions that are specifically manifested as lysis of the protoplast. Such a delineation does not infer that other stresses which could result in other potentially lethal lesions do not exist or are not significant. Our objective is merely to examine the potentially lethal stress barriers presented by contraction and expansion and the resultant plasma-membrane-associated manifestation, lysis, which intuitively is most likely to result. The possibility and likelihood of other more subtle lesions resulting from these or other stresses and manifested as metabolic dysfunctions of the membrane (32) are not precluded.

B.   Osmometric Behavior During a Freeze-Thaw Cycle

Protoplasts behave as ideal osmometers and exhibit characteristic Boyle van't Hoff behavior in solutions between 0.35 and 2.75 osmolal, the limits of the range examined (48). A typical plot for rye protoplasts is presented in Figure 3. Volume varies linearly (r^2 = 0.94) with osmolality^{-1}. The minimum volume is equal to the osmotically-inactive volume of the cell, V_b. The volume at any particular osmolality may be described by the equation:

$$\frac{V}{V_o} = \frac{P_o}{P} \ (1-V_b) + V_b$$

where V = volume, P = osmolality, V_b = osmotically inactive volume. The volumetric behavior is largely determined by the osmotically-active internal solute content.

During the course of extracellular ice formation and cellular dehydration, protoplasts will contract with the minimum volume achieved a function of the lowest temperature encountered. Subsequently, they will expand to an extent determined by the osmolality of the suspending medium. The osmolality of the partially frozen solution bathing the cells at any given subzero temperature can be approximated as:

$$m \ \tilde{} \ - \frac{T}{1.86}$$

The cell volume at any given subzero temperature can be approximated by assuming that the cells behave as osmometers and applying the Boyle-van't Hoff law (3). Thus, the intensity of the freezing stress can be expressed as a function of dehydration rather than temperature. This would be especially important if dehydration rather than temperature is the primary stress resulting in injury and most important if the extent of cellular contraction is of immediate concern.

The above description of volumetric changes during freezing is a greatly simplified portrayal and at best can only provide a qualitative description of cellular volumetric changes during a freeze-thaw cycle. Numerous models have been presented (11, 12, 13, 14, 21, 22, 24, 25, 34, 39) and have considered various parameters which will influence the fundamental behavior, most notably the water permeability of the plasma membrane, solution non-ideality, diffusivities of solutes in the intra- and extra-cellular solution, and the surface area to volume ratio of the cell. Mazur (25) acknowledges that a major source of error can be attributed to the incorrect estimation of hydraulic water pemeability and its temperature coefficient. Until these problems are resolved, the volumetric responses of cells in disequilibrium at subzero temperatures can only be qualitatively predicted, however, the predictions are valid under equilibrium conditions.

Using a recently developed cryomicroscope (16) we have directly observed the volumetric changes in protoplasts during a freeze-thaw cycle. Video recording of the sequence allows for the accurate and precise determination of cellular volumetric changes during the freeze-thaw cycle. Resolution of 0.2 μm is possible as 2.5 mm on the video monitor represents 1.0 μm in actual dimensions. Routinely, cereal

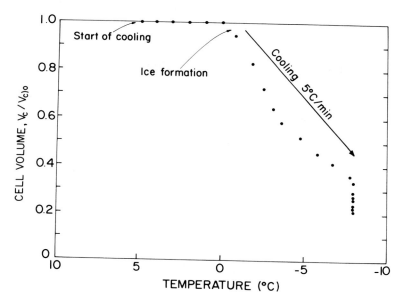

FIGURE 4. Changes in cell volume of an isolated rye protoplast during freezing to -8°C at 5°C/minute.

protoplasts are in the range of 20 to 30 μm in diameter at 0.54 osmolal. Although protoplasts subjected to osmotic manipulation at room temperature contract spherically in concentrations up to 2.75 osmolal, contraction during freezing may be irregular, especially when temperatures lower than -5° are imposed. Beginning as spheroids, protoplasts may become contorted in the ice field, and an ellipsoidal model is then required for the calculation of the volume. Such transformations, however, are only required after the majority of contraction has already occurred. Fractional volumetric changes in a rye protoplast frozen to -8°C at 5°C/min are shown in Figure 4.

C. Plasma Membrane Lesions Induced by Cellular Volumetric Changes

Previously, we have demonstrated simulation of a freeze-thaw cycle may be achieved by osmotic manipulation of protoplasts at room temperature (48). The extent of injury (percent lysis) incurred by a population of cells subjected to a freeze-thaw cycle can be quantitatively accounted for by the extent of contraction and expansion the protoplasts undergo. Additionally, the kinetics of freeze-thaw injury were as predicted by osmotic manipulation. Thus, both the extent and kinetics of injury in protoplasts subjected to a freeze-thaw cycle are similar to those subjected to osmotic manipulation. These facts strongly suggest that injury to protoplasts during a freeze-thaw cycle and manifested as cell lysis is due to the same stresses of contraction and expansion that result from osmotic manipulation in the absence of ice or low temperatures. However, other stresses, most notably electrolyte concentration, could have been responsible for the injury observed. This possibility (with respect to lysis as the manifested injury) is diminished, however, because similar results were obtained whether the osmotic manipulations or the freeze-thaw cycle were conducted in an ionic ($CaCl_2$ + NaCl) or non-ionic osmoticum (sorbitol or mannitol). Additionally, if the cells were exposed to varied degrees of contraction, either by varied concentrations of osmoticum or varied freezing temperatures, the extent of lysis could be modified by altering the extent of expansion--by varying the osmolality of the suspending medium during expansion or thawing. We submit that the stresses of contraction and expansion, in themselves, result in potentially-lethal strains or membrane lesions manifested as lysis of the cell. Previously, Scarth *et al.* (38) suggested that injury to the plasma membrane was related to plasmolysis and deplasmolysis which occurred during freezing and thawing and that lysis occurred during deplasmolysis.

Meryman (28) has proposed a "minimum critical volume" hypothesis to account for freeze-induced lysis in a wide variety of cell types. This hypothesis proposed ". . .that membrane injury from freezing is the result of osmotic stress produced by the freezing-out of water and the associated increase in osmolality of the suspending solution leading to a reduction in cell volume beyond a tolerated minimum" (29). Such a minimum cell volume may be interpreted in two ways: either the occurrence of injury is simply correlated with the attainment of a certain cell volume or that there is, in fact, a physical minimum volume at osmolalities less than infinity beyond which the cell cannot shrink, and that the attainment of this physical minimum volume creates stresses which lead to hemolysis. For the latter interpretation to be valid, deviations from ideal Boyle-van't Hoff osmometric behavior should occur at osmolalities which result in injury (with lysis subsequently occurring on dilution and expansion). Protoplasts do not exhibit any such deviant behavior in the region of osmolalities which result in subsequent lysis. Recent analysis of earlier data also suggests that human erythrocytes do not exhibit such deviant behavior (51). On this basis, we consider that the "minimum critical volume" hypothesis should be viewed from the perspective that injury is simply correlated with the attainment of a certain cell volume.

Several proposals have been put forth to account for the fact that injury occurs following exposure to hypertonic conditions when the cell is simultaneously experiencing high solute concentrations (strictly electrolytes in some cases) and volume reductions. Most explictly state or infer that lysis will occur during cell expansion when the cells are returned to isotonic (18, 19, 52) or hypotonic (33) conditions. Lovelock (18) proposed that red blood cells become permeable to Na^+ and the uptake of solutes causes the cells to subsequently plasmolyze if resuspended in "isotonic conditions." Zade-Oppen (52) proposed that cells took up electrolytes in hypertonic solutions and burst upon return to isotonic conditions due to the fact that the cell exceeded a maximum tolerable volume. Similarly, Meryman (29) concluded that an influx of solute during hypertonic exposure increased the intracellular solute concentration and that the cell will therefore ". . .reach hemolytic volume at a higher osmolality." Mazur (26) invoked a transient hydrostatic tension which produces a driving force for a net influx of solutes.

The proposals differ in some respects, most notably, Lovelock proposes that the altered permeability is attributable to the increase in electrolyte concentration; Meryman attributes it to physical stresses associated with a minimum critical volume; Mazur indicates that it appears related to isotropic cell shrinkage. However, regardless of the reason, all of the proposals suggest that solute influx occurs during the hypertonic exposure and lysis is attributed to increased volumes which occur on subsequent exposure to hypertonic solutions. Implicit is the notion of a fixed maximum volume which is merely achieved at higher osmolalities. If this were the case, the cells would exhibit altered osmometric behavior during dilution. To determine whether a "loading" of protoplasts occurs under hypertonic conditions, protoplast volumes were determined in various hypertonic solutions before and after varying extents of dilution from these solutions (Fig. 5). Had an irreversible

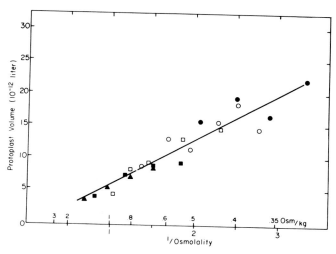

FIGURE 5. Boyle van't Hoff plot of protoplast volume after 10-min. contraction in salt (equiosmolar $CaCl_2$ + NaCl) followed by immediate addition of H_2O to induce protoplast expansion and obtain final osmolalities shown in the figure. Contraction was achieved in salt concentrations of 0.536 osmolal (●), 0.803 osmolal (O), 1.071 osmolal (◻), 1.338 osmolal (■), and 1.607 osmolal (▲). Regression coefficient (r = .951) is significant at the 1% level.

influx of solutes occurred, protoplast volumes would deviate from a linear van't Hoff plot. If more solute influx was to have occurred in protoplasts exposed to 1.6 osmolal solutions than in those exposed to 1.0 osmolal, the volume of the former, following dilution to 0.8 osmolal, would be greater than that of the latter, which were also diluted to 0.8 osmolal. The linearity of the volumetric response and coincidence of the volumetric responses for both contraction and expansion excursions demonstrate that a solute influx and subsequent bursting upon dilution cannot explain the lysis of protoplasts observed following dilution.

Most proposals relate injury as a function of volumetric responses. Red blood cells are thought to possess a maximum critical volume (33). It should be emphasized, however, that although osmometric behavior is best characterized as a function of volume and extensive data suggest that a reduction in cell volume *per se* is a significant stress imposed on a cell during hypertonic treatment; lysis is the result of a dissolution of the plasma membrane when a critical surface area of the plasma membrane is exceeded. Thus, while the cellular response to hypertonic

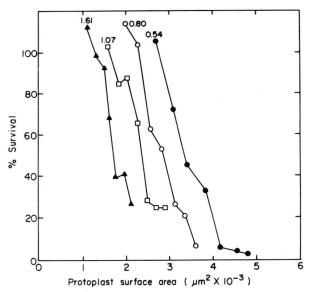

FIGURE 6. Protoplast survival plotted as a function of surface area of the protoplasts when contracted in varying salt concentrations and then expanded to varying extents by dilution of the osmoticum, as in Fig. 5. Numbers above each line represent the highest osmolality to which protoplasts were exposed before dilution. Surface areas were calculated from volumes given by the least squares line in Figure 5; osmolalities are represented by the same symbols as in Fig. 5.

treatment is a decrease in cell volume, the repercussions are manifested as changes in the surface area over which the plasma membrane must be extended. The change in surface area in response to volumetric changes will depend on the geometrical shape of the cell. Red blood cells are concave discoids which are able to double in volume with little or no change in surface area (6). Plant protoplasts are spherical and must increase in surface area during volumetric expansion.

Whether protoplast lysis is or is not associated with the attainment of a maximum size as a function of surface area can be determined by observing survival as a function of plasma membrane surface area (Fig. 6). The spherical geometry of protoplasts permits accurate determinations of surface area from diameter measurements. Protoplasts were isolated in 0.54 osmolal osmoticum ($CaCl_2$ + NaCl) and subsequently exposed to hypertonic solutions before dilution of the osmoticum. Figure 6 demonstrates that 50 percent of the protoplasts exposed to 0.54 osmolal solutions and diluted to varying extents had lysed before they achieved a surface area of approximately 3400 μm^2. Additionally, 50 percent of the protoplasts contracted in 1.61 osmolal solutions before dilution lysed before they achieved a surface area of approximately 1700 μm^2---a surface area less than that which they possessed in 0.54 osmolal solutions. Protoplasts which had been contracted to intermediate values lysed when they were expanded to surface areas intermediate between these two values. The data demonstrate that protoplasts do not possess a fixed maximum surface area at which lysis occurs. Rather, the surface area at which lysis occurs is decreased if protoplasts have previously been subjected to contraction.

The similarity of slopes of the curves further suggests that survival is a function of the absolute surface area increment achieved during expansion, regardless of the other conditions to which they were exposed. For instance, 50 percent survival occurred when protoplasts incurred an absolute surface area increment of approximately 900 μm^2 (Fig. 7a)-- regardless of whether they were initially exposed to 1.61 osmolal and diluted to 0.81 osmolal or exposed to 0.54 osmolal followed by dilution to 0.36 osmolal. When survival is plotted as a function of volume change (Fig. 7b), it becomes apparent that the cellular volume at which lysis of 50 percent of the protoplasts occurs varies with the degree of contraction. The data suggest that although osmometric behavior of protoplasts is best described as a function of volume, the lytic lesion resulting from volumetric changes is best described as a function of surface area of the plasma membrane during expansion.

On the basis of these observations we have proposed that freeze-thaw injury to isolated protoplasts is the result of two major strains: a freeze- or contraction-induced membrane alteration which decreases the maximum critical surface area of the plasma membrane and a thaw- or expansion-induced dissolution of the plasma membrane which occurs when the maximum critical surface area is exceeded. These two strains interact during a freeze-thaw cycle and result in lysis of the cell. Thus, cellular volumetric changes during a freeze-thaw cycle result in injury

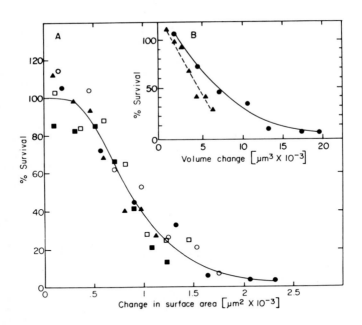

FIGURE 7. *Dependence of protoplast survival on the absolute increase in surface area or volume of the protoplasts. Conditions and symbols were as listed in Figures 5 and 6.*

due to an alteration in the protoplast resilience--the capability of a strained body to recover its size and shape after deformation caused especially by a compressive stress. While the altered resilience is the result of an alteration in the plasma membrane which occurs during contraction, it is not manifested until the protoplasts are induced to expand during thawing--when disruption of intermolecular forces in the membrane causes protoplast lysis.

As early as 1939, Tornava (46) observed protoplast diameters of several plant species at the moment of hypotonic lysis. The general conclusion was that protoplasts burst when the surface area was approximately double the initial surface area, although it was added that ". . .in many experiments (the surface area at the moment of lysis, as a percentage of the original surface area) has been 110 percent or less, and in several trials, where the protoplast first has been made to contract itself or been plasmolyzed, not even the original surface area has been reached." Scarth *et al.* (38) also described this phenomenon in a qualitative manner: ". . .the limit (of expansion) varies greatly with such simple treatment as plasmolysis. The point of bursting of cells which have been strongly plasmolyzed beforehand is lower than those weakly plasmolyzed." The results presented in this paper and an earlier report

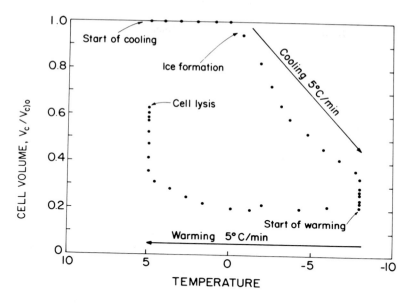

FIGURE 8. *Changes in cell volume of an isolated protoplast (the same cell as in Figure 4) during a freeze-thaw cycle. Minimum temperature attained was $-8^{\circ}C$, cooling and warming rates were $5^{\circ}C/min$.*

(48) are in complete agreement with these earlier observations. Recently, direct measurements of cellular volumetric changes during a freeze-thaw cycle support the conclusion that cells which have undergone contraction will undergo lysis at a size smaller than the original size (Fig. 8). The observations further support the contention that protoplasts do not possess a single unalterable surface area at which lysis occurs. In previous proposals regarding the mechanism of injury, the basis for injury was ascribed to altered volumetric relations due to solute influx. In protoplasts, we propose that contraction directly alters the membrane resilience and the need for a solute influx to provide a driving force for lysis is not required.

As was stated earlier, Meryman (28, 29) has proposed the "minimum critical volume" hypothesis to account for freeze-thaw induced lysis which can be interpreted in two ways. For the hypothesis to be applicable to plant protoplasts, the interpretation that injury is correlated with the attainment of a certain cell volume rather than the attainment of a physical minimum volume must be made. Furthermore, while volume/surface area reduction does lead to some membrane alteration which limits the membrane expansion potential, this contraction-induced alteration is a continuous (or nonresolvable discrete) function of contraction rather than the existence of some minimum critical volume.

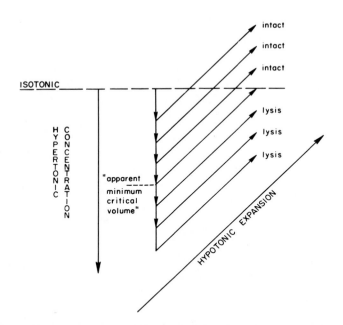

FIGURE 9. A schematic representation of the "apparent" minimum
critical volume of a protoplast.

As a result, any inferred minimum critical volume is only an "apparent"
volume, the manifestation of which depends on the extent of contraction
relative to the extent of subsequent expansion. In other words, the
manifestation of injury resulting from contraction depends on the extent
of expansion during subsequent hypotonic conditions. The "apparent"
nature of a minimum critical volume is illustrated in Figure 9.

    1.   Contraction-Induced Lesion. When a protoplast is exposed to a
hypertonic solution and contracts in a uniform spherical configuration,
components of the plasma membrane should become more tightly
appressed. It is reasonable to expect that this would be a reversible
process and that the membrane should be able to return to its original
surface area when the suspending medium is diluted. However, as was
previously demonstrated, when a protoplast is osmotically induced to
expand from a contracted state, lysis can occur before it regains its
original surface area. Such observations suggest that a contraction-
induced membrane alteration limits the expansion potential of the
plasma membrane. There are two obvious alternatives to account for
this possibility: a qualitative alteration in the ductility of the membrane
or a quantitative reduction in membrane material. Siminovitch and
Levitt (40) concluded that the surface membrane of protoplasts stiffened

when dehydrated osmotically and, as a result, ruptured more easily when subjected to tension. Meryman *et al.* (31) have suggested that lipid material is lost during plasmolysis.

To directly test this latter possibility, the plasma membrane of protoplasts was labelled with fluorescein conjugated Concanavalin A (Con A) (49). In order to minimize autofluorescence by chlorophyll, protoplasts were isolated from the achlorophyllous tissue of the crown region of wheat seedlings. If the protoplasts were subsequently contracted, fluorescence from the Con A-fluorescein appeared aggregated and internalized in the cytoplasm. During expansion the fluorescent material remained in the interior of the cell and there was no observable reincorporation of the material into the plasma membrane. Such observations suggest that invagination and endocytotic vesiculation of the plasma membrane resulted during contraction. Preliminary experiments with thin section electron microscopy indicate that vesiculation of the plasma membrane occurs in contracted protoplasts. Furthermore, this process did not appear to be reversible, as suggested by Meryman *et al.* (31). Evidence that the contraction-induced alteration is irreversible or only slowly reversible was obtained by varying the rate of expansion. If contracted protoplasts are expanded by direct osmotic manipulation, the amount of survival is similar in those which are expanded to the same extent in a stepwise manner with 30 minute equilibration periods between successive dilutions. Thus, the extent of injury is the same, whether expansion is immediate or sequential. The irreversible physical deletion of components from the plasma membrane during contraction may, in part, account for the contraction-induced lesion which limits the expansion potential of the plasma membrane.

2.   *Expansion-Induced Lesion.* Cellular expansion during thawing is a potentially lethal stress which is usually considered from the perspective of altered osmometric characteristics. Tolerance of the membrane to the resultant strain, which is manifested as lysis of the cell, is usually considered constant on the basis of the assumption that the expansion potential will remain constant. The expansion-induced dissolution of the plasma membrane which occurs when the maximum critical surface area is exceeded is a primary manifestation of freezing injury in protoplasts (Fig. 8). It is clearly a temporally separable strain and the environment during thawing could influence the extent to which this strain may be tolerated. Several observations suggest that biochemical influences on the plasma membrane may have a significant influence on freeze-thaw survival of protoplasts.

Both osmotic manipulation and freeze-thaw induced lysis of spinach protoplasts are influenced by the type of monovalent ions present in the suspending medium (45). The sensitivity of spinach protoplasts to a freeze-thaw cycle follows the series: $Li^+ = Na^+ < K^+ = Rb^+ = Cs^+$ and $Cl^- < Br^- < I^-$. Further work has shown that the absolute surface area increment of the plasma membrane that resulted in lysis was significantly less in the presence of K^+ than when Na^+ was present.

Additionally, in the presence of an ionic osmoticum ($CaCl_2$+NaCl or $CaCl_2$+KCl), osmotically-induced lysis gradually increased as the pH decreased below 6.0. Also, when protoplasts are frozen in solutions of varying ratios of $CaCl_2$+NaCl: sorbitol, an optimum in survival was observed. Other work (50) indicates that specific aspects of the lytic lesion and its interaction with the chemical environment varies in protoplasts isolated from different species. Unlike protoplasts from spinach, lysis of wheat leaf protoplasts is not strongly influenced by the monovalent cations present in the osmoticum or by pH changes. Also, specifically in wheat protoplasts, expansion-induced lysis, but not ehe contraction-induced alteration, is temperature dependent. A given extent of expansion is more injurious at 20°C than at 10°C or 4°C. Such a temperature influence was not observed in spinach protoplasts.

The observations strongly suggest that the chemical nature of the environment during a freeze-thaw cycle may influence the tolerance of the plasma membrane to the strains incurred during expansion. Such interactions depend on the species in question and are presumably due to differences in plasma membrane composition. In the instances where such chemical interactions are observed, the physical stresses of contraction and expansion remain as the primary stresses responsible for lysis, but tolerance of the cell to these stresses may be quantitatively altered by the chemical environment. In some cases their influence may be minimal, as Meryman *et al.* (31) have indicated that they ". . .find no evidence that biochemical mechanisms are involved in either freezing injury or cryoprotection. . ." Clearly, further work is required to elucidate the molecular nature of both the contraction-induced alteration and the expansion-induced dissolution lesions to precisely delineate the interaction of the membrane with its environment.

D.   *Proposed Hypothesis of Freezing Injury*

To reiterate, freeze-thaw injury to isolated protoplasts is the result of two major strains: a freeze- or contraction-induced membrane alteration which decreases the maximum critical surface area of the plasma membrane and a thaw- or expansion-induced dissolution of the plasma membrane which occurs when the maximum critical surface area is exceeded. These two strains interact during a freeze-thaw cycle and result in lysis of the cell. Specifically, protoplasts do not possess a single maximum surface area at which lysis occurs (Fig. 6). The precise surface area at which lysis occurs is a function of two factors: the extent of contraction incurred and the absolute magnitude of change in surface area which can be tolerated before lysis occurs.

1.   Practical Considerations. In a population of cells, the absolute magnitude of change in surface area that results in lysis of 50 percent of the cells appears to be constant and independent of the extent of contraction (Fig. 7a). We refer to this constant as the Tolerable Surface

Area Increment and use $TSAI_{50}$ to denote the tolerable surface area increment where 50 percent of the cells lyse. Thus, the maximum critical surface area at which 50 percent of the population will lyse is equal to the minimum surface area plus the $TSAI_{50}$. Protoplasts will contract during freezing at slow cooling rates, with the minimum surface area a function of the lowest temperature attained. Subsequently, during thawing they will expand as a function of the osmolality of the suspending medium. Absolute volumes and corresponding surface areas will be a function of the characteristic osmometric behavior and the initial cell sizes. If the extent of contraction is equal to or less than the $TSAI_{50}$, 50 percent or more of the population will be able to subsequently expand to the original surface area without lysing. Alternatively, if the extent of contraction equals or exceeds the $TSAI_{50}$, 50 percent or more of the cells will lyse before regaining their initial (pre-contraction) surface area. By determining the osmometric behavior and $TSAI_{50}$, survival of isolated spinach protoplasts can be quantitatively accounted for by the absolute magnitude of change in surface area which the plasma membrane undergoes during the freeze-thaw-induced contraction and expansion events (48). Further corrections for temperature influence on expansion-induced lysis are required for similar predictions for wheat protoplasts (50).

It is readily apparent that $TSAI_{50}$ is a powerful parameter for assessing the capacity of a population of cells to survive a given freeze-thaw cycle. This value varies among species examined to date: $TSAI_{50}$ = 900 μm^2 for spinach protoplasts (48), 400 μm^2 for wheat protoplasts (49), and 500 μm^2 for rye protoplasts (43). However, one cannot infer a ranking of hardiness based on $TSAI_{50}$ alone. Because of the absolute nature of $TSAI_{50}$, a particular $TSAI_{50}$ value must be viewed in relation to the characteristic osmometric behavior of the cells in question. This behavior will be influenced by the internal solute concentration and the original cell size. Smaller cells or cells with higher internal solute concentrations or smaller osmotically active fractional cell volumes will undergo smaller changes in surface area per unit change in external osmolality or freezing temperature. Thus, both the $TSAI_{50}$ and the osmometric behavior are required to predict the cells capacity to withstand a given freeze-thaw cycle. In such an analysis, the osmometric behavior will describe the volumetric response of the cells to a given freeze-thaw cycle--or the intensity of the strain imposed, while the $TSAI_{50}$ will describe the tolerance of the cells to an imposed strain. Thus, the analysis affords the opportunity to assess the contribution of factors which influence the cells' sensitivity separately from those which influence the actual response to a given freeze-thaw cycle.

2.   *Theoretical Considerations.* The foregoing discussion has been based largely on observations of population averages of cells exposed to osmotic manipulation or freeze-thaw cycles. The observations and interpretations are useful to describe the responses of a <u>population</u> average of cells and the discussion has solely addressed population averages. In order to provide insight into the specific nature of the

plasma membrane alterations and the mechanism of injury, these observations must be projected to individual cells. Unfortunately, isolation of protoplasts from leaves results in a population of cells of varying sizes. Thus, we are able to infer the behavior of individual cells within a population from population averages only by making several assumptions regarding the behavior of the plasma membrane components in cells of different size ranges exposed to compressive and tensile stresses. Derivation of assumptions that are physically reasonable requires us to delineate the physical and/or biochemical processes which might occur during contraction and expansion, and which lead to membrane dissolution. At the present time, we can only attempt to speculate on the molecular mechanism of lysis based on a relatively limited knowledge of the intermolecular interactions of membrane molecules.

First we make the simplest assumption, that intensive mechanical properties of the plasma membrane of initially small cells are identical to the intensive mechanical properties of the plasma membrane of initially larger cells. We use the term "intensive properties" in the thermodynamic sense - properties which are independent of the amount of substance considered. In the case of the situation presently being considered, this assumption will hold true only if the composition and topological arrangement of membrane constituents is identical in initially large and initially small cells. If experimental findings are contrary to this assumption, our model will necessarily be quantitatively incorrect. However, qualitatively our model could still serve useful for the derivation of parameters for individual classes of cells within a mixed population (4).

Identity of intensive properties would further mean that the population surviving a contraction-expansion event would have a size distribution consistent with the original population distribution. In addition, extensive mechanical properties would vary with cell size. One obvious example of an intensive property in the model given later would be the energy of attraction between adjacent membrane components, or the intermolecular spacing between components. An example of an extensive property will then be the $TSAI_{50}$ value. That is, $TSAI_{50}$ would vary as a direct function of initial protoplast size. These aspects of the problem will be approached later as consequences of the model.

It should be noted at this point that all volumes and surface areas in the studies reported were calculated on the basis of the average diameter of the sampled population. That is, the diameters of sampled protoplasts were measured, an average was taken and from this average volumes and surface areas were calculated. This procedure was used because protoplast diameters followed an apparently normal distribution. Since n=25 or 30 in these studies, the Central Limit Theorem (42) was not applicable to surface area or volume averages. Hence, it was deemed appropriate to use the diameter average to calculate the average surface area and volume of the samples so that the resulting means could be evaluated with conventional statistical techniques and residual errors would on the average equal 0. This manipulation of the data is similar in principle to the transformation of skewed data to approach a paranormal distribution, e.g., the use of a geometric mean rather than an arithmetic mean (42).

The fact that protoplasts sufciciently contracted are unable to regain their initial surface area without lysing suggests that some alteration occurs such that the expansion potential of the plasma membrane is decreased. Since the surface area of the plasma membrane can be reduced by more than 50% of the original area during contraction (which translates to an average 29% decrease in the linear distance between adjacent molecules), it is reasonable to assume that this alteration is located within the plasma membrane itself.

Considering membrane structure from a purely mechanistic perspective, it is believed that in the "native" state the membrane is structured such that the global free energy is at a minimum (41), or at least at a local minimum or saddle point (9, 10, 20). The global free energy represents the sum of individual inter- and intramolecular energies of attraction and repulsion. Luke and Kaplan (20) have calculated the family of spherical solutions for the free energy of vesicles as a function of area and have shown it to be an apparent single-minimum energy structure devoid of saddle points. That is, both compression and expansion will result in an increase in the global free energy, which may be unfavorable from an energetic standpoint. Some membrane alteration in structure and/or composition which reduces this increase in free energy would be favorable. Such an alteration could consist of either a rearrangement of membrane molecules intra-or intermolecularly, a deletion of membrane molecules, or both. Results of fluorescent labeling of the plasma membrane (i.e., the apparent internalization of plasma membrane-bound fluorescein-Concanavalin A following contraction) are consistent with a postulated deletion of molecules. It appears that a membrane alteration does occur during contraction, and this alteration consists at least in part of a deletion of membrane components. The possibility exists that this contraction-induced membrane alteration is sufficient to lower the global free energy of the contracted membrane to its previous isotonic value. This would then explain some of the phenomena observed following protoplast expansion.

Consider the following model, which attempts to describe the phenomenon of protoplast lysis as a result of contraction followed by expansion. We present this model here for several reason: 1) to aid the reader in intuitively understanding the lytic phenomenon, and 2) to formally present one interpretation of the data which, although hypothetical, is testable.

A fundamental principle of vesicle (or membrane) mechanics is the relationship:

$$dw = T \, dA$$

where T is membrane tension, A is surface area and w is the work done on the vesicle by the environment (7). In protoplasts, T must be relatively constant during both contraction and expansion in the range of osmolalities tested (0.35 to 2.75 osmolal) since protoplasts follow the Boyle-van't Hoff law. If T increased with an increase in cell surface area, the membrane would become more resistant to stretching and would not follow the Boyle-van't Hoff law (27). Generally, the value of T in an unstrained state has been assumed to be 0 (7), although no direct evidence exists. Rand (35) has calculated the existence of a small, but non-zero, tension. Assuming that T has a relatively constant, non-zero value in protoplasts, dw will be directly proportional to dA. The work done on the membrane is directly related to the increase in its kinetic energy (2). These relationships lead to the concept that the disruption of intermolecular forces in the membrane cause protoplast lysis. One can conceptually envision the process as follows. A membrane is composed of many distinct components whose association is the result of characteristic energies of attraction (37). As the plasma membrane expands, the above relationships demand that the kinetic energy of the membrane increase. Experimentally it has been shown that at least some of this kinetic energy is entropically translated into increased membrane fluidity of plant protoplasts as well as liposomes (1). We speculate that when this increased kinetic energy exceeds the energy of association of some particular site(s), the association(s) will no longer be stable and membrane denaturation (lysis of the cell) will follow.

Such a concept of membrane disruption can be used to consider the observation that, upon dilution, lysis in a population of protoplasts is correlated with the average increase in surface area (48). This implies that disruption of the membrane of a cell of a given initial size requires the same amount of total work (in a given type of osmoticum) regardless of the surface area of the cell at the time of expansion. That is, regardless of the extent to which the protoplast had been contracted, the total increase in kinetic energy of the membrane of a protoplast at the time of lysis remains constant. Thus, we tentatively suggest that protoplast lysis occurs when the kinetic energy dissipated into the membrane during expansion exceeds the magnitude of the weakest of the intermolecular forces joining the membrane together (i.e., the weak link). This suggestion has relevant consequences to the nature of the contraction-induced membrane alteration.

The speculation made above is a generalization of the more specific, but easier to conceptualize, idea that lysis is related to the intermolecular distance between adjoining molecules. Since this latter concept is easier to envision, the following discussion will center about it. However, in an effort to stress the point that this discussion is purely hypothetical, and a gross simplification of true membrane properties, dimensionless numbers will be used. In addition, we are forced to limit discussion to relatively large, homogeneous lipid vesicles. Such vesicles do contract uniformly, remaining as spheres, when exposed to hypertonic solutions, and they do lose their semipermeability when the osmolality is lowered (36). Assume that vesicle lysis will occur at an intermolecular spacing $\delta_{50}=1.225$ and that an "isotonic" vesicle has $\delta=1$. Emperically we observe that 50% of the vesicles lyse when the membrane undergoes a surface area increase $TSAI_{50}=10$. Now let us follow the hypothetical vesicle through a freeze-thaw cycle that results in 50% lysis. The vesicle with an initial "isotonic" surface area of 20, is forced to contract to a surface area of 10 during freezing. After thawing, the vesicle attempts to re-expand to its initial surface area of 20. If all the membrane molecules remained in the vesicle, the intermolecular distance in the contracted state would be 0.707 and following expansion the intermolecular distance would return to 1. However, it was noted that 50% lysis occurred, and this corresponds to $\delta_{50}=1.225$. For the intermolecular distance in this case to return to 1.225, δ in the contracted state would have to be 0.866. In other words, roughly 33% of the original vesicle components have been deleted or intramolecularly altered so as to change their volume, and the remaining or altered components have become somewhat compressed. It is useful to make the analogy with a spring which breaks when its coils are a fixed distance apart. In order to shorten the spring we remove some of the coils and compress those remaining. Although this is a very simplified view of as complex an entity as the plasma membrane, it is intuitively pleasing when this conceptual model is extrapolated back to the original, abstract supposition that lysis occurs at a fixed kinetic energy of the membrane. In an energetic sense, the two systems may not be very different.

Next consider the case of a vesicle with initial surface area of 20 which is contracted to 5. In this hypothetical case 75% of the vesicles lysed, which corresponds to $\delta_{75}=1.4$. In the contracted state $\delta_c=0.70$, so that this contracted vesicle contains only 51% of the initial amount of vesicle material. In this case lysis of 50% of the vesicles should occur following expansion to a surface area of 15 (i.e., when $\delta_c=\delta_{50}=1.225$).

The values of $\delta_\%$ as a function of % lysis can be derived in the following manner. The equations for the above example are:

$$\delta_\% = \delta_{50} \left[\frac{A_i}{A_c + TSAI_{50}} \right]^{1/2} \tag{1}$$

and [see Levin et al., (17)]

$$\%S = 100 \exp(-2.27\,\Delta A^{1.71}) \tag{2}$$

where $\delta_\%$ is the relative distance between adjacent molecules in survivors of a population that underwent a given % lysis, δ_{50} is the distance at 50% lysis (=1.225 in this example), A_i is initial vesicle area (20 in this example), A_c is contracted vesicle area, $TSAI_{50}$ is the surface area increment at 50% survival, %S is % survival and ΔA is

$$\frac{A_i - A_c}{A_i}$$

The distance between adjacent molecules in the contracted state is

$$\delta_c = \delta_\% \left[\frac{A_c}{A_i} \right]^{1/2} \tag{3}$$

The percentage of the initial amount of vesicle material remaining in the contracted vesicle is

$$\% \text{ material remaining} = \frac{100 A_c}{A_i \, r_c^{\,2}} \tag{4}.$$

In this manner we quantitate how much membrane material must be deleted (or how much the partial molar volume of the individual components must be reduced) at any given extent of contraction, in order to facilitate testing of the model. It should be emphasized that at the present stage in the development of the model, the above relationships are not necessarily quantitatively accurate, especially since the plasma membrane is not homogeneous. Nevertheless, they do provide a basis for devising new experiments and for extending the theory.

A corollary of the above conceptual model may also be presented. That is, since more molecules/membrane are present in the contracted state in the first case ($A = 20 \rightarrow 10$) than in the second ($A = 20 \rightarrow 5$), and since the absolute surface area increment at 50% survival is constant, it apparently takes a lower amount of kinetic energy/molecule to cause 50% of the membranes to lyse in the first case. This does, in fact, appear to be the case in the above model since it should take less energy to spread the molecules from $\delta = .866$ to $\delta = 1.225$ than to spread them from $\delta = .70$ to $\delta = 1.225$.

The above model makes no assumptions concerning the behavior of cells within a given percentile range of a population. The simplest possible behavior of cells would be the case of a single %S vs. $\delta_\%$ surface valid for all percentiles fo the population. That is, as mentioned earlier, $\delta_\%$ is an intensive property and identical for all cells in the population. This would mean that $TSAI_{50}$ would vary with the original size of the protoplast, and the population distribution of survivors of a freeze-thaw cycle should be identical to the original distribution. Preliminary data, with limited numbers of observations, are consistent with this type of behavior (47). An aternative type of behavior would be that the %S vs. $\delta_\%$ surface varies with the initial size of the cells, perhaps because of nonhomogeneity of cell type in the population. For instance, if initially large cells had a larger δ_{50} than cells that were initially small, the population distribution after a freeze-thaw cycle would be skewed to the right of the original distribution. The magnitude of the difference would be dependent upon the magnitude of the differences in δ_{50}. Likewise, if initially small cells had a larger δ_{50} than cells that were initially large, the resulting distribution of survivors would be skewed to the left of the original distribution. In these cases equation 4 would be a sum of the individual characteristics of each class of the population. Further work with large numbers of observations of the initial population and of survivors of a freeze-thaw cycle could resolve the behavior of individual cells within a population. Currently, this approach and direct observations of individual cells exposed to a freeze-thaw cycle are being implemented.

II. REFERENCES

1. Borochov, A. and Borochov, H. *Biochim. Biophys. Acta 550,* 546–549 (1979).
2. Cromer, A. H. "Physics for the Life Sciences." McGraw-Hill, N.Y. (1974).
3. Dick, D. A. T. *In* "International Review of Cytology" (G. H. Bourne and J. F. Danielli, eds.), Vol. 8. pp. 387–448. Academic Press, New York (1959).
4. Dick, N. P., and Bowden, D. C. *Biometrics 29,* 781–790 (1973).
5. Dowgert, M. F., and Steponkus, P. L. *Plant Physiol. 63,* (Abstract) (1979).
6. Evans, E. A., and LeBlond, P. F. *Biorheology 10,* 393–404 (1973).
7. Evans, E. A., and Waugh, R. *J. Coll. Interf. Sci. 60,* 286–298 (1977).
8. Garber, M. P., and Steponkus, P. L. *Plant Physiol. 57,* 673–680 (1976).
9. Helfrich, W. *Z. Naturforsch 28C,* 693–703 (1973).
10. Jenkins, J. T. *SIAM J. Appl. Math. 32,* 755–764 (1977).
11. Levin, R. L., Cravalho, E. G., and Huggins, C. E. *Cryobiology 13,* 415–429 (1976).
12. Levin, R. L., Cravalho, E. G., and Huggins, C. E. *J. Heat Transfer Trans. ASME 99,* 322–329 (1977a).
13. Levin, R. L., Cravalho, E. G., and Huggins, C. E. *J. Biomechanical Engineering Trans. ASME 99,* 65–73 (1977b).
14. Levin, R. L., Cravalho, E. G., and Huggins, C. E. *Biochim. Biophys. Acta 465,* 179–190 (1977c).
15. Levin, R. L., Ferguson, J. F., Dowgert, M. F., and Steponkus, P. L. *Plant Physiol. 63,* (Abstract) (1979a).
16. Levin, R. L., Steponkus, P. L., and Wiest, S. C. *Agronomy Abstracts* p. 80 (1978).
17. Levin, R. L., Wiest, S. C., and Steponkus, P. L. Proc. 10th Ann. Modeling and Stimulation Conference. Pittsburgh, (in press). (1979b).
18. Lovelock, J. E. *Biochim. Biophys. Acta 10,* 414–426 (1953).
19. Lovelock, J. E. *Proc. Roy. Soc. 147,* 427–433 (1957).
20. Luke, J. C. and Kaplan, J. I. *Biophys. J. 25,* 107–111 (1979).
21. Mansoori, G. A. *Crybiology 12,* 34–45 (1975).
22. Mazur, P. *J. Gen. Physiol. 47,* 347–369 (1963).
23. Mazur, P. *Ann. Rev. of Plant Physiol. 20,* 419–448 (1969).

24. Mazur, P. *Science 168*, 939-949 (1970).
25. Mazur, P. *Cryobiology 14*, 251-272 (1977a).
26. Mazur, P. *In* "The Freezing of Mammalian Embryos" pp. 19-48. CIBA Foundation Symp. London (1977b).
27. Mela, M. J. *Biophys. J. 7*, 95-110 (1967).
28. Meryman, H. T. *Nature 218*, 333-336 (1968).
29. Meryman, H. T. *Cryobiology 8*, 488-500 (1971).
30. Meryman, H. T. *Ann. Rev. Biophys. 3*, 341-363 (1974).
31. Meryman, H. T., Williams, R. J., and Douglas, M. St. J. *Cryobiology 14*, 287-302 (1977).
32. Palta, J. P., and Li, P. H. *In* "Plant Cold Hardiness and Freezing Stress-Mechanisms and Crop Implications." (P. H. Li and A. Sakai, eds.), pp. 93-115. Academic Press, New York (1978).
33. Ponder, E. *Protoplasmatologia 10 (Pt. 2)*, 1-123 (1955).
34. Pushkar, N. S., Etkin, Y. A., Bronstein, V. L., Gordiyenko, E. A., and Kozmin, Y. V. *Cryobiology 13*, 147-152 (1976).
35. Rand, R. P. *Biophys. J. 4*, 303-316 (1964).
36. Reeves, J. P., and Dowben, R. M. *J. Cell Physiol. 73*, 49-60 (1969). (1969).
37. Salem, L. *Can. J. Biochem. Physiol. 40*, 1287-1298 (1962).
38. Scarth, G. W., Levitt, J., and Siminovitch, D. *Cold Spg. Harbor Sym. 8*, 102-109 (1940).
39. Silvares, O. M., Cravalho, E. G., Toscano, W. M., and Huggins, C. E. *J. Heat Trans. ASME 97*, 582-588 (1975).
40. Siminovitch, D., and Levitt, J. *Can. J. Res. 19*, 9-20 (1941).
41. Singer, S. J., and Nicolson, G. L. *Science 175*, 720-731 (1972).
42. Snedecor, G. W., and Cochran, W. G. *Statistical Methods* Iowa State Univ. Press, Ames, Iowa (1967).
43. Steponkus, P. L., Dowgert, M. F., and Levin, R. L. *Plant Physiol. 63*, (Abstract) (1979).
44. Steponkus, P. L, Garber, M. P., Myers, S. P., and Lineberger, R. D. *Cryobiology 14*, 303-321 (1977).
45. Steponkus, P. L., Wiest, S. C., and Levin, R. L. *Agronomy Abstracts* p. 85-86 (1978).
46. Tornava, S. R. *Protoplasma 32*, 329-341 (1939).
47. Wiest, S. C. Ph.D. Thesis. Cornell Univ., Ithaca, N.Y. (1979).
48. Wiest, S. C., and Steponkus, P. L. *Plant Physiol. 62*, 599-605 (1978a).
49. Wiest, S. C., and Steponkus, P. L. *Plant Physiol. 61 (Supple.)*, .32 (1978b).
50. Wiest, S. C., and Steponkus, P. L. *Agronomy Abstracts*, p. 88 (1978c).
51. Wiest, S. C., and Steponkus, P. L. *Cryobiology 16*, 101-104 (1979).
52. Zade-Oppen, A. M. N. *Acta Physiol. Scand. 73*, 341-364 (1968).

MEMBRANE STRUCTURAL TRANSITIONS:
PROBABLE RELATION TO FROST DAMAGE
IN HARDY HERBACEOUS SPECIES[1]

C. Rajashekar, L. V. Gusta[2] and M. J. Burke[3]

Department of Horticulture
Colorado State University
Ft. Collins, Colorado

I. INTRODUCTION

There is ample evidence in these proceedings that temperature
dependent membrane structural transitions occur in a wide variety of
chilling sensitive plants. These structural transitions lead to altered
activities of membrane associated functions that eventually cause plant
injury. It is an attractive hypothesis that temperature dependent
membrane structural transitions also occur at subfreezing temperatures
in intermediately frost hardy plants, and that such temperature
dependent membrane structural transitions are responsible for frost
injury.

Before discussing the similarities between chilling and frost injury
mechanisms it must be noted that there are marked differences between
chilling and frost injury. Chilling injury in plants occurs at temperatures
above $0^{\circ}C$, it does not involve the presence of ice and the tissue is not

[1]These studies were supported in part by grants to M. J. Burke from
the Petroleum Research Fund (PRF 9702-AC1,6), the Research
Corporation, the Horticultural Research Institute, and the Colorado
Experiment Station and by grants to L. V. Gusta from the National
Research Council of Canada (A-9661). Scientific Journal Series Paper
#2456 of the Colorado Experiment Station.
[2]Visiting Scientist from the Crop Development Centre, Crop
Science Department, University of Saskatchewan, Saskatoon,
Saskatchewan S7 NOWO.
[3]Present address, Department of Fruit Crops, University of Florida,
Gainesville, Florida 32611.

severely dehydrated due to the growth of ice crystals. Chilling often results from metabolic dysfunctions which are likely to be very minimal in frozen and dehydrated tissues, and therefore of negligible importance in frost injury. Chilling sensitive plants must often be given prolonged exposure below their critical temperatures for damage to develop and frequently recover from the damage if exposed to warm temperatures. In contrast, injury during freezing stress develops after brief exposure below a critical temperataure. It is cataclysmic and the direct damage is irreversible and lethal.

There are many similarities, however, between chilling and frost injury. In both types of injury it is clear that the membranes are of pivotal importance. Chilling injury has been associated with membrane structural transitions (16). After frost injury has occurred, one of the first signs of damage is the loss of membrane semipermeability (15). In both cases there is a critical temperature involved and injury is observed only after exposure below this critical temperature. For frost hardy Kharkov winter wheat crowns, no injury is observed above -22°C, but a brief exposure below this temperature leads to frost killing (10). For the chilling sensitive tomato, prolonged exposure below 10°C is required for visible symptoms of injury to develop, whereas exposure above 10°C causes no injury.

There are several difficulties in studying frost injury. Conventional biochemical and biophysical methods are not easily applied to systems encased in ice and at low temperatures. These experimental problems make it difficult to determine the moment or time the frost injury is initiated in a sample. This is a classical question discussed in detail by Levitt (14, 15). To determine the mechanism of frost injury, the moment in time that injury is initiated during freezing or thawing must be known. There is only scant evidence whether injury is initiated during freezing, immediately passing below a critical temperature, during thawing or after thawing. The experimental results suggesting at what point during the freeze-thaw cycle injury initiates are based on qualitative evidence [for more discussion see Levitt (14)] , primarily from microscopic observations (19), or from fluorescent changes in cells under frozen conditions (13). These results suggest that freezing injury is sometime initiated immediately on passing below a critical temperature, the frost injury temperature.

The objective of this study is to establish the moment and site of freezing injury. The results conclusively establish in the systems studied here that injury occurs during freezing immediately upon passing through a critical temperature and suggest that the primary site of injury is the plasma membrane. Herbaceous plants with well-defined killing temperatures (lower than -15°C) were chosen as model systems for the study. At the killing temperature of these species (winter wheat, Kentucky bluegrass and cactus), little or no change in cell dehydration occurs. Thus the effect of dehydration can better be separated from low temperature effects.

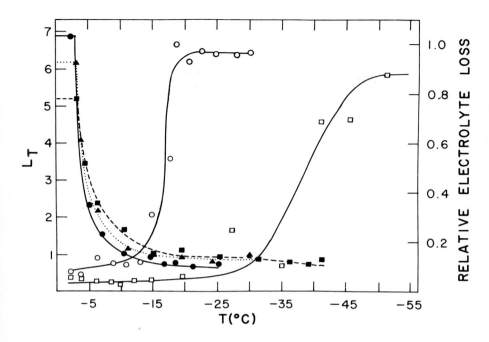

FIGURE 1. Freezing curves (closed symbols) and electrolyte loss curves (open symbols) for Kharkov winter wheat (Triticum aestivum L.) leaf segments (●, O). Kentucky bluegrass (Poa pratensis L., Merion) leaf segments (■, □) and cactus (Opuntia polyacantha Haw.) stem (▲). L_T is the liquid water content at temperature T in gm liquid water per gm dry tissue. On the relative electrolyte loss scale, a value of 1 is the electrolyte loss for a sample held at $80°C$ for 30 minutes. All electrolyte loss measurements were made on the thawed samples that had been slowly cooled to and warmed from the temperature indicated on the ordinate. Note absence of anomalies in the freezing curves at the inflections in the electrolyte loss curves. The nmr procedures for measuring liquid water content are described by Gusta et al. (9).

Winter wheat was grown in a greenhouse for 3 weeks prior to moving to a cold chamber at $3°C$ and 12 hour dark period where it was held for 4 additional weeks or more. Kentucky bluegrass plugs were obtained from the field during December as and when required. Leaf segments were 1 cm in length and were taken from the middle portion of leaf blades. Cactus was grown at $23°C$ and cylindrical plugs 1 cm long and 0.7 cm diameter from stem blades were used for the experiment. The cooling rate was $3°C$/hour or less.

II.   FREEZING OF WATER AT THE KILLING TEMPERATURE

In several tissues studied, water freezes like an ideal salt solution and usually does not supercool appreciably (5, 9, 11). Under slow freezing conditions, most tissue water freezes above $-10^{\circ}C$ (Fig. 1) resulting in no major change in frost dehydration below this temperature. Nevertheless, many herbaceous plants are killed well below $-10^{\circ}C$ as evidenced by the sharp increase in electrolyte leakage for wheat leaves at $-17^{\circ}C$ and for Kentucky bluegrass leaves at $-38^{\circ}C$ (Fig. 1). The inflections in the electrolyte loss curves are sharply defined over the narrow temperature range. Unfortunately, electrolyte loss curves do not provide information as to the moment when the electrolyte leakage is initiated during the freeze-thaw cycle. Each point on the electrolyte leakage curve represents the fractional leakage measured after the sample was cooled to the stress temperature (indicated on the abscissa in Fig. 1) and thawed. From such results, it is not known if the electrolyte leakage was initiated during freezing to the test temperature, at the test temperature or during thawing from the test temperature.

The freezing curves in figure 1 are obtained using nuclear magnetic resonance (nmr) relaxation time differences between liquid water and ice as described in detail by Gusta *et al.* (9). The curves obtained during freezing and thawing are identical. No anomalous behavior in the freezing or thawing curve could be detected in the temperature region of the major electrolyte loss. Furthermore, differential thermal analysis could not detect anomalous heats near the temperatures of the electrolyte loss inflections. In these experiments, freezing in leaves of wheat and Kentucky bluegrass was initiated around $-3^{\circ}C$ with ice nucleation and differential thermal analysis was carried out using a cooling rate of $1^{\circ}C/min$. Results from these and from other extensive nmr analyses (3, 5, 9) suggest absence of freezing points associated with the frost killing temperature.

III.   NUCLEAR MAGNETIC RESONANCE STUDIES

One of the few direct means for characterization of water in a partially frozen sample is to study the nmr relaxation properties of water protons. The two relaxation processes can be described by the longitudinal (T_1) and transverse (T_2) relaxation times and they have been used extensively to study water in plant tissues (1, 2, 4, 22). These relaxation times are sensitive to interactions between water and various cellular contents and therefore reflect, to some degree, the environment the water experiences.

The nmr relaxation properties of wheat leaf segments in figure 2 are typical of those found in plant tissues (4, 21) and in model protein systems (12). There are two exponentially decaying components for T_2

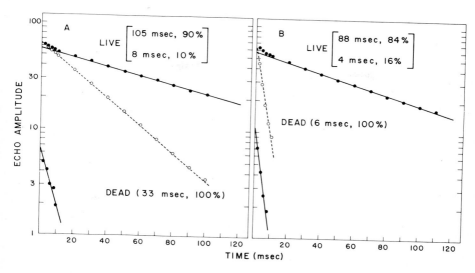

FIGURE 2. Echo amplitude decay in a Carr-Purcell-Meiboom-Gill experiment for measuring T_2 nmr relaxation. See Farrar and Becker (6) for more details on the nmr methods. All measurements were made at $25^{\circ}C$ on wheat leaf segments. The relaxation for live samples (●) is analyzed using two first order reaction rate processes. For wheat leaf segments (A) the major component, comprising 90% of the nmr signal, had a time constant, T_2 of 105 msec and the minor component had a time constant of 8 msec. The frost killed sample (O), which had been slowly cooled and warmed from $-70^{\circ}C$, had only one component with a T_2 of 33 msec. Experiments in B are identical to those in A except that in part B, the leaf segments were equilibrated in a 0.031 molar $MnSO_4$ solution for about 30 minutes prior to the nmr measurements.   Note the drop in T_{2+} relaxation after frost death, which is more pronounced in the Mn^{2+} treated sample.

relaxation, one accounting for 90% of the water protons with a T_2 of 105 msec and the other faster relaxing component with a T_2 of 8 msec. The T_1's for these two components are 276 msec (85%) and 39 msec (15%), respectively. There is usually about a 5% discrepancy in the proportion of the two fractions depending on if they are measured by the T_2 or T_1 process. These two relaxation times are only slightly affected by the addition of the Mn^{2+}, (Fig. 2B), suggesting that the Mn^{2+} does not penetrate to the bulk of the tissue water. Paramagnetic ions like Mn^{2+} when added to water cause a substantial reduction in both T_2 and T_1 relaxation times. If wheat leaf samples are killed by slow cooling to $-70^{\circ}C$, two major changes in the nmr relaxation times are observed. First, the T_2's shorten significantly. The shortening of T_2 is much more pronounced in the Mn^{2+} containing samples (Fig. 2B). This indicates that

C. Rajashekar *et al.*

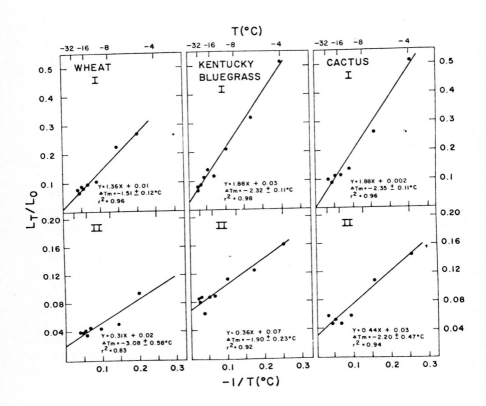

FIGURE 3. *Freezing curves for the two components of T_2 relaxation described in figure 2. Components I and II are the major and minor fractions of tissue water respectively. The data are plotted as suggested by Gusta et al (9). L_T/Lo is the liquid water of component I or II at temperature T divided by the total water of the sample. In the equations for linear regression, Y is L_T/Lo and X is $1/T$ (^oC). ΔTm is the melting point depression of the tissue solution when thawed and is obtained by dividing the slope of the line by the total water of the component. r^2 is the coefficient of determination. The results suggest that both T_2 nmr components freeze as ideal solutions, with melting point depressions between -1.5^o and -3.1^oC.*

TABLE I. Nmr Relaxation Times for Live and Frost Killed Kharkov Winter Wheat Leaf Segments, Kentucky Bluegrass Leaf Segments and Cactus Stem. The Killing Temperature Was 17^0C for Wheat, -38^0C for Kentucky Bluegrass and -24^0C for Cactus.

Sample	Treatment[a] (lowest temperature in cooling-warming cycle) (^0C)	Relaxation time (msec)			
		T_2		T_1	
		Long (I)	Short (II)	Long (I)	Short (II)
Wheat (measurements at -9.6^0C)	-9.6	11.0 (0.72)[b]	2.5 (0.28)	1750 (0.64)	30 (0.36)
	-14	11.0 (0.69)	2.4 (0.31)	1600 (0.66)	34 (0.34)
	-20	6.0 (0.66)	1.5 (0.34)	1490 (0.38)	45 (0.62)
	-70	4.7 (0.71)	1.3 (0.29)	1580 (0.31)	55 (0.69)
Kentucky bluegrass (measurements at -15^0C)	-15	7.2 (0.66)	1.4 (0.34)	2100 (0.71)	40 (0.29)
	-30	7.3 (0.63)	1.4 (0.37)	2035 (0.68)	50 (0.32)
	-40	4.1 (0.69)	0.3 (0.31)	2300 (0.38)	47 (0.62)
	-70	3.8 (0.68)	0.2 (0.32)	2020 (0.40)	40 (0.60)
Cactus (measurements at -4.8^0C)	-4.8	131.0 (0.77)	12.0 (0.23)	1200 (0.66)	110 (0.34)
	-23	126.0 (0.75)	12.0 (0.25)	1220 (0.64)	116 (0.36)
	-29	48.0 (0.70)	2.0 (0.30)	1300 (0.52)	130 (0.48)
	-70	43.0 (0.71)	1.8 (0.29)	1340	

[a] The treatments consist of a slow cooling to the test temperature (-9.6^0C, wheat; -4.8^0C, Kentucky bluegrass; -15^0C, cactus) where the first relaxation time measurements were made. The samples were then cooled to the next temperature indicated and then warmed back to the test temperature and relaxation times again measured. This process was repeated on the same sample for each succeeding temperature.

[b] The value in parenthesis refers to the proportion of total nmr signal decaying with this relaxation time.

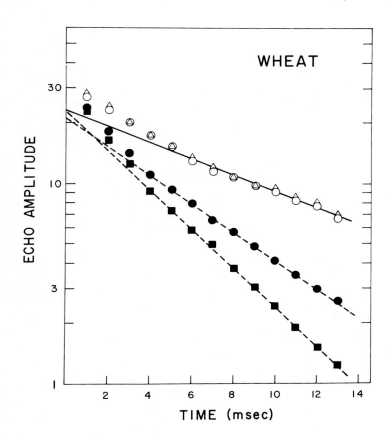

FIGURE 4. Echo amplitude decay in a Carr-Purcell-Meiboom-Gill experiment for T_2 determination. All measurements were made at -9.6°C. First, the echo amplitudes were obtained for wheat leaf segments slowly cooled to -9.6°C (O). The same sample was cooled subsequently to -14°C (3°C above the killing temperature) and warmed to the test temperature, -9.6°C, again to measure the echo amplitudes (△). This sample was then cooled to -20°C (3°C below the killing temperature) and rewarmed to -9.6°C to make a similar measurement (●). Final echo amplitude measurement (■) was made on the sample cooled to -70°C and rewarmed to -9.6°C. These changes in echo amplitude decay occur in a sample not warmed above -9.6°C. Similar results for Kentucky bluegrass and cactus are presented in table I.

the Mn^{2+} readily penetrates the bulk of the tissue water after frost injury. Second, the long and short relaxing components become one component after frost injury (Fig. 2) in wheat leaf segments. Similar results are observed with Kentucky bluegrass leaf segments and cactus stem.

The two nmr relaxing components are also observed at subfreezing temperatures and freezing of each of these components can be monitored (Fig. 3). The data in figure 3 are plotted as suggested by Gusta *et al.* (9) and show the typical ideal freezing behavior observed with herbaceous and some woody plants. The freezing curves can be described with two parameters, the melting point depression of the aqueous component, ΔT_m, obtained from the slope of these lines and a parameter b, the intercept of these lines, which is the fraction of the component that is not freezable. These parameters are given in the figure 3.

Various attempts have been made to identify the two nmr relaxing components in plant and model protein systems (12, 21). In frozen protein solutions and in protein crystals, the two exponentially decaying components of nmr relaxation are attributed to the magnetization transfer between protein and water and a distribution of rotational correlation times for the unfrozen aqueous fraction of the frozen protein systems. Interestingly, Stout *et al.* (22) have proposed that the two nmr relaxing components come from two physically separated aqueous fractions of the plant tissues. They concluded that the component with a short T_2 relaxation time (component II here) is associated with the extracellular water. The component with long T_2 relaxation time (component I here) is associated with the intracellular water, presumably the vacuolar water which constitutes the largest fraction of cellular water and has a minimal interaction with macromolecules. However, for these tissues it has not been possible to determine the origin of the two components of nmr relaxation times.

The T_2 and T_1 relaxation times were measured at subfreezing temperatures (Table I). Typical data for wheat leaf segments are in figure 4. In this experiment, wheat leaf segments were slowly cooled to $-9.6^{\circ}C$. T_2 was determined from the echo amplitude from a CPMG pulse sequence. If the sample was subsequently cooled slowly to $-14^{\circ}C$ ($3^{\circ}C$ above the killing temperature) and warmed to $-9.6^{\circ}C$ there was no change in the T_2 values. However, when the same leaf segments were cooled to $-20^{\circ}C$ ($3^{\circ}C$ below their killing temperature) and warmed again to $-9.6^{\circ}C$, a considerable decrease in T_2 was observed. This T_2 hysteresis is further increased if the sample is cooled to $-70^{\circ}C$ and rewarmed to $-9.6^{\circ}C$. These results strongly suggest that an irreversible change in T_2 occurs only on passing below the killing temperature and in the absence of any appreciable thawing of the tissue water. Similar results are observed in Kentucky bluegrass leaf segments and cactus stem, which are killed at $-38^{\circ}C$ and $-24^{\circ}C$ respectively.

To establish the moment the hysteresis initiates, T_2 was measured during cooling and warming in wheat and Kentucky bluegrass leaves and cactus stem tissue (Fig. 5). The T_2 hysteresis is only observed on warming the sample from below the killing temperature and is generally initiated in the absence of significant thawing of the frozen sample.

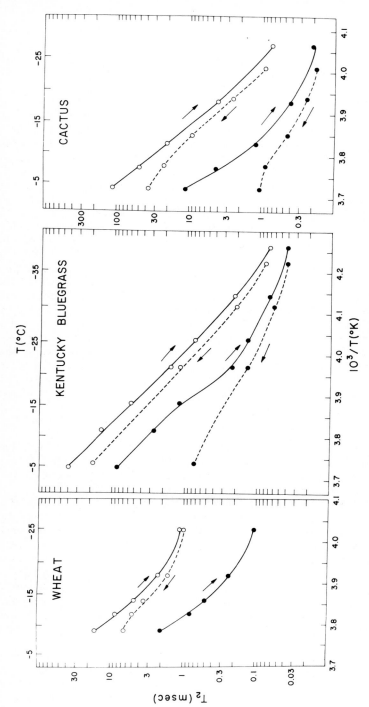

FIGURE 5. *Temperature dependence of T_2 during slow cooling and warming. The samples used are similar to those described in figure 1. The arrows indicate cooling and warming curves. All samples were cooled at the rate of 2^0C/hour just below their respective killing temperatures and warmed at the rate of 3^0 to 4^0C/hour. The hysteresis generally became greater with progressive warming for component I (O) and II (●).*

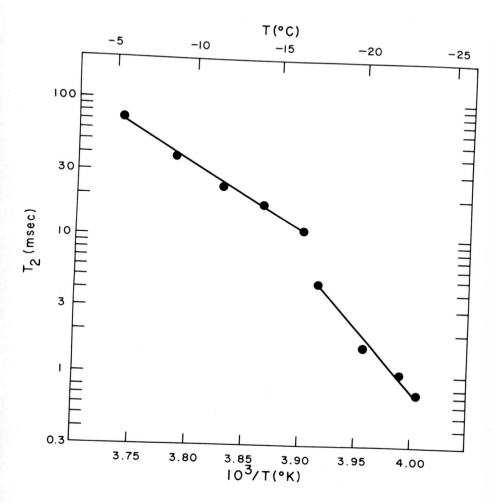

FIGURE 6. Temperature dependence of T_2 during cooling of Kharkov winter wheat leaf segments containing Mn^{2+}. The sample was equilibrated in 0.031 molar $MnSO_4$ solution for about 15 minutes prior to the nmr measurements. The T_2 for slow relaxing component (component I) is presented here. Note the anomalous shift in the curve at about $-17°C$, the killing temperature of these leaf segments. In repeated experiments it was noted that addition of Mn^{2+} decreased the values of T_2 for unfrozen samples, but during freezing they were relatively higher than those for samples untreated with Mn^{2+}.

It was noted in figure 2 that addition of Mn^{2+} in low concentrations to wheat leaf segments made T_2 relaxation times much more sensitive to changes induced by frost damage. For instance, live wheat segments had a T_2 of 88 msec for the major water fraction but when freeze killed to $-70^{o}C$, the T_2 value dropped to 6 msec, a 14 fold reduction. The decrease in T_2 suggests that frost injury results in loss of water compartmentalization.

TABLE II. Time Constants (t) for Live and Freeze Killed Kharkov Winter Wheat Crowns Supercooled to $-4^{o}C$. The Freezing Rates Followed First Order Reaction Kinetics. The Plants Used Here are Described in Figure 1 with the Tender Samples Obtained Directly from the Greenhouse (Killing Temperatures, $-23^{o}C$ for Hardened and $-5^{o}C$ for Tender).

Tissue	Treatment	t (min)
Hardened	Alive	14.0
	Liquid N_2 [a]	11.8
	Slow Freeze $(-60^{o}C)$ [b]	12.0
Tender	Alive	9.0
	Liquid N_2	5.5
	Slow Freeze $(-60^{o}C)$	5.4

[a] In this treatment samples were plunged directly into liquid nitrogen.
[b] In this treatment samples were slowly cooled ($3^{o}C$/hr) to $-60^{o}C$.

TABLE III. Dependence of Time Constants for Freezing on Degree of Supercooling. Tissues Used were Tissue Culture Cell Aggregates of Manitou Spring Wheat and Kharkov Winter Wheat.

Temperature (^{o}C)	t (min)			
	Manitou		Kharkov	
	Tender	Hardened	Tender	Hardened
-3	12.5	12.2	12.2	10.6
-5	8.2	7.8	7.1	7.0
-7	4.7	4.7	4.7	4.7
-10	2.5	2.4	2.5	2.7

T_2 values can readily be determined for tissue in the frozen state. A twofold decrease in the long T_2 for wheat leaf tissue incubated in Mn^{2+} was observed the moment the tissue was cooled below its critical temperature (Fig. 6). This supports the above finding that Mn^{2+} gains access to intracellular unfrozen water in the tissue precisely at the killing temperature. This establishes that injury of the tissue occurs during the freezing process and not during the thawing process.

IV. FREEZING AND THAWING RATE STUDIES

The kinetics of freezing and thawing in leaf segments, crowns and tissue culture cell aggregates of winter cereals have been measured using nmr techniques. The experiments are of two general types. In the first, supercooled samples are inoculated with ice at different subfreezing temperatures and the rate of freezing measured under isothermal conditions (Table II and III). In the second type of experiments, samples are cooled to $-2^{\circ}C$, inoculated with ice and allowed to reach equilibrium overnight. The samples are cooled slowly to either $-3^{\circ}C$ or $-4^{\circ}C$, ensuring equilibrium freezing. Generally, over 60% of the sample water is frozen at $-4^{\circ}C$. The freezing and thawing rates are then measured during a rapid temperature change, between $-4^{\circ}C$ and $-8^{\circ}C$ for crowns and $-3^{\circ}C$ and $-5^{\circ}C$ for cell cultures. Freezing and thawing rates can be obtained for samples slowly cooled and warmed from various temperatures above and below the killing temperature without warming above $-3^{\circ}C$. Typical results are in figure 7 and table 3.

Freezing of supercooled samples always followed first order kinetics and could be described by a single time constant, t (i.e., the reciprocal of the first order rate constant). On the average, the time constant for freezing of hardened winter cereal crowns supercooled to $-4^{\circ}C$ was approximately 14 minutes as compared to 9 minutes for tender crowns (Table II). The time constants for frost killed crowns were reduced slightly. The difference in freezing rates for hardened and tender crowns is reduced substantially if the crowns are vacuum infiltrated with water. There is no difference in the freezing rates of hardened and tender cereal cell suspensions supercooled to -3°, -5°, -7° and $-10^{\circ}C$. However, the rate of freezing did show strong dependence on the temperature of ice initiation (Table III). Overall there are no dramatic differences between living and frost killed, or hardy and tender samples. The rate of freezing in supercooled samples is dependent upon many things including at least the following: the rate of water efflux (membrane permeability), the degree of supercooling, the rate of ice front growth, and the rate of heat removal. The role of membranes in freezing can be best determined in samples prefrozen to $-3^{\circ}C$ or $-4^{\circ}C$. If the temperature is changed quickly, the primary factor limiting freezing would be the contribution of the membrane.

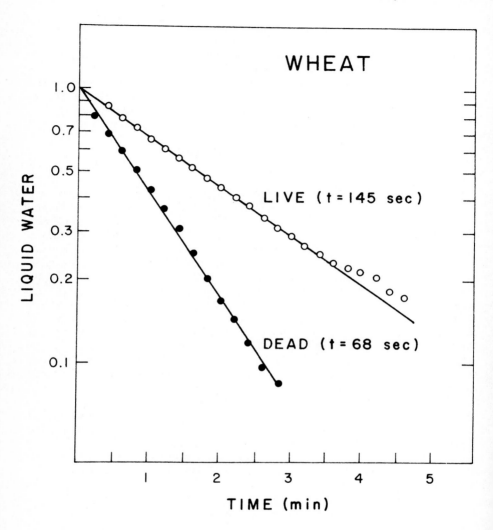

FIGURE 7. *Freezing rates for Kharkov winter wheat leaf segments vacuum infiltrated with water. The sample was slowly cooled to -4°C and allowed to reach equilibrium. The sample temperature was then dropped to -8°C with the following of liquid water content as a function of time (Live tissue, 0). The liquid water follows first order kinetics with time constant t. The sample was then slowly cooled to -25°C (8°C below the killing temperature) and rewarmed to -4°C. The freezing rate was again determined for the frost killed tissue (●) as described above.*

In experiments where the fast temperature change procedure described above was used, there were substantial differences in the rate of freezing and thawing between living and frost killed samples (Fig. 7). In the case of wheat leaf segments the time constant for freezing between $-4°C$ to $-8°C$ was 145 seconds. If the sample was slowly cooled to $-25°C$ ($8°C$ below the killing temperature), and slowly warmed to $-4°C$, the time constant for freezing between $-4°C$ to $-8°C$ was reduced to 68 seconds. In control experiments where the samples were cooled only to $-14°C$ ($3°C$ above the killing temperature), the freezing time constant remained unchanged. Similar behavior between live and frost killed leaf segments was observed for the thawing time constants between $-8°C$ and $-4°C$. The time constant for thawing of frost killed samples was about a third as large as that of living samples.

TABLE IV.   The First Order Time Constants of Freezing and Thawing of Puma Rye Crowns and Brome Grass Tissue Culture Cell Aggregates.

Species	Killing Temperature	5(sec)	
		Freezing	Thawing
Winter Rye		$-4°C$ to $-8°C$	$-8°C$ to $-4°C$
Control	$-30°C$	160	241
SC $-40°C$[a]		85	100
FC $-60°C$[b]		75	93
Heat killed[c]		70	87
Brome Grass		$-3°C$ to $-5°C$	$-5°C$ to $-3°C$
Tender	$-11°C$	135	165
SC $-40°C$		75	120
Hardened	$-22°C$	146	210
SC $-60°C$		75	116

[a] Sample slowly cooled and warmed from $-40°C$ ($2°C$/hour).
[b] Sample rapidly cooled and warmed from $-60°C$ ($60°C$/min).
[c] Heat killed samples were warmed to $80°C$ for 30 minutes prior to making measurements.

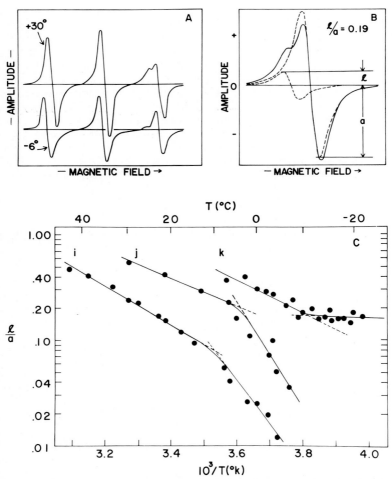

FIGURE 8.   A) Electron spin resonance spectra of TEMPO in
macerated potato leaf at two temperatures.   The partitioning of TEMPO
between the lipid and aqueous leaf fractions is determined using the high
magnetic field band on the right.   B) The calculated high magnetic field
band in a TEMPO spectrum resolved into two lines.   The amplitude of the
line attributed to TEMPO in lipid is indicated by a 1 and the comparable
aqueous TEMPO line by an a.   See text for further description.   C) van't
Hoff plots for the amplitude ratio, (1/a) of macerated tomato leaf, i;
macerated potato leaf, j; and whole wheat leaf, k. The bends for tomato,
potato and wheat are at +11°, + 3° and -11°C, respectively.   In these
experiments 1/a was raised to near 0.5 for the highest test temperature
by controlled sample dehydration.   Further experiments were conducted
on wheat leaf and no additional bends were found up to 30°C, the highest
test temperature with wheat. In a given experiment, measurements were
made proceeding from high to low temperatures.

In live crown tissue, two fractions of water were observed during freezing and thawing. However, if the tissue was frost killed only one fraction of water was observed, which had a marked time constant reduction for both freezing and thawing (Table IV). Freezing rates in fast temperature change experiments, such as these, are thought to be limited by the rate of water efflux from within the cells to extracellular sites containing ice. The plasma membrane is likely to be a major resistance to the movement of water. The results suggest that the resistance to water movement is reduced significantly after the samples are cooled below their killing temperatures. This might be caused by permanent rupture of the plasma membrane and/or formation of ice within the cells.

V. ELECTRON SPIN RESONANCE STUDIES

The electron spin resonance spin label, TEMPO (2,2,6,6-tetramethylpiperidine-1-oxyl) has been used extensively in model systems and in lipids extracted from biological tissues (8, 17, 20). Changes in temperature dependent partitioning of TEMPO provide evidence of membrane structural transitions. The advantage of using TEMPO in living systems is its solubility in both lipids and water which allows its distribution throughout the living tissues. TEMPO was incorporated into macerated tomato and potato leaf tissue and into whole Cheyenne winter wheat leaves (frost killing temperature -12°C). The details of TEMPO labeling results are reported elsewhere (7).

Figure 8A shows the three band TEMPO spectrum for potato leaf with typical temperature dependence of the high field band. The bands are somewhat broader than found in model membrane studies (8, 17, 20). Consistent with other investigations (17, 20), the high field band is the sum of two lines. Of the two lines the lower field line is from TEMPO dissolved in lipid or possibly other hydrophobic materials, and the higher field line is from TEMPO in an aqueous environment. In this analysis, the amplitude of each of the two lines is assumed to be proportional to the TEMPO concentration in the lipid or aqueous environment. The amplitudes were obtained from the measured spectra by a curve fitting procedure. The lipid and aqueous lines in the high field band were calculated as Lorentzian curves (Fig. 8B) and for convenience the line widths for the two curves were assumed equal and allowed to vary with temperature. The observed spectrum was then matched with the calculated spectra obtained by varying the line widths, and the line amplitudes. The best fit was chosen by visual comparison. The temperature dependence of amplitude ratios (i.e., 1/a in Fig. 8B) obtained in this way are plotted in figure 8.

The equilibrium constant for partitioning of TEMPO between lipid and aqueous leaf fractions K_{eq}, is the ratio of the TEMPO concentrations in the lipid and aqueous fractions and follows:

$$K_{eq} = \left(\frac{i'}{a'}\right)\left(\frac{A}{L}\right)$$

where i' and a' are the concentrations of TEMPO in the lipid and aqueous environments, respectively, and L and A are the volumes of the lipid and aqueous environments of the leaf, respectively. In the absence of freezing, A/L is a constant and i'/a' is expected to be proportional to the measured amplitude ratio, i/a. Therefore, K_{eq} should be proportional to the amplitude ration, i/a. For this reason figure 8C is a van't Hoff plot.

van't Hoff plots obtained have bends near $11^{\circ}C$ in tomato, $3^{\circ}C$ in potato and $-11^{\circ}C$ in wheat leaves. Similar anomalous behavior of TEMPO partitioning in other membrane systems has been interpreted to suggest membrane phase separations (8, 17, 20). In addition, chilling sensitive tomato is known to have a phase separation near $11^{\circ}C$ (18). This further lends validity to this technique for finding membrane phase separations. The bend near $3^{\circ}C$ in potato is consistent with the occurrence of storage injury to potato tubers at this temperature (23) and suggests that a membrane phase separation may be responsible. The change of slope for the supercooled wheat leaf is opposite to that of the other plants and is near $-11^{\circ}C$. Supercooled samples of these leaves survive exposure to below $-11^{\circ}C$; however, when frozen the killing temperature of these samples was near $-11^{\circ}C$. The bend in the van't Hoff plot for wheat suggests that a membrane structural change occurs at the killing temperature. Possibly such membrane structural change at $-11^{\circ}C$ in combination with the severe frost dehydration that occurs when such tissues freeze, leads to frost injury observed at $-12^{\circ}C$ in these leaf segments.

VI. CONCLUSIONS

The plant tissues studied here have frost killing temperatures where very little or no freezing of water takes place. Therefore, it is unlikely that the freeze induced dehydration alone can account for the injury observed. These tissues have sharply defined killing points as evidenced by the rapid electrolyte loss at these temperatures. It is very typical in tissues, with a high to moderate water content (such as leaves) to lose most of the cellular electrolytes as a result of cell lysis following frost injury. The question arises as to the mechanism that could account for the lysis of cells over a narrow range of temperature. This could be approached by establishing the possible site and time or moment the frost injury is initiated.

The inflection in the electrolyte loss curves shows damage or lysis of plasma membrane on passing below the lethal temperatures in a freeze-thaw cycle. Unfortunately, the results do not show when during the freeze-thaw cycle the lysis occurs. Neither freezing anomolies at the killing temperature nor hysteresis in the temperature dependent freezing and thawing could be found using either nmr or differential thermal analysis. The results demonstrate that these tissues freeze like an ideal solution. During a slow freeze-thaw cycle, nmr T_2 relaxation times are found to exhibit hysteresis which is initiated after passing below the killing temperature and is observed well before any significant tissue thawing occurs. This strongly suggests irreversible cellular changes at the killing temperature while the tissue is still frozen. In addition, an anomalous drop in T_2 could be observed precisely at the killing temperature in samples containing Mn^{2+} ions. This occurs during cooling and suggests the penetration of Mn^{2+} ions through the plasma membrane at this lethal temperature. The access of Mn^{2+} ions to the liquid water of the frozen tissue could either be due to ice penetration of the cells or possible membrane lysis. Consistent with results implicating membrane damage are the freezing and thawing rates which substantially increase on frost injury. It is assumed here that a resistance to freezing is the plasma membrane, in that the cell water must pass through the plasma membrane before it can freeze in the extracellular spaces. The increased freezing and thawing rates are due to the membrane damage. Direct evidence to the temperature dependent membrane structural transitions similar to those observed in chilling sensitive plants has been found by electron spin resonance methods associated with the killing temperature in winter wheat leaves.

These studies suggest that a membrane structural transition, causing membrane damage at the killing temperature, results in a change in the environment of unfrozen cellular water, allows Mn^{2+} ion penetration into the cells and reduces resistance to water movement during freezing and thawing. It is clearly demonstrated that little or no thawing is necessary for observing these effects.

VII. REFERENCES

1. Burke, M. J., Bryant, R. G., and Weiser, C. J. *Plant Physiol.* 54, 392-398 (1974).
2. Burke, M. J., George, M. F., and Bryant, R. G. In "Water Relations of Foods" (R. B. Duckworth, ed.), pp. 111-135. Academic Press, New York (1975).
3. Chen, P. M., Burke, M. J., and Li, P. H. *Bot. Gaz.* 137, 313-317 (1976).

4. Chen, P. M., Gusta, L. V., and Stout, D. G. *Plant Physiol. 61,* 878–882 (1978).
5. Chen, P. M., Li, P. H., and Burke, M. J. *Plant Physiol. 59,* 236–239 (1977).
6. Farrar, T. C., and Becker, E. D. "Pulse and Fourier Transform NMR." Academic Press, New York (1971).
7. Fey, R. L., Workman, M., Marcellos, H., and Burke, M. J. *Plant Physiol.* (In Press) (1979).
8. Grant, C. W., Wu, S. H., and McConnell, H. M. *Biochem. Biophys. Acta 363,* 151–158 (1974).
9. Gusta, L. V., Burke, M. J., and Kapoor, A. C. *Plant Physiol. 56,* 707–709 (1975).
10. Gusta, L. V., and Fowler, D. B. *Can. J. Plant Science 57,* 213–219 (1977).
11. Harrison, L. C., Weiser, C. J., and Burke, M. J. *Plant Physiol. 62,* 899–901 (1978).
12. Hsi, E., Mason, R., and Bryant, R. G. *J. Phys. Chem. 80,* 2592–2597 (1976).
13. Krasavtev, O. A. *Fiziol. Rastenii 9,* 359–367 (1962).
14. Levitt, J. "The Hardiness of Plants." Academic Press, New York (1956).
15. Levitt, J. "Responses of Plants to Environmental Stresses." Academic Press, New York (1972).
16. Lyons, J. M. *Ann. Rev. Plant Physiol. 24,* 445–466 (1973).
17. McConnell, H. M. *In* "Spin Labeling: Theory and Applications" (L. J. Berliner, ed.), pp. 525–560. Academic Press, New York (1976).
18. Raison, J. K. *In* "Mechanisms of Regulation of Plant Growth" (R. L. Bieleski, A. R. Ferguson, and M. M. Cresswell, eds.), Bull. 12, pp. 487–497. The Royal Society of New Zealand, Wellington (1974).
19. Salcheva, G., and Samygin, G. *Fiziol. Rastenii 10,* 65–72 (1963).
20. Shimshick, E. J., and McConnell, H. M. *Biochem. 12,* 2351–2360 (1973).
21. Stout, D. G. "A Study of Plant Cell Permeability and of the Cold Acclimation Process in Ivy Bark." Ph.D. Thesis, Cornell University, Ithaca (1976).
22. Stout, D. G., Steponkus, P. L., and Cotts, R. M. *Plant Physiol. 62,* 636–641 (1978).
23. Workman, M., Kerschner, E., and Harrison, M. *Amer. Potato J. 53,* 191–204 (1976).

POSSIBLE INVOLVEMENT OF THE TONOPLAST LESION IN CHILLING INJURY OF CULTURED PLANT CELLS

S. Yoshida, T. Niki and A. Sakai

The Institute of Low Temperature Science
Hokkaido University
Sapporo, Japan

I. INTRODUCTION

A great deal of information has been accumulated so far concerning the mechanism of chilling injury in plant cells, mainly from biochemical and biophysical point of view (7). A vast number of experiments have been focused on the impairment (24, 25, 26, 27) or depression of oxidative activity of mitochondria during chilling (1, 8, 13, 14, 21). On the other hand, only limited studies (5, 6, 12, 20) have been reported on the ultrastructural responses associated with chilling injury.

The present study was designed to improve our understanding of the mechanism of chilling injury in plants by utilizing cultured cells, which were highly susceptible to chilling. Emphasis was placed on the primary ultrastructural responses of the cells to chilling stress and special attention was focused on the possible involvement of the tonoplast lesion in the mechanism of chilling injury in cultured plant cells.

II. MATERIALS AND METHODS

A. Cell Culture and Evaluation of Cell Viability

The callus derived from cambial area of twig pieces of *Cormus stolonifera* were used as the materials. They were subcultured at $26^{\circ}C$ in the dark on the agar medium of Murashige-Skoog containing NAA (3mg/1) with a slight modification. The flasks containing callus on the 10th day of culture were cooled to $0^{\circ}C$ and kept there for different lengths of time. The cell viability after chilling was evaluated both with the TTC reduction test and the regrowth test. For the determination of the regrowth rate after chilling, cell culture was continued on fresh medium at 26°C for fifteen days and then the increase in fresh weight was determined.

B.   Electron Microscopy

For transmission electron microscopy, callus was sampled at the desired time of chilling and fixed in 3% glutaraldehyde in 1/15 M potassium-phosphate buffer, pH 7.2 for 2 hr at 0°C or 26°C. The specimens were then post-fixed in 2% osmium tetroxide solution for 2 hr at 0°C before dehydration in an ethanol series and finally in n-butylglycidylether. The embedded specimens in Spurr's epoxy resin (19) were sectioned and stained with saturated uranyl acetate and then with Reynolds' lead solution (15).

For freeze-fracture electron microscopy, callus was fixed in glutaraldehyde as described above.   The pre-fixed specimens were successively treated in 10, 20, 30, and 35% glycerol each for one hour after a brief washing in potassium-phosphate buffer, pH 7.2.   The glycerol-treated specimens were placed on an aluminum sample holder and rapidly frozen in Freon 22 at -156°C. The frozen specimens were fractured at -96°C in 2-4 x 10^{-5} mm Hg in a freeze-etching apparatus, JEOL-4. After minimum etching, the fractured surface was shadowed with both platinum and carbon at -92°C for a few seconds. The replicas were cleaned with a chromic acid solution and viewed with an electron microscope, JEM-100 C.

C.   Preparation of Crude Mitochondria

Ten g each of unchilled or chilled callus was ground in a mortar and a pestle with 10 ml of the grinding medium, 4 g of sea sand and 1.5 g of Polyclar AT. The grinding medium contained 100 mM Tris-HCl buffer, pH 7.8, 0.5 M sorbitol, 5 mM EGTA, 5 mM $MgCl_2$, 10 mM KCl, 0.5% BSA and 20 mM cysteine-HCl.   The cell-free extracts were successively centrifuged at 300, 1,000 and 14,000 g for 5, 10 and 15 min, respectively. The 14,000 g pellets were washed once with the grinding medium and suspended in the reaction medium.

D.   Sucrose Density Gradients

Ten g of unchilled or chilled callus were ground in a motar and a pestle with 10 ml of the grinding medium, 4 g of sea sand and 1.5 g of Polyclar AT at 0°C for 100 seconds. The grinding medium contained 150 mM tris-HCl buffer, pH 7.8, 0.5 M sorbitol, 5 mM EGTA and 20 mM cysteine-HCl. After passing through two layers of gauze and one layer of Miracloth, the cell-free extracts were centrifuged at 300 g for 5 min to remove cell debris and then the supernatants were subjected to centrifugation at 189,000 g for 30 min. The 189,000 pellets were suspended and loaded onto a linear sucrose gradient (15-55%, w/w) in 5 mM Tris-HCl buffer, pH 7.6 and 1 mM EGTA, and centrifuged at 96,000 g for 2 hr. Antimycin A-insensitive NADH cyt c reductase activity was assayed with an aliquot of the fractionated samples (1.2 ml each). The reaction mixture contained 100 μ moles potassium phosphate buffer, pH 7.2, 0.5 mg of cyt c (oxidized form), 5 μ moles of NaN_3, 2 μ M of antimycin A and an aliquot of the enzyme preparations. Reactions were performed at $25^{\circ}C$ and followed the reduction of cyt c at 550 nm.

E. Determination of Respiratory Activities in vitro and in vivo

The oxidative activity of mitochondria in vitro was measured as oxygen uptake by oxygen electrode. The reaction mixture consisted of 0.5 M mannitol, 20 mM potassium phosphate buffer, pH 7.23, 1 mM $MgCl_2$ and 0.1% BSA. The rate of oxidation of substrate (20 mM succinate) was determined in the presence (state 3) or absence (state 4) of 100 μ M ADP with 1.0 mg of mitochondrial protein in 2.0 ml of the reaction medium.

Respiratory activity in vivo was assayed in 3 ml of the basal culture medium with the oxygen electrode. The oxidation of added dopamine in vivo was also determined by measuring the increased uptake of oxygen by oxygen electrode.

III. RESULTS

A. Effects of Temperature on Cell Growth and Cell Lesion

The optimum temperature for the growth of the callus utilized in the present study lay in the very narrow range between 20 and $26^{\circ}C$. No growth was observed below $10^{\circ}C$ or above $28^{\circ}C$ during culture. When the callus on the 10th day of culture at $26^{\circ}C$ were subjected to temperatures below $10^{\circ}C$, they suffered injury and lost their viability depending both on the temperatures and the time duration. The lower the temperature, the more severe the injury within a short period. As presented in Fig. 1A, TTC reduction rate of the callus subjected to $0^{\circ}C$ for 24 hr decreased to nearly sixty percent of the unchilled control. However, only a slight decrease was observed in the TTC reduction rate in the callus chilled for 12 hr at $0^{\circ}C$. The callus subjected to $0^{\circ}C$ for different lengths of time were transferred to fresh medium and culture was continued at $26^{\circ}C$ to determine the capability for regrowth (Fig. 1B). The evaluation of cell viability with the TTC reduction test was correlated with the results from the regrowth test. Thus, the callus were observed to be very susceptible to chilling and injured within a relatively short time of chilling at $0^{\circ}C$.

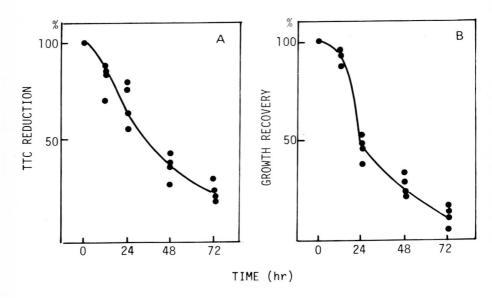

FIGURE 1. TTC reduction rates (A) and regrowth rate (B) of callus subjected to 0°C for various lengths of time. Rates of TTC reduction and regrowth are expressed as percent of the control values. Regrowth was performed on fresh medium at 26°C for 15 days after chilling.

B.   Changes in Cell Permeability during Chilling

As presented in Fig. 2, only a partial release of amino acids was observed in the callus after chilling for more than 48 hr, whereas little or no leakage was observed within 24 hr of chilling. Abrupt permeation and oxidation of added dopamine *in vivo*, however, were observed in the callus chilled for more than 48 hr (Fig. 2). The oxidation of the added dopamine by the callus was strongly inhibited by the addition of phenylthiourea which is a potent inhibitor of polyphenoloxidases. The polyphenoloxidase was determined to be located in the callus in various cellular fractions (Table 1).   About fifty percent of total activity, however, was located in the soluble supernatant.   When protoplasts prepared from unchilled callus were preincubated with dopamine and then washed, no uptake of the substrate was observed. Little or no leakage of the enzyme outside the cells was also observed even in the cells chilled for 72 hr at 0°C. These results may indicate that plasma membranes retain their normal functions until cells are severely injured in the later stages of chilling and the lesion of the membranes *per se* is by no means the primary reason for the chilling injury of the callus, but follows the secondary cellular events.

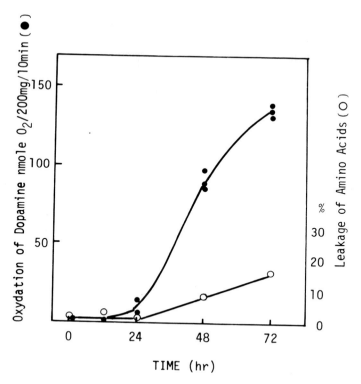

FIGURE 2. *Amino acids leakage and oxidation of added dopamine after chilling for various lengths of time. The chilled cells were leached in distilled water at 26°C for one hr and the amino acids released were determined. The values are expressed as the percent of the killed cells. The permeation and oxidation of added dopamine was measured at 26°C with an oxygen electrode.*

C. Changes in Respiratory Activity in vivo and in vitro

As presented in Fig. 3, respiratory activity *in vivo* was slightly reduced at the time of 12 hr of chilling. However, it returned almost to the normal level upon prolonged chilling up to 24 hr. After 24 hr, the respiratory activity was decreased further with the time lapse of chilling. The enhanced respiration obtained by the addition of uncoupler (FCCP 1 μ M) *in vivo* was also reversibly diminished at the time of 12 hr. Nearly the same results were obtained with respiratory activity in crude mitochondria isolated from chilled callus. The respiratory control declined temporarily after 12 hr of chilling and then was restored nearly to the normal level upon prolonged chilling up to 24 hr. After 48 hr of chilling, a marked decline of the respiratory control was observed, suggesting an irreversible dysfunction of the respiratory control systems.

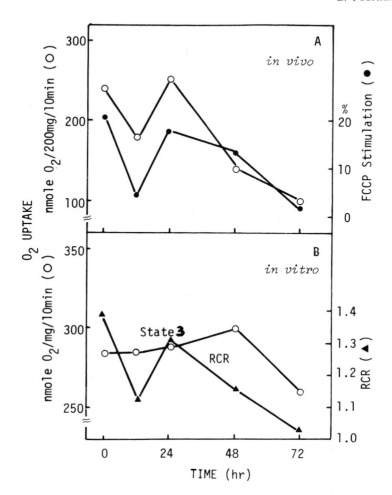

FIGURE 3. Changes in respiratory activities in vivo (A) and in vitro (B) after chilling. The crude mitochondrial fractions were prepared from callus chilled for the indicated lengths of time. The oxygen uptakes both in vivo and in vitro were measured with an oxygen electrode at $26°C$. Succinate (20 mM) was used as the substrate and 115 μM of ADP was used for the measurement of state 3 respiration in vitro. FCCP was added in a final concentration of 1 μM.

TABLE I.   Cellular Localization of Polyphenoloxidase in Callus of Cornus stolonifera.
Ten g of callus on 12th day of culture was ground as described in text and the cell-free homogenate was subjected to differential centrifugations. The pellets were suspended in 5 mM of Tris-HCl buffer, pH 7.3. The reaction mixture consisted of 50 mM potassium phosphate buffer, pH 6.5 an aliquot of the enzyme preparations and 2 mM dopamine in a final volume of 2 ml. The oxidation of added dopamine was measured at $26^\circ C$ with an oxygen electrode. The blank values without addition of the substrate were subtracted.

Fraction	Protein (mg)	Total activity O_2 nmoles/ 10 min	(%)	Specific activity O_2 nmoles/mg/ 10 min
300-1,000 g	6.7	549.1	20.0	81.9
1,000-14,000 g	12.1	549.0	20.0	45.3
14,000-189,000 g	5.9	287.0	10.9	48.6
189,000 g Sup	22.8	1248.0	47.3	54.7

Thus, observed changes in the respiratory activity *in vivo* during chilling may be ascribable to a reversible or an irreversible alteration of the mitochondrial functions.

D.   Ultrastructural Changes in Cells Associated with Chilling Injury

The ultrastructures of unchilled control cells on the 10th day of culture are presented in Fig. 4A, B in different magnifications. In these cells a large central vacuole, rough ER and irregularly invaginated nuclei (Fig. 4A) were characteristically observed.   Proplastids showed an ellipsoidal form containing osmiophilic globules. Many ribosomes were observed in the cytoplasm (Fig. 4B). Ultrastructural changes were first detected in proplastids in the cells chilled for 6 hr at $0^\circ C$ (Fig. 4C). Proplastids showed an increase in the electron density of the matrix and somewhat stretched and constricted shape which was accompanied by folding. The osmiophilic globules in the proplastids were scarcely visible in this early stage of the chilling treatment. In the ultrastructure of the other cell organelles, little or no visible change was detected within 6 hr of chilling. As shown in Fig. 4D, on further prolonged exposure to $0^\circ C$ up to 12 hr, a partial dilation followed by microvesiculation of the rough ER membranes and release of ribosomes from the membranes were clearly observed (Fig. 4D, arrows in the lower left hand).   The normal form of

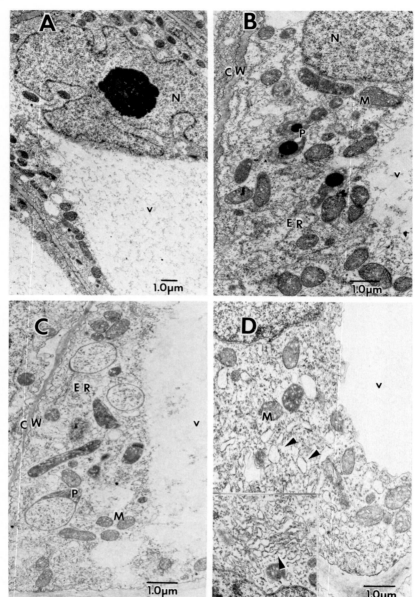

FIGURE 4A-D.   Ultrastructures of unchilled and chilled cells.
Unchilled control cell in lower magnification (A), x 4,000, and in higher
magnification (B), x 8,000; cells chilled for 6 hr (C) and 12 hr (D), x 8,000.
Inset in D shows the initiation of dilated structures of rough ER (Arrow)
at a higher magnification, x 10,000. P, proplastids contained osmiophilic
globules; M, mitochondria; ER, rough endoplasmic reticulum; V, central
vacuole; N, nuclei; CW, cell wall.

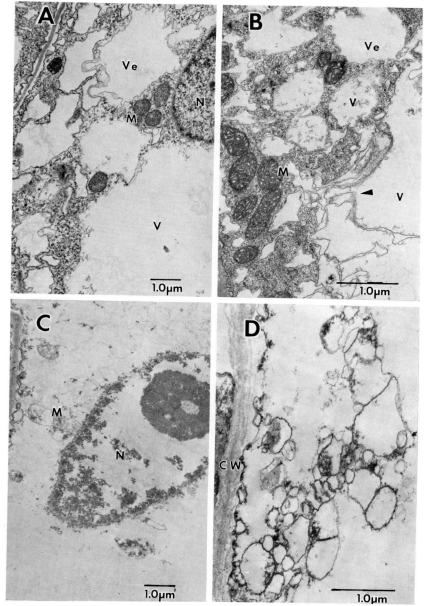

FIGURE 5 A-D. Ultrastructures of chilled cells. Chilled for 24 hr (A, x 8,000; B, x 16,000), 48 hr (C, 8,000) and 72 hr (D, 16,000). Ve, vacuolated endoplasmic reticulum. Arrow in B indicates a partial disruption of tonoplast.

rough ER was hardly detected in this stage of chilling. No detectable change, however, was noted in the ultrastructures of nuclei, mitochondria and plasma membranes. Invagination and infolding of the tonoplasts was occasionally observed in this stage. Numerous ribosomes were still visible in the cytoplasm.

In the cells exposed to $0^{\circ}C$ for 24 hr, more striking ultrastructural changes became evident (Fig. 5A,B). The cytoplasm was occupied by vacuolated vesicles. The large vacuolated vesicles appeared to be further developed from the micro-vesiculated ER observed in the 12 hr stage (Fig. 4C, arrows). In the 24 hr stage, a remarkable structural change also appeared in the tonoplasts. Tonoplasts often exhibited irregularly infolded and engulfed structures. Partial disruption of the tonoplasts was conspicuously visible in about thirty percent of these cells (Fig. 5B, arrow). Most of the mitochondria still retained the normal structure, but quite a few became swollen and less dense. Proplastids showed somewhat less dense structure, but nuclei and plasma membranes still remained unchanged. As clearly shown in Fig. 5C, the cells exposed to $0^{\circ}C$ for 48 hr displayed an autolytic degradation of the cytoplasm. Most of the cell organelles such as mitochondria, Golgi apparatus and proplastids had disappeared. The tonoplast was completely disrupted and disappeared. Degraded nuclei and mitochondria still remained recognizable. After 72 hr of chilling, most of the cells had structures similar to those in Fig. 5D.

The observed changes in the ultrastructure of ER during the early stage of chilling, i.e., transformation of rough ER into vacuolated smooth ER vesicles, were also confirmed by the changes in the sucrose density profiles of antimycin A-insensitive cyt c reductase as the presumed marker enzyme for ER membranes. As presented in Fig. 6, 24 hr of chilling resulted in a marked shift of the major activity peaks to a lighter portion, suggesting the transformation of rough ER to a vacuolated smooth one. Upon further chilling up to 48 hr, the activity peak with lighter density was shifted again to the heavier portion, suggesting further deteriorative changes in those membranes.

Thus, sequential changes in the ultrastructures of cell organelles were observed to be closely related to the chilling injury of the callus subjected to $0^{\circ}C$ for different lengths of time.

To give some insights into the intramembranous structures of cellular membranes, experiments were performed utilizing the freeze-fracture technique (11). No detectable change in the number and the distribution patterns of the intramembranous particles on both faces P and E in several membranes could be detected within 6 hr of chilling (data not shown). On further chilling up to 12 hr, however, remarkable changes in the distribution of the particles were observed in the P face of tonoplast in considerable number of cells, while no detectable change was observed in plasma membranes in this stage (Fig. 7B). This fact may suggest that in the 12 hr chilled cells, changes in the nature of the tonoplasts precede the changes in the plasma membranes. Upon further prolonged chilling up to 24 hr, an aggregation of the particles on the P

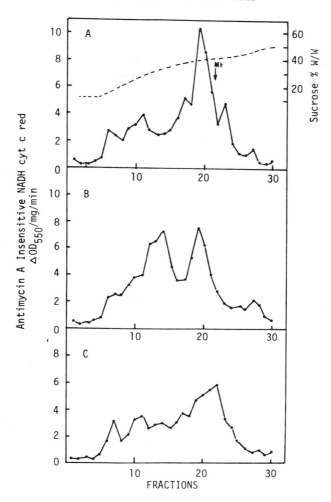

FIGURE 6. Changes in the sucrose density profiles of membranes associated with antimycin A-insensitive NADH cyt c reductase. A, B and C correspond to unchilled, 24-hr chilled and 48-hr chilled callus, respectively. Mitochondria were banded at the indicated position (Mt).

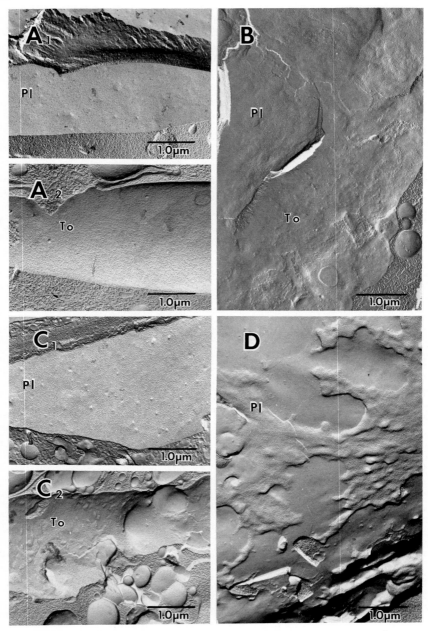

FIGURE 7.   Freeze-fracture electron micrographs of plasma
membranes and tonoplasts in unchilled control and chilled cells.
Fracture faces of the half membranes left frozen to the cytoplasm (P
faces) are indicated. Unchilled control cells ($A_{1,2}$). Chilled for 12 hr (B),
24 hr ($C_{1,2}$) and 48 hr (D). Pl, plasmamembrane; To, tonoplast x 15,000.

face of plasma membranes was initiated in a small number of the cells, but most of the cells gave the normal structure of plasma membranes as shown in Fig. 7C. A marked reduction in the particle numbers was also detected on the E face of tonoplast and severe aggregation of the particles was observed on the P face of the tonoplast (Fig. 7C$_2$).

The alteration in the intramembranous structures of plasma membranes became quite common and more striking on further prolonged chilling up to 48 hr. The particles on the P face of plasma membranes were severely aggregated and large particle-free areas, forming protuberant portions, were formed in most of the cells (Fig. 7D). As already mentioned, in this stage of chilling, almost all of the cytoplasmic components were degraded and the cell viability declined severely.

IV. DISCUSSION

One of the most drastic cellular events observed in the chilled callus was the abrupt disruption of cell organelles, including nuclei, mitochondria, proplastids and ribosomes, after 48 hr of chilling at 0°C. This was apparently followed by observable changes in tonoplasts, i.e., invagination, infolding and a partial disruption, and a concomitant development of the vacuolated smooth ER after 24 hr of chilling. These facts may suggest that the crucial event leading to an irreversible cell decay under chilling is the rupturing of the tonoplasts. A comparable sequence of cellular events regarding the tonoplast has been reported in late senescence of plant cells (2, 16). Vacuoles have been shown to be a lysosomal apparatus in plants, compartmentalizing a variety of lytic enzymes (9, 10). In the callus utilized in the present study, an abrupt decrease in DNA content occurred after 24 hr of chilling. This fact also suggests that the compartmentation of DNAase in the cells has been markedly disturbed during the later stages of chilling.

The ultrastructures of callus that had been moderately altered by 12 hr of chilling were almost totally restored within 12 hr of warming at 26°C. Even in the cells chilled for 24 hr, the severely altered ultrastructures partially or completely restored after rewarming, unless the tonoplast had been severely changed (Niki, T., 1979, in preparation).

The current theory explains chilling injury in terms of a temperature-induced phase transition of lipids from gel to liquid crystal in cellular membranes (see Lyons in this volume) and/or a temperature-dependent alteration in the hydrophobic nature of membrane proteins (24, 25, 26, 27), especially in mitochondrial membranes. If it were also the case in our materials, then we could expect to observe some visible changes in either the distribution or the number of the intramembranous particles in any sort of cellular membranes immediately after lowering the environmental temperature sufficiently below the critical temperature to induce the gel-liquid crystal transition of membrane lipids, with exclusion of the particles by the ordered lipid areas. That is, intramembranous particles may be excluded from the ordered lipid domains, the lateral displacement into still fluid regions producing particle aggregation (4, 17, 22) or vertical displacement producing an apparent loss of particle numbers (18, 23). In chilled callus, however, no visible change was detected both in the particle distribution and the particle number in any sort of membranes immediately after chilling to 0°C and even after 6 hr of chilling. No detectable change in the particles in various membranes, except for tonoplasts, was observed until 24 hr of chilling. In tonoplast, an aggregation and reduction in the number were detected early on both P and E faces after 12 hr of chilling. After 24 hr, an aggregation of particles on the P face of plasma membranes was initiated in a limited number of cells. Further chilling up to 48 hr produced a severe aggregation of particles on plasma membranes in almost all of the cells. These changes in plasma membranes were coincident with the abrupt permeation and oxidation of the added dopamine, and a partial release of amino acids in the cells chilled for 48 hr. These changes in plasma membranes observed in the later stage of chilling, where most of the cells are severely injured, may follow a secondary cellular event. Suggesting this notion, the severely aggregated membrane-particles on plasma membranes from 48 hr chilled cells were found to be completely relocated to the normal state by simply preincubating with potassium phosphate buffer, pH 7.2 at 0°C immediately before fixation in glutaraldehyde (11). Accordingly, the particle aggregation on plasma membranes during the later stage of chilling might be attributed to the altered nature of the membrane *per se* and/or reduction of pH in the cytosol as a result of the lesion of the tonoplasts.

No ultrastructural change in mitochondria was determined with 24 hr of chilling. A temporary depression of the respiratory activity, however, was noted in the 12 hr of chilling and then it returned nearly to the normal state upon prolonged chilling up to 24 hr. Considering the observed changes in the intra-membranous structures of tonoplasts and the concomitant dilation and vacuolation of ER membranes in the early stage of chilling, the reversible changes in the respiration may be ascribable to a reversible disturbance in the cytosolic environment, i.e., changes in the pH, salt concentration, etc., presumably caused by a permeability change in tonoplast. In *Frankenia* salt gland cells, the vesiculation of rough ER has been reported to be related to the salt accumulation (3). If it were the case in the callus utilized in the present study, we could speculate that the ER membranes have, more or less, the similar function to balance the cytoplasmic concentration of salts and/or proton and so forth.

Further experiments are clearly needed to understand entirely the mechanism of chilling injury in cultured plant cells. At the moment it seems possible to conclude that the lesion of the tonoplast in the primary step in chilling injury of the cultured plant cells, in a reversible or an irreversible way, may subsequently lead to irreversible alteration of cell functions. The reason for the tonoplast lesion *per se* under chilling remains unsolved. Some experiments are under investigation in our laboratory by utilizing isolated vacuoles from the callus.

Finally, our hypothesis for the mechanism of chilling injury in cultured plant cells is schematically presented in Fig. 8.

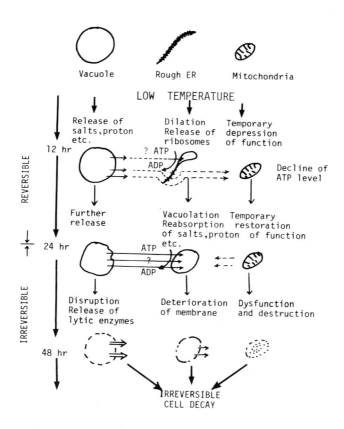

FIGURE 8.   A hypothetic mechanism of chilling injury of cultured plant cells (Cormus stolonifera). Some kinds of ATPase are presumed to take part in the process of salt-uptake and the vacuolation of ER in the reversible stage of chilling.

290   S. Yoshida *et al.*

V.   REFERENCES

Breidenbach, R. W., Wade, N. L., and Lyons, J. M. *Plant   Physiol.*
     54, 324–327 (1974).
2.   Butler, R. D. *J. Expt. Bot. 18,* 535–543 (1967).
3.   Campbell, N. and Thomson, W. W. *Ann. Bot. 40,* 681–686 (1976).
4.   Fukushima, H., Martin, C. E., Iida, H., Kitayama, Y., Thompson, G.
     A. and Nozawa, T. *Biochim, Biophys. Acta 431,* 165–179 (1976).
5.   Ilker, R., Waring, A. J., Lyons, J. M., and Breidenbach, R. W. *Pro-
     toplasma 90,* 229–252 (1976).
6.   Kimball, S. L., and Salisbury, F. B. *Amer. J. Bot. 60,* 1028–1033
     (1973).
7.   Lyons, J. M. *Ann. Rev. Plant Physiol. 24,* 445–466 (1973).
8.   Lyons, J. M. and Raison, J. K. *Plant Physiol. 45,* 386–389 (1970).
9.   Matile, P. and Moor, M. *Planta 80,* 159–175 (1968).
10. Matile, P. *The Lytic Compartment of Plant Cells,* pp. 18–38.
     Springer-Verlag, Wien-New York (1975).
11. Niki, T. Doctor Thesis, Hokkaido University, Japan (1979).
12. Platt-Aloia, K. A. and Thomson, W. W. *Cryobiology 13,* 95–106
     (1976).
13. Raison, J. K., Lyons, J. M., Mehlhorn, R. M., and Keith, A. D. *J.
     Biol. Chem. 246,* 4036–4040 (1971).
14. Raison, J. K., Lyons, J. M., and Thomson, W. W. *Arch.   Biochem.
     Biophys. 142,* 83–90 (1971).
15. Reynolds, E. S. *J. Cell Biol. 17,* 208–212 (1963).
16. Shaw, M. and Manocha, M. S. *Can. J. Bot. 43,* 747–755 (1965).
17. Shimonaka, H., and Nozawa, Y. *Cell Struct. and Function, 2,* 81–
     89 (1977).
18. Speth, V. and Wunderlich, F. *Biochim. Biophys. Acta 291,* 621–628
     (1973).
19. Spurr, A. R. *J. Ultrastruct. Res. 26,* 31–43 (1969).
20. Taylor, A. D. and Craig, A. S. *Plant Physiol. 47,* 719–725 (1971).
21. Wade, N. L., Breidenbach, R. W., Lyons, J. M., and Keith, A. D.
     Plant Physiol. 54, 320–323 (1974).
22. Wunderlich, F., Speth, V., Batz, W., and Kleining, H. *Biochim.
     Biophys. Acta 298,* 39–49 (1973).
23. Wunderlich, R., Batz, W., Speth, V., Wallach, D. F. H. *Biochim,
     Biophys. Acta 373,* 34–43 (1974).
24. Yamaki, S. and Uritani, I. *Plant and Cell Physiol. 13,* 67–69 (1972).
25. Yamaki, S. and Uritani, I. *Plant Physiol. 51,* 883–888 (1973).
26. Yamaki, S. and Uritani, I. *Agr. Biol. Chem. 37,* 183–186 (1973).
27. Yamaki, S. and Uritani, I. *Plant and Cell Physiol. 15,* 669–680
     (1974).

FREEZING STRESS IN POTATO

P. H. Li
J. P. Palta
H. H. Chen

Laboratory of Plant Hardiness
Department of Horticultural Science and L.A.
University of Minnesota
St. Paul, Minnesota

The potato[1] is an ideal crop to exploit in an effort to meet the increasing food demand of the world because of its economic value among the world major food crops (9), high energy production per unit of land (36), and high nutritive value compared with other food crops (18). It provides not only a good source of carbohydrates, but also a high quality of proteins (19), minerals and vitamins. More than 100 countries at present grow potatoes. However, it has been primarily cultivated in the Temperate Zone of North America and Europe, and tropical Andean highlands of South America where frost is often a major factor in reducing production or resulting in crop failure (22). Breeding for better adapted clones to frost is promising by genetic manipulation of existing potato germplasms (8). A better understanding of potato frost resistance (2), its hardening characteristics (1, 5), and the establishment of a laboratory screening method (34) have contributed to the recent successful breeding efforts (8).

Furthermore, the development of frost tolerant clones will greatly expand to the areas much of which is currently marginal due to low temperature. By the same token, the potato production can be increased in the presently cultivated areas by simply extending the growing season.

---

[1]*'Potato' in this paper is referred to the tuber-bearing Solanum species in addition to the S. tuberosum potato.*

I.   FROST HARDINESS AND FREEZING INJURY

A.   *Frost Hardiness*

Frost hardiness is defined as the resistance of foliage to a freezing temperature above which no injury occurs in the tissue. During freezing, water moves from the cells (mainly vacuole) to the intercellular ice that results in cell dehydration. As freezing proceeds, ices cause cellular contraction. Thus, during freezing, besides the low temperature, a plant experiences mainly three types of stresses: a dehydration stress due to extracellular ice formation, an osmotic stress due to removal of water from vacuole and a mechanical stress due to ice accumulation and cell contraction.   The mechanisms of frost hardiness in herbaceous plants have been reviewed by Olien (24) and Levitt (21). Levitt (21) concluded that the tolerance of dehydration and other stresses listed above due to extracellular freezing and the avoidance of intracellular freezing are mechanisms of frost hardiness. In general, herbaceous plants cannot withstand temperatures below $-20^{\circ}C$.

The difference in frost hardiness between resistant and sensitive potatoes is only about 3 to $4^{\circ}C$. In some plant species, a small variation in tissue water content (g H_2O/g dry weight) could result in such a difference in a plant's capacity to survive. A high cell sap concentration may increase slightly the ability of tissue to supercool; thus, the tissue could survive greater stress by avoiding freezing. However, the tissue water content and the cell sap concentration have not correlated to the frost hardiness in potato leaves (2).

Using freezing curve analysis, Sukumaran and Weiser (35) found that the leaf tissue of *S. acaule* potato could tolerate a greater amount of ice formation than the sensitive species, *S. tuberosum*. Nuclear magnetic resonance spectroscopic studies have shown that the resistant potato species can tolerate more freeze-induced dehydration than the susceptible ones (2). For example, at the killing temperatures the amount of unfrozen tissue water averaged to about 20 and 45% of total liquid water in *S. acaule* and *S. tuberosum,* respectively.   Thus it appears that the major difference in frost hardiness between resistant and susceptible potatoes is due to the ability of the hardy potato to tolerate more frozen water than the susceptible one.

B.   *Freezing Injury*

Freezing injury has been thought to involve cell membranes. One of the most common signs of freezing injury is the intercellular space infiltered with water, leading to a soaked appearance and loss of cell turgor. Freezing injury also often causes leakage of ions from the cells. The efflux of ions and the infiltration with water following a freezing injury have been assumed due to the breakdown of semipermeable properties of cell membrane (21, 35). Recently, Palta *et al.* (26, 27) have shown that during the progress of freezing injury (reversible or irreversible), semipermeable properties (membrane lipids) remain intact whereas the active transport properties (membrane intrinsic proteins) are damaged.

TABLE I.   Percent Ionic Leakage[a] of Two Potato Species at Two Time Intervals after Freezing-thawing (T_1 = 12 hr., T_2 = 60-120 hr.).

Freezing Temperature (°C)	0		Days in cold hardening conditions[b]			
			7		14	
	T_1	T_2	T_1	T_2	T_1	T_2
			Solanum tuberosum			
-2.0	40.5	20.2	-	-	-	-
-2.5	89.0	57.7	45.2	24.7	16.7	14.7
-3.0	90.9	68.1	59.0	38.0	33.1	37.4
-3.5	92.5	93.0	54.2	59.5	55.4	68.1
Control	11.3	8.3	13.7	7.1	5.6	4.1
			Solanum acaule			
-4.0	23.5	13.7	10.0	5.6	-	-
-5.0	54.0	41.3	34.6	21.6	12.5	9.5
-6.0	93.0	92.9	90.8	92.6	52.5	39.1
-7.0	94.0	92.3	93.0	93.8	59.3	70.6
-8.0	-	-	-	-	93.5	91.9
Control	9.5	6.1	7.1	5.4	5.9	4.6

[a]These data were presented at the Society for Cryobiology Annual Meeting (28).
[b]Cold hardening conditions (1).

1. *Reversibility of Membrane Damage after Freezing.* Plants of *S. acaule* and *S. tuberosum* were subjected to cold hardening conditions (for details see Ref. 1). At 0, 7, 14 and 21-day intervals, leaflets of both species were frozen from $-2^{\circ}C$ to $-9^{\circ}C$ at a cooling rate of $1^{\circ}C/hr$. They were then thawed slowly over ice. During the 60 to 120 hours following the thaw, measurements were made on the conductivity of the effusate. In the unhardened plants (0 day, Table I) the leaflets frozen to -2.0 and $-2.5^{\circ}C$ for *S. tuberosum,* and -4 and $-5^{\circ}C$ for *S. acaule* recovered from freezing injury during the post-thaw period, while at the lower temperatures, injury increased or remained unchanged with time until final death of the tissue. After 14 days of hardening, the leaflets frozen to $-3^{\circ}C$ in *S. tuberosum* and $-6^{\circ}C$ in *S. acaule* also recovered during the post-thawing periods (Table I). In both species, a recovery in injury was followed by disappearance of the infiltration with water, increase in turgidity and a decrease in the conductivity of the effusate (Table I). The freezing temperatures at which injury was reversible or irreversible varied by 0.5 or $1.0^{\circ}C$ depending upon the age of plant materials. These results suggest that the initial freezing injury may be to the active transport systems of the cell membranes because only the recovery of such process can lead to an active uptake of effusates, such as K^{+}, against the cell gradient. These observations are in agreement with earlier reports (26, 27).

2. *Microscopic Observations of Freezing Injured Cells.* As discussed, a freeze injured cell may or may not have the ability to recover completely. In order to further study the nature of injury, microscopic observations were carried out on injured cells (22, 29). Leaflets of both resistant *(S. acaule)* and sensitive species *(S. tuberosum)* were slowly frozen down to $-7^{\circ}C$ from $-2^{\circ}C$. Immediately after slowly thawing, segments from the middle portion of the leaflets were collected and fixed for ultrastructural studies. Although the tissue became soaked and flaccid various organelle appeared normal at a low degree of freezing stress. Abnormalities at the subcellular level seemed to start with swelling of the protoplasm (29). With increasing stress, swelling of the mitochondria and chloroplast was apparent. Separation of plasma membrane and cell wall (frost plasmolysis) and coagulation of the protoplasm were observed in dead cells (22). In these cells coagulation of the protoplasm and disruption of the tonoplast and plasma membrane were observed. Possible sequence of freezing injury in potato leaf cells appears to initiate with some disturbances in the cell membrane leading to swelling of protoplasm, and then swelling of mitochondria and chloroplasts, followed by breakdown of the tonoplast and plasma membrane system and coagulation of protoplasm resulting in cell death.

3. *Membrane Permeability after Freezing Injury.* In spite of leakage of ions and loss of turgor due to freezing injury, microscopic observations revealed that injured yet living cells were visually intact in cell membrane. The leakage of ions appears due to alterations in the cell

membrane properties rather than the membrane rupture. Furthermore, freeze injured cells could be plasmolyzed in a hypertonic mannitol solution (0.8 osm), and they remained plasmolyzed for several days similar to the unfrozen control cells (29). Even the irreversibly injured cells that showed a swelling of the protoplasm could be plasmolyzed in hypertonic mannitol solution just like the unfrozen control or reversibly injured cells (29). These observations indicate that the semipermeable properties of these cell membranes are still intact.

During the post-thaw period, an injured yet living cell may have the ability to recover from freezing injury (reversible injury) or may eventually die (irreversible injury). In order to examine the nature of alterations in membrane properties, the transport of urea, methyl-urea and potassium across the cell membrane were studied in freezing injured yet alive cells (29). A plasmometric method described by Stadelmann (33) was used in measuring the cell membrane permeability. In these experiments, the plasmolyzed protoplasts were allowed to expand in response to the passive uptake of permeable solute (equimolar solution of KCl, urea and methyl-urea). The protoplast expansion was measured directly with a microscope with an eyepiece micrometer.

The unfrozen cells did not take up K^+ in two hours and remained plasmolyzed at a constant volume. The injured cells, on the other hand, expanded immediately upon transfer to KCl solution and gradually deplasmolyzed. The rates of K^+ uptake varied for individual cells in the same tissue because the extent of injury was different in different cells. Although freeze injury resulted in a rapid increase in the rate of K^+ transport across the cell membranes, no change in rates was detected for nonelectrolytes such as urea and methyl-urea. The permeability of a nonelectrolyte has long been known as a direct function of its lipid solubility (6). Therefore, a change in the nonelectrolyte permeability should reflect the alterations in the lipid component of the membrane. No change in the permeability constants of these nonelectrolytes indicates that the freezing injury does not alter the lipid portion of the cell membranes. The finding that water permeability constants remain unchanged during freezing injury further supports this conclusion (26).

The increase in K^+ permeability provides evidence that the protein components of the membrane are possible targets of freezing injury. Membrane proteins are of two types, intrinsic and peripheral. Some of the intrinsic proteins pass entirely through the bilayer (31). This type of intrinsic proteins has been proposed as being sites for active transport of ions (32). Sublethal freezing temperature can lead to the denaturation of membrane proteins resulting in an inactivation of the active transport systems. When inactivated, these intrinsic proteins could serve as channels for passive ion transport, giving very high K^+ permeability values and swelling of the protoplasm. A large passive efflux of ions could also occur through these channels in the direction of the concentration gradient from the vacuole to the extracellular solution. A repair of the inactivation leads the tissue to recover from freezing injury (reversible) and failure to do so leads ultimately to the death of cells (irreversible).

II.   COLD HARDENING

Many herbaceous plants undergo increase in frost hardiness when subjected to a specific environmental condition such as low temperatures, short daylength, changes in quality and quantity of light, reducing water supply, etc. Some species can harden (increase in frost hardiness) extensively in response to these environmental changes, while others will harden only a few degrees and some do not harden at all.

A.   *Species, Varieties and Tissues*

Prior to 1976, there was some dispute as to whether the tuber-bearing *Solanum* potatoes can be cold hardened. Hayden *et al* (14) indicated that potatoes possess a stable frost hardiness level and do not harden. Richardson and Estrada (30) reported that 2 to 3 weeks of cool temperatures could differentiate hardy from nonhardy potatoes. Chen and Li (1) confirmed that some noncultivated species like *Solanum acaule, S. chomatophilum, S. commersonii and S. multidissectum* could harden, while the cultivated *S. tuberosum* (varieties of "Red Pontiac," "Kennebec," "Norland" and "Norchip") failed to harden.

Recently, additional tuber-bearing *Solanum* species were screened for their frost hardiness and cold hardening ability. Based on existing information, we can classify them into four groups in terms of their frost hardiness and cold hardening (Table II). These groups are (I) the species which possess frost hardiness (survive to -4.0°C or colder) and are able to cold harden, (II) the species which possess frost hardiness but are unable to cold harden, (III) the species which possess no frost hardiness (survive to -3.0°C or warmer) but are able to cold harden, and (IV) the species which possess no frost hardiness and are unable to cold harden. Example species for each group are listed in Table II. *Solanum Tuberosum,* the commonly grown potatoes, falls into the last category.

The potato is a cool season grown crop and is usually chilling resistant. We observed chilling injury in *S. trifidum* (PI 25541) when plants were grown in $2/2^{\circ}$C, day/night, regime. It is possible that additional tuber-bearing *Solanum* species, which are chilling sensitive, exist.

Stem-cultured plants of *S. tuberosum* and *S. commersonii* grown in agar medium showed similar hardening patterns as the potted plants (5). Leaf callus tissues of *S. tuberosum,* "Atlantic," "Kennebec" and "Red Pontiac," showed no hardening ability and were killed at -3°C. The leaf callus tissues of *S. acaule,* however, can be hardened to -9°C after 15 days at 3°C in darkness, a 3°C increase in frost hardiness (5). These results agreed with observations from potted plants previously reported (1).

TABLE II.   Classification of Tuber-bearing Solanum Potatoes in Terms of Frost Resistance and Cold Hardening

Categories	Species (examples)	Killing Temp (C^o)	
		Before Treatment[a]	After Treatment[b]
Group I: frost resistance and able to cold harden	S. acaule (Oka 3885)[c]	-6.0	-9.0
	S. commersonii (Oka 5040)	-4.5	-11.5
Group II: frost resistance but unable to cold harden	S. sanctae-rosae (Oka 5697)	-5.5	-5.5
	S. megistacrolobum (Oka 3914)	-5.0	-5.0
Group III: frost sensitive but able to cold harden	S. oplocense (Oka 4500)	-3.0	-8.0
	S. polytrichon (PI 184773)	-3.0	-6.5
Group IV: frost sensitive and unable to cold harden	S. tuberosum	-3.0	-3.0
	S. stenotomum (PI 195188)	-3.0	-3.0

[a] Plants were grown in a regime of 20/15°C day/night, 14 hrs.
[b] Plants were grown in a regime of 2°C day/night, 14 hrs, for 20 days.
[c] Identification number at Potato Introduction Station, Sturgeon Bay, Wisconsin.

B.   Optimum Hardening Conditions

Low temperatures and short days are the two major environmental factors in inducing frost hardiness. Some species can be hardened by low temperatures with a sufficient amount of light, regardless of the photoperiod (20). Potatoes belong to this category. Low temperature conditioning for maximum hardiness varies from species to species. For example, winter rape (17) and cabbage (20) required freezing temperature during the hardening to achieve the maximum hardiness. The optimum hardening temperatures for Solanum potatoes are about 1 to 2°C above freezing.

Cold hardening of potato can be achieved by directly exposing plants to constant day/night low temperatures (5) or stepwise lowering the temperature (1). The hardiness level which can be achieved is dependent upon the temperatures used. Lower temperatures result in greater

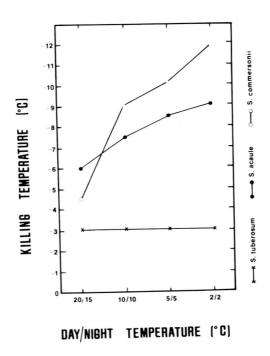

FIGURE 1. *Changes in frost hardiness of three Solanum species at constant low day/night temperatures with 14 hr light for 2 weeks.*

degrees of hardiness (Fig. 1). Under constant low temperatures, plants appeared to reach their maximum hardiness at about 15 days, while plants that were subjected to stepwise hardening conditions continued to harden even after 20 days (5).

C. Growth Regulators

The application of growth regulators to induce frost hardiness has long been considered but the results have been contradictory (21). Most studies were conducted by exogenous application to the foliage or the soil. Foliar applications raise certain problems because leaf surfaces have a tendency to restrict the uptake of exogenously applied chemicals. Microbiological degradation causes problems with the soil application. For studying the effect of growth regulators on potato hardiness, *Solanum* stem-cultured plants grown in the agar medium were therefore used (5). Preliminary results indicated that the exogenously applied abscisic acid (ABA) can increase potato frost hardiness. No increase in frost hardiness was found with gibberellic acid treatments. The increase and change in hardiness with ABA treatments are the same in both warm and low temperature regimes. It appears that the exogenous ABA may substitute the functional role of the low temperature in inducing frost hardiness. However, different levels of ABA produced the same results. As suggested by Kacperska-Palacz (17), ABA may serve as a driving force to shift the metabolic pathway in favoring cold hardening. Possibly higher levels of ABA have an effect similar to the lower levels once the endogenous ABA reaches the functional level. Apparently, ABA can induce hardiness not only in those species which are able to harden but also in *S. tuberosum* which is a species that is unable to harden at low temperature.

D.   Biochemical Changes

Many studies have been reported regarding biochemical changes in herbaceous plants during cold-hardening (7, 11, 12). We also examined and compared some of the biochemical changes in leaves of *S. acaule* and *S. tuberosum* during cold hardening (4). The results are summarized below. Since DNA in matured leaf cell is maintained at a constant level during growth (25) metabolite changes during hardening would be better expressed on unit DNA than on dry weight basis. Following discussions relative to the changes in sugars, starch, nucleic acids, proteins and lipids are therefore based on unit DNA (same numbers of cells).

1.   *Sugars and Starch.* Sugar increase was found in both *S. acule* and *S. tuberosum* during hardening. Total sugar content more than doubled in both species after 15 days hardening. Levitt (21) reviewed the protective role of sugars against frost injury. However, such a role in potato frost hardiness is questionable because *S. tuberosum* fails to harden even when its sugar content increases more than twofold during hardening. Huner and Macdowall (16) found that the cold adaptive rye plants increased the stability of RuBP carboxylase in chloroplasts under cold stress. This would enable plants to maintain photosynthetic ability under cold hardening condition. Therefore, sugars can accumulate in leaf tissues.

Under hardening condition, leaf starch content was also increased in both *S. acaule* and *S. tuberosum*. Hatano (13) reported an increase of starch in *Chlorella* during cold hardening under light. Heldt *et al* (15) reported that isolated spinach chloroplasts from hardy winter materials showed higher starch synthesis rate than from nonhardy winter materials. The decrease in starch content during cold hardening was almost always found in nonphotosynthetic tissues such as bark (23) and root (10) which possess very little or no photosynthetic activity. The cold hardening is a process requiring energy. The energy source in nonphotosynthetic tissues seems to be the stored food such as starch. On the other hand, leaves which can maintain their photosynthetic activity at low temperature should be able to directly supply photosynthate as the energy source for hardening. This may be an explanation for both starch and sugar increases in potato leaves during cold hardening.

2.   *Ribonucleic Acids.* In *S. acaule,* plants grown under hardening environment always maintained at a higher level of rRNA than plants grown under unhardening environment. In *S. tuberosum,* there was no different change in rRNA content between hardened plants and controls. It has been reported that the temperature had a marked influence on rRNA metabolism while the photoperiodic response was not as great in potato plants (25). The higher level of rRNA in *S. acaule* plants during hardening is paralleled with higher level of soluble protein and the increase of hardiness. The difference in rRNA levels between *S. acule* and *S. tuberosum* in response to hardening suggests that the hardening process likely initiates at the transcription and/or translation levels.

3.   *Proteins.* In *S. acaule,* soluble proteins were maintained at a much hjgher level in cold hardened plants than in controls. The soluble protein fraction also increased at a much higher rate in the former than in the latter during hardening. No significant protein changes were observed in both controls and hardened plants of *S. tuberosum,* which failed to harden. Since the increase in hardiness is always parallel with the increase in soluble protein content, several researchers have discussed the role of soluble protein in cold hardening (17, 21). Cox *et al* (7) concluded that only plants that were able to conduct active protein synthesis at low temperature had the capability to harden. This conclusion was supported by Hatano's work (13) in which cyclohexamide, an inhibitor for protein synthesis, could inhibit the hardening in *Chlorella.*

4.   *Lipids.* During hardening, total lipid increased in *S. acaule* and a more or less constant level was maintained in *S. tuberosum.* The increase in lipid in *S. acaule* during hardening supported the EM observations of lipid bodies accumulations in hardened chloroplasts (3). The phospholipid content in hardened *S. acaule* plants was much higher than in controls, while lower level of phospholipid was observed in *S. tuberosum* hardened plants than in controls during hardening. Cold hardening alters membrane properties and increases its stability during freezing as suggested (17). The development of hardiness in *S. acaule* is probably associated with alterations in membrane properties as evidenced in phospholipid increase.

E.   *Electron Microscopic Observations*

Comparative ultrastructural studies were undertaken between hardy *(S. acaule)* and tender *(S. tuberosum)* potatoes grown under warm and cold temperature regimes. In less than 10 days during cold hardening, the hardy species of *S. acaule* showed a drastic increase in starch grains in chloroplasts (Fig. 2B). Such an increase, however, was

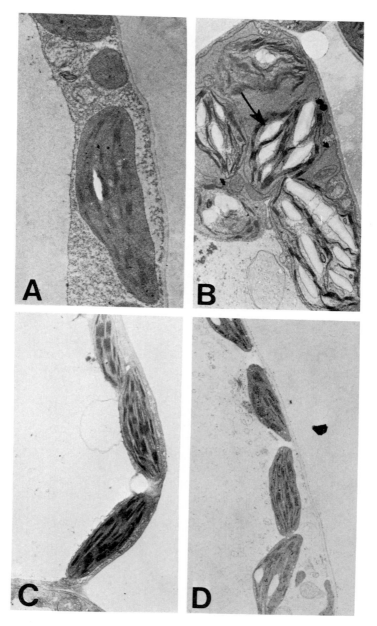

FIGURE 2. Electron microscopic observations of leaf cross-sections showing the accumulation of starch grains (→) in chloroplasts after 8 days cold hardening (5/2°C day/night temp., 14 hr light). A,B: Solanum acaule; C,D; Solanum tuberosum; A & C: before hardening; B & D: after hardening.

not observed in chloroplasts of the tender species of S. *tuberosum*
(Fig. 2D) grown under the same regime. It appears that energy sources
for cold hardening in potato came directly from photosynthate rather
than from starch to sugar transformation. Chen *et al* (3) have reported the swelling and irregularity of thylakoid membranes and large
patches of stroma in chloroplasts of hardened S. *acaule,* while no significant changes were observed in cold treated S. *tuberosum* chloroplasts.

III. SUMMARY

   The difference in frost hardiness between a resistant and sensitive
type of potato is about 3 to 4°C. This difference is not due to the
avoidance of intracellular freezing because cell sap concentration and
tissue water content have shown no correlation to the variations of frost
hardiness among different species; rather, it is due to the tolerance of
freeze-induced dehydration (extracellular freezing). During the initial
stage of freezing injury, semipermeable properties of the cell
membranes remain intact. Freezing injury, due to extracellular ice
formation, results in alterations of membrane transport properties, most
likely the intrinsic membrane proteins which involve in active transport
systems.   Freezing injury can be reversible (leading to complete
recovery) or irreversible (leading to death). The mechanisms of frost
hardiness and cold hardening appear to be independent. Increases in
ribosomal RNA, soluble proteins and phospholipids are associated with
the increase of frost hardiness during cold hardening.

IV. REFERENCES

1.   Chen, P. M., and Li, P. H. *Botanical Gaz. 137,* 105–109 (1976).
2.   Chen, P. M., Burke, M. J., and Li, P. H. *Botanical Gaz. 137,* 313–317 (1976).
3.   Chen, P. M., Li, P. H., and Cummingham, W. P. *Botanical Gaz. 138,* 276–285 (1977).
4.   Chen, H. H. Master Plan B Paper, University of Minnesota, Saint Paul (1978).
5.   Chen, H. H., Gavinlertvatana, P., and Li, P. H. *Botanical Gaz. 140,* June Issue (1979).
6.   Collander, R., and Barlund,H. *Acta. Bot. Fenn. 11,* 1–114 (1933).
7.   Cox, W., and Levitt, J. *Plant Physiol. 44,* 923–928 (1969).
8.   Estrada, R. N. *In* "Plant Cold Hardiness and Freezing Stress," (P. H. Li and A. Sakai, eds.). pp. 333–334, Academic Press, New York (1978).
9.   Food and Agricultural Organization. Production Year Book 1974 (1974).

10. Glier, J. H., and Caruso, J. L. *Cryobiology 10*, 328-330 (1973).
11. Greathouse, G. A., and Stuart, N. W. *Plant Physiol. 12*, 685-702 (1937).
12. Green, D. G., and Ratzlaff, C. D. *Can. J. Bot. 53*, 2198-2201 (1975).
13. Hatano, S. *In* "Plant Cold Hardiness and Freezing Stress," (P. H. Li and A. Sakae, eds.). pp. 175-196, Academic Press, New York (1978).
14. Hayden, R. E., Dionne, L, and Fensom, D. S. *Can. J. Bot. 50*, 1547-1554 (1972).
15. Heldt, H. W., Chon, C. J., and Maronde, (1965). 1155 (1977).
16. Huner, N. P. A., and Macdowall, F. D. H. *Biochem. Biophys. Res. Comm. 73*, 411-420 (1976).
17. Kacperska-Palacz, A. *In* "Plant Cold Hardiness and Freezing Stress," (P. H. Li and A. Sakae, eds.). pp. 139-152. Academic Press, New York (1978).
18. Kaldy, M. S. *Econ. Bot. 26*, 142-144 (1972).
19. Kapoor, A. C., Desborough, S. L., and Li, P. H. *Potato Res. 18*, 469-478 (1975).
20. Kohn, H., and Levitt, J. *Plant Physiol. 40*, 476-480 (1965).
21. Levitt, J. Responses of Plants to Environmental Stresses, Academic Press, New York (1972).
22. Li, P. H., and Palta, J. P. *In* "Plant Cold Hardiness and Freezing Stress," (P. H. Li and A. Sakai, eds.). pp. 49-71, Academic Press, New York (1978).
23. Marvin, J. W., and Morselli, M. *Cryobiology 8*, 339-344 (1971).
24. Olein, C. R. *Ann. Rev. Plant Physiol. 18*, 387-408 (1967).
25. Oslund, C. R., Li, P. H., and Weiser, C. J. *J. Amer. Soc. Hort. Sci. 97*, 93-96 (1972).
26. Palta, J. P., Levitt, J., and Stadelmann, E. J. *Plant Physiol. 60*, 393-397 (1977).
27. Palta, J. P., Levitt, J., and Stadelmann, E. J. *Plant Physiol. 60*, 398-401 (1977).
28. Palta, J. P., and Li, P. H. Society for Cryobiology Annual Meeting, Abs. 25 (1977).
29. Palta, J. P., and Li, P. H. *In* "Plant Cold Hardiness and Freezing Stress," (P. H. Li and A. Sakai, eds.). pp. 93-115, Academic Press, New York (1978).
30. Richardson, D. G., and Estrada, R. N. *Amer. Potato J. 48*, 339-343 (1971).
31. Singer, S. J., and Nicolson, G. L. *Science 175*, 720 (1972).
32. Singer, S. J. *In* "Cell Membrane." pp. 35-44, H. P. Publishing Company, New York (1975).
33. Stadelmann, E. J. *Methods in Cell Physiology 2*, 143-216 (1966).
34. Sukumaran, N. P., and Weiser, C. J. *Hort. Sci. 7*, 467 (1972a).
35. Sukumaran, N. P., and Weiser, C. J. *Plant Physiol. 50*, 564-567 (1972b).
36. van der Zaag, D. E. *Potato Res. 19*, 37-72 (1976).

FLUORESCENCE POLARIZATION STUDIES OF MEMBRANE PHOSPHOLIPID PHASE SEPARATIONS IN WARM AND COOL CLIMATE PLANTS

Carl S. Pike[2,3], Joseph A. Berry, and John K. Raison[4]

Department of Plant Biology
Carnegie Institution of Washington
Stanford, California

I. INTRODUCTION

The search for a primary locus of plant responses to low temperature has focused on the physical properties of the membrane lipids. At low temperature membrane lipids may undergo a phase transition from a fluid, liquid-crystalline state to a more solid, gelled state. When a lipid mixture is cooled, the lipids with the highest melting points solidify first and undergo a lateral phase separation from the remaining fluid lipid (19). The temperature dependence of the physical state of membrane lipids (as studied with electron spin resonance probes in intact membranes or vesicles prepared from isolated lipid fractions) has been compared to the temperature dependence of various biochemical processes catalyzed by membrane-bound proteins. These studies, based largely on crop plants, have suggested a strong correlation between the lipid phase separation temperature and an increase in the activation energy of various reactions (13, 14). An abrupt change in the temperature dependence of spin label motion in mung bean lipids correlated well with the lower limit of seedling growth (16). Lipid phase changes in a series of *Passiflora* species corresponded to the species' known temperature responses for

[1] CIW/DPB Publication Number 660
[2] Permanent address: Department of Biology, Franklin and Marshall College, Lancaster, PA 17604
[3] Supported in part by an NSF Science Faculty Professional Development Award and by Franklin and Marshall College
[4] Permanent address: Plant Physiology Unit, CSIRO Division of Food Research and School of Biological Sciences, Macquarie University, North Ryde, N.S.W., 2113

growth (11). The physical state of the membrane lipids is thus suggested to be a major determinant of the lower temperature limit for a plant's survival or growth.

Recent work by Sklar *et al.* (23, 24, 25) has shown the usefulness of a pair of naturally-occurring fluorescent polyene fatty acids, *cis-* and *trans*-parinaric acid (Fig. 1), as probes of membrane lipid properties. When biosynthetically incorporated into membrane lipids or merely added to a membrane or lipid vesicle preparation, these molecules would be expected minimally to perturb the lipid orientation, compared to a typical ESR probe bearing a bulky 2.2-dimethyl-N-oxyloxizolidine ring attached to a fatty acyl chain (2). The probes are very rapidly incorporated into lipid bilayers when added to a liposome preparation. Measurements of the temperature dependence of the fluorescence intensity can provide information on lipid phase transitions, as shown in studies of single-component phospholipid vesicles (25), bacterial membranes and extracted phospholipid vesicles (26), and mammalian membrane and phospholipid preparations (17). These probes have been used to demonstrate the substantial difference in phase transition temperatures in membrane lipids from an *E. coli* fatty acid auxotroph fed fatty acids differing in chain length and saturation (26). The two isomers differ in their partitioning behavior between solid and fluid phases in a mixed system: the *trans* form has a strong preference for the solid phase, while the *cis* form has a very slight preference for the fluid phase. As a result, *trans*-parinaric acid is sensitive to a few percent solid, but becomes insensitive to the formation of solid lipid above about 50% (29).

The fluorescence polarization behavior of *trans*-parinaric acid-labeled lipid preparations can provide information on fluidity behavior, in the same manner as ESR probes [See Shinitzky and Barenholz (20) for review of fluorescence polarization]. The polarization ratio (fluorescence intensity with emission and excitation polarizers oriented parallel to one another divided by the intensity with the polarizers in perpendicular orientation) is used as an expression of fluidity, with a higher ratio corresponding to lower fluidity (25).

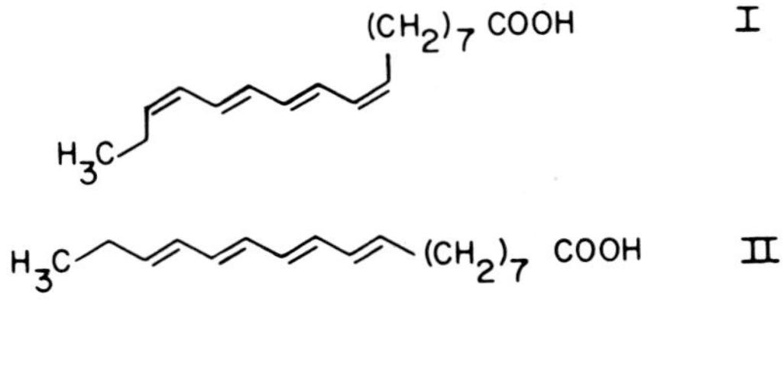

FIGURE 1. The structures of cis-parinaric acid (I) and trans-parinaric acid (II).

Most studies on chilling sensitivity and resistance have dealt with crop plants, which have undoubtedly been extensively modified by artificial selection. We have chosen to emphasize investigations of wild plants, since the temperature responses of their lipids might more closely reflect the constraints of natural selection operating in the native environment. Plants native to Death Valley and other areas of the Mojave Desert have been used for much of this study. This region is a temperate desert with substantial seasonal temperature fluctuations (the mean daily maximum and minimum are 45C/31C in July and 18C/3C in January). Several annuals, considered to be primarily cool- or warm-season active, were investigated. Whenever possible, plants were grown from seed at the same temperature, to allow us to focus on genetic, rather than environmentally-induced, differences in lipid physical properties.

II.   MATERIALS AND METHODS

The temperature conditions during growth of each species are indicated in the tables. For plants grown from seed, controlled environment chambers were used. Within the groups of grasses and dicots, chamber-grown plants received identical temperature and light regimes; chilling sensitivity (as assessed by growth) is dependent on these factors (2). A few plants were collected in Death Valley; the mean daily maximum and minimum temperatures for the months in which the collections were made are indicated below.

Approximately 5 gm of leaf tissue was extracted for 5 min in 50 ml of boiling methanol containing 1 mg of of butylated hydroxytoluene. Then 100 ml of chloroform was added, the tissue was ground in a Vir-Tis homogenizer, and the extract was filtered through miracloth. The extract was partitioned (6) 4 times with 0.55 M KCl, once with water, and once with 0.06 M KCl. The chloroform layer was dried over Na_2SO_4, concentrated, and loaded onto a Bio-Sil A (Bio-Rad Laboratories) column which was sequentially eluted with chloroform, acetone, and methanol. The phospholipid-rich methanol fraction was used in the fluorescence studies. The presence of substantial amounts of pigments in the polar lipid fraction (galactolipids plus phospholipids, the methanol eluate obtained following washes with chloroform and 10% acetone in chloroform) interfered with the fluorescence measurements, so the phospholipid fraction was used. Aliquots were dried onto the walls of a glass vial under N_2, held in a vacuum desiccator, and then dispersed by gentle sonication in 0.1M Tris-HCl buffer, pH 7.2, containing 0.005M Na_2 EDTA.

For fluorescence studies, 400 μg of lipid vesicles was added to buffer containing either 25% or 33% ethylene glycol (a concentration without effect on phase separation temperatures). *Trans*-parinaric acid (0.7 μg in ethanol) was added and the sample stirred for 15 min at room temperature The fluorimeter (Perkin-Elmer MPF-3L) contained a temperature-regulated cuvette holder and a magnetic stirrer; temperatures were measured with a copper-constantan thermocouple. The excitation beam (320 nm) was passed through a polarizing prism (Karl Lambrecht and Co.); the emitted light was passed through a plastic polarizer (Edmund Scientific) and a 350 nm cut-off filter, and the monochromator was set at 420 nm. All temperature scans were made in the ascending direction; there were at least 2 replicates of each measurement. Phase transitions of replicate samples usually did not differ by more than 1C degree.

III. PHASE SEPARATION PROCESSES

We will first examine data from corn (Zea mays) as a general illustration. Figure 2 shows a plot of the fluorescence intensity measurements and Figure 3, the polarization ratio. The change in slope of the intensity curve occurs at 9C, in good agreement with ESR determinations (3). Likewise, the sharp discontinuity in the polarization ratio occurs at 10C; in fact, in all preparations studied the two procedures indicate changes within about 1C degree of one another. The abrupt change in intensity is associated with a change in fluidity, and the fluidity change is more specifically shown by the polarization ratio. Although we have continued the measurements to 50C, we have not detected the change usually seen around 25-35C with some ESR probes (14, 16).

At temperatures above the polarization ratio slope change, the low ratio indicates a fluid probe environment (25). Analysis of model systems has shown that this probe is sensitive to the appearance of a few percent solid (25). Thus we interpret the slope change at 9C in corn as the beginning of the appearance of detectable solid as the temperature is lowered   This interpretation is supported by studies in which the phase separation process of *Anacystis nidulans* membranes was observed using freeze-fracture electron microscopy and the properties of extracted phospholipids were studied with *trans*-parinaric acid (1, 15). A similar correlation is seen in studies of *E. coli* (7, 26). The idea that chilling sensitivity correlates with the first appearance of solid lipid as the temperature is lowered is in contrast to previous ESR work which suggested that this point was the end of solidification (16). That assignment had been made by analogy to model systems (28). At present we do not know what sort of physical change is detected by certain ESR probes at 25-35C; in model systems Sklar *et al.* (25) found that some structural order may persist 10C degrees or more above the transition temperature.

In Figure 3 the levelling off of the polarization ratio of the corn lipids at about 2.0 (seen at about -5C) does not indicate the completion of the solidification process. In model system studies Sklar *et al.* (25) showed that levelling-off occurred at about 50% solid; thus, the *trans*

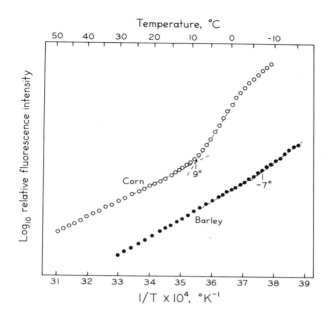

FIGURE 2. Logarithmic plots of trans-parinaric acid fluorescence intensity of corn and barley phospholipid vesicles. The phospholipids were dispersed by gentle sonication in 0.1M Tris-HCl buffer, pH 7.2, containing 5 mM Na_2 EDTA and 25% or 33% (V/V) ethylene glycol. A suspension containing 400µg lipid per 3 ml of buffer was labeled with 0.7 µg of trans-parinaric acid. Polarized light at 320 nm was used to excite the sample and fluorescence was measured at 420 nm, with the excitation polarizer parallel to the emission polarizer.

isomer is not suitable to detect the completion of solidification. The *cis* isomer would be a better indicator of that point, which may not occur in corn phospholipids until below -20C.

The intensity (Fig. 2) and polarization ratio (Fig. 3) plots for barley *(Hordeum vulgare)* phospholipids are markedly different from the corn plots. There are rather slight discontinuities at about -6C. The polarization ratio shows that even below that point the lipids are still not very ordered, compared to the ratio found for corn. We have compared two legume crops, chilling-sensitive beans *(Phaseolus vulgaris)* and chilling-resistant peas *(Pisum sativum),* and found the phase separations at 9C and -4C, respectively.

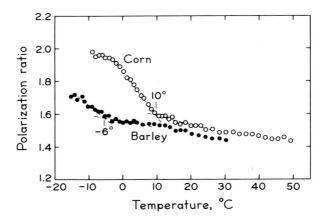

FIGURE 3. *Trans-parinaric acid fluorescence polarization of corn
and barley phospholipid vesicles. Fluorescence emission was measured
with the polarizer parallel $(I_{||})$ and perpendicular $(I_{\perp})$ to the orienta-
tion of the excitation polarizer. The polarization ratio is $I_{||}/I_{\perp}$.*

IV. PHASE SEPARATION TEMPERATURES IN WARM AND COOL CLIMATE PLANTS

A. *Grasses*

Table I shows the separation temperatures for several grasses
generally viewed as cool- and warm-temperature active. The warm
temperature species show separations several degrees higher than the
cool-temperature plants. In the case of the crop plants, the separation
temperatures correlate with observations on chilling sensitivity (corn)
and resistance (barley, oats).

B. *Desert Dicots*

Mojave Desert plants, chiefly annuals, were selected to exemplify
the winter and summer-active flora (10, 22). As shown in Table II, the
winter-active species exhibited phospholipid phase separations at 3C or
less. These plants would seem well adapted to the typical temperature
regime in the Death Valley area, where temperatures are rarely below 0C
(the mean daily minimum temperature in the coolest month is 3C). The
summer-active species present a somewhat more complex pattern. *Ti-*

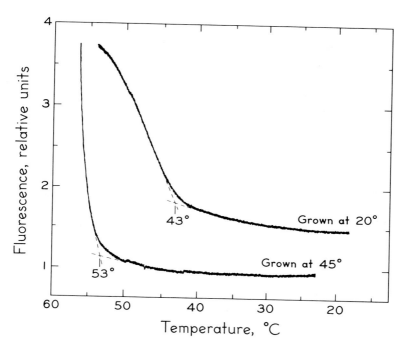

FIGURE 4.   Change in the relative intensity of chlorophyll
fluorescence in whole, attached Nerium oleander leaves as a function of
temperature.   Leaves from plants grown at the indicated temperatures
were heated at about 1C degree/min. The fluorescence from excitation
by extremely low intensity light was measured continuously.   The
temperature of the fluorescence increase was determined as the point of
intersection of the lines extending the two linear portions of the curves
as shown.   From Raison et al. (15).   Reprinted by permission of John
Wiley and Sons, Inc.

destromia oblongifolia is a winter-dormant perennial which shows
optimal growth at very high temperatures (about 45C/32C day/night). In
a 16C/11C regime, there is no net growth although the plants are not
killed (3). Plants grown from seed at 45C/32C showed a phase separation
at 12C (cf. corn, beans, and Boerhaavia coccinea). Also, some leaves
were collected in Death Valley in December, when there was probably
little net growth. The phase separation occurred at 7C; the plants were
routinely experiencing much lower temperatures. Typically, Tidestromia loses its leaves in winter. The low separation temperature for
Atriplex elegans elegans is an exception, and may indicate that it is not
necessary for a summer-active species to have a high phase separation
temperature.   Although this plant is part of the summer flora in the

TABLE I.   Phospholipid Phase Separation Temperatures
             for Annual Grasses[a]

Species	Separation temperature[b] C
Cool Climate	
Avena fatua	-9
Avena sativa	-11
Bromus rigidus	-10
Hordeum vulgare	-6
Warm Climate	
Chloris virgata	4
Digitaria sanguinalis	8
Panicum texanum	7
Zea mays	9

[a] All plants were grown at constant 27C and a 16 hr photoperiod.
[b] Determined as the average value from trans-parinaric acid fluorescence intensity and polarization ratio plots of data from 2 samples.

Mojave Desert, it can also be found during the winter (22). Its phase separation behavior may reflect a genetic adaptation to the coolest conditions the plant may experience. We have not observed any winter-active plants with high phase separation temperatures, so we suggest that there is a direct relationship between the occurrence of low temperature and natural selection for lipids with a low phase separation temperature.

Some Death Valley plants are active through much of the year. We investigated the phospholipids from 4 such species collected in January. In all cases the separation occurred at 1C or below. Thus, there seems to be a strong correlation between the ability of a plant (either an ephemeral or a perennial) to grow at low temperatures and the possession of a low phase separation temperature. The desert species we have studied co occur, yet in general, natural selection has apparently yielded lipid compositions with quite different phase separation temperatures in plants with different seasonal preferences.

A plant with a relatively high phase transition temperature seems restricted from growth at chilling or freezing temperatures. In the case of wild species, we base this statement on the known seasonal activity of the plants For crop plants (such as corn and beans), direct information

TABLE II. Phospholipid Phase Separation Temperatures for Desert Dicots

Species	Growth[a] conditions	Separation[b] temperature C
Cool Climate		
Boerhaavia annulata	DV 19C/3C	2
Camissonia claviformis	DV 22C/6C	-3
Cryptantha angustifolia	L 28C/21C	2
Eriogonum inflatum	DV 22C/6C	-4
Lepidium lasiocarpum	L 28C/21C	-1
Perityle emoryi	L 28C/21C	3
Warm Climate		
Atriplex elegans elegans	L 28C/21C	-1
Boerhaavia coccinea	L 28C/21C	12
Pectis papposa	L 28C/21C	13
Tidestromia oblongifolia	L 45C/32C	12
	DV 19C/3C	7
Active throughout the year		
Atriplex hymenelytra	DV 18C/3C	None detectable to -10C
Heliotropium curassavicum	DV 18C/3C	-2
Larrea divaricata	DV 18C/3C	-6
Psathyrotes ramosissima	DV 18C/3C	1

[a] For laboratory-grown (L) plants, the growth chamber day and night temperatures are given (16 hr photoperiod). For plants collected in Death Valley (DV), the mean daily maximum and minimum temperatures for the month of collection are given.
[b] Determined as in Table I.

on growth responses is available. Although the precise reason for this restriction is not yet established, our studies suggest that the complex of phenomena known as chilling injury (9) relates to the appearance of solid phase lipid. A plant growing in warm conditions need not have a high phase separation temperature; we do not know how common the Atriplex elegans elegans situation might be.

It is important to realize that fluorescent probe analysis, like other biophysical techniques, can only provide an average measure of the bulk lipid properties (20). If some subfraction of the lipids, such as protein boundary layers (27), is critical to chilling sensitivity or resistance, then these studies would not specifically investigate its properties. However, our results and others (11) provide at least qualitative agreement between bulk properties and plant temperature preference.

IV. ADAPTATION TO HIGH TEMPERATURE

Many of the cold climate plants we have studied are unable to grow at high temperature. The fluorescence measurements do not indicate any change in lipid physical properties above the phase separation temperature as the basis for this limitation. Studies conducted on *Nerium oleander,* a C_3 evergreen capable of acclimating to a wide range of temperatures, provide information on the relation of lipid properties and activity at high temperature.

The acclimation of cloned oleander plants to growth in chambers at 45C/32C (day/night) and 20C/15C is indicated by the differences in photosynthetic temperature optima: between 35 and 40C for the warm-grown plants and between 25 and 30C for the cool-grown. Also, the onset of reversible thermal inhibition of photosynthesis is at 47 and 40C, respectively (4). At any temperature between 14C (the lowest studied) and 40C, the photosynthetic rate of the cool-grown plants exceeds that of warm-grown plants. Lipid phase separation temperatures of about 7C and -3C for warm- and cool-grown plants were determined with *trans*-parinaric acid. Does this change in lipid properties seen at low temperature have any relation to the differences in photosynthetic function at high temperature?

The intactness of the chloroplast membranes can be revealed by examining the fluorescence yield of chlorophyll. In intact, undamaged lamellae, chlorophyll fluorescence is highly quenched by excitation energy transfer to the reaction centers. An increase in F_o chlorophyll fluorescence suggests a disruption of this transfer (18). Measurements on intact oleander leaves from warm- and cool-grown plants showed an increase in fluorescence at 53C and 43C, respectively (Figure 4), temperatures which are quite close to the point of irreversible inhibition of whole-leaf photosynthesis (4). This result suggests a 10C degree difference in membrane thermal stability as a result of differences in growth temperature (15).

Chloroplast membrane polar lipid vesicles from these plants were studied to see if there is a relationship between lipid physical properties and membrane thermal stability. Figure 5 shows the temperature dependence of the motion (expressed as rotational time) of the spin label membrane probe 12NS [3-oxazolidenyloxy-2-(10- carbmethoxydecyl)-2 hexyl-4, 4-dimethyl] . The motion decreases as temperature increases. There are no changes in the temperature coefficient of motion over the

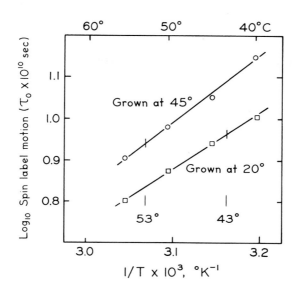

FIGURE 5. *The motion of a spin label incorporated into the polar lipids of chloroplast membranes from Nerium oleander as a function of temperature. Plants were grown at either 45C/32C day/night temperatures or 20C/15C. The lipids were suspended as described for Figure 2 and labeled with 12NS[3-oxazolidenyl-2-(10 carbmethoxydecyl)-2-hexyl-4,4-dimethyl]. Motion was calculated as described by Raison and Chapman (16) from 5 spectra at each temperature. The standard deviation for motion at 20C was ± 0.2 x 10^{-10} sec and at 50C, ± 0.05 x 10^{-10} sec for both samples.*

range 20C to 50C, showing that there is no change in the molecular ordering of the lipids in the region monitored by this probe (15). The same conclusion was reached using *trans* parinaric acid. The lipids from both plants showed a change in lipid order with the probe 5N10 at about 30C.

At any given temperature the motion of 12NS in the polar lipids from the cool grown oleander is faster than the motion in the lipids from the warm-grown plant. Thus, since lipid viscosity is related to spin label motion (8), the lipids from the cool-grown plant are more fluid than those from the warm-grown plant. If membrane thermal stability is related to membrane lipid viscosity, then the increase in chlorophyll fluorescence should occur at a temperature which corresponds to the same viscosity in the two lipid samples. This is indeed the case (Fig. 5); for lipids from a plant grown at 20C/15C the motion at 43C is 8.7 x 10^{-10} sec and for lipids from a plant grown at 45C/32C the motion at 53C is 8.8 x 10^{-10} sec (15). We conclude that the thermal stability of these membranes is related to the physical properties of their lipids.

The oleander experiments suggest that acclimation to widely different growth temperatures results in at least two changes in the membrane lipids. First, there is a shift of about 10C degrees in the low temperature phase separation point. We have not explored the chilling responses of oleander to see if there is a corresponding shift in the sensitivity of any physiological function. Second, the threshold for thermal damage shifts by about 10C degrees, and this loss of functional integrity occurs at a common lipid viscosity, although there is no accompanying discontinuity in the temperature dependence of spin label motion. Studies of other widely-tolerant species should indicate if these two sets of changes generally occur together, which would suggest a common basis in a changed lipid composition.

V. CONCLUSIONS

Fluorescence intensity and polarization measurements with *trans*-parinaric acid are convenient for investigating plant phospholipid physical properties. Because the probe partitions preferentially into solid phase lipid, it can reveal very sharply the first appearance of solid. If the physical properties of lipids are of interest, they should be measured directly, since an analysis of the fatty acid composition is not always a sure indication of changes in physical properties (14). The utility of this probe in surveys of plant lipids should be explored further.

Based on our work with *trans*-parinaric acid the phase separation seen at about 10C in various typical chilling-sensitive plants appears to represent the beginning of the appearance of ordered (gel-phase) lipids, a process which is not completed until well below 0C. These processes are shifted to lower temperatures in cool-climate species, including typical chilling-resistant plants. This proposed assignment of order-disorder transitions ought to be verified by other techniques.

Our survey results to date suggest that growth at low temperature is accompanied by (and may require) a low phospholipid phase separation temperature. Growth at high temperature need not always require a higher phase separation temperature. Further investigations of highly adaptable plants under warm and cool conditions are in progress.

The order-disorder transition extends over many degrees, because plant membranes are complex mixtures. Our results suggest that plants normally growing in a certain environment synthesize a mixture of membrane phospholipids which will not phase separate in the usual thermal regime.

It is not known what property of the membrane lipids is acted on by natural selection. Present techniques can only study the bulk lipid properties. There is no evidence to determine if temperature preference depends on the maintenance of a proper fluidity in the bulk phospholipids or if the bulk phase behavior is a chance reflection of some other properties of some crucial fraction. Indeed, different fractions might set the lower limit for growth in different plants or at different developmental stages in a given plant. If the phase separation sets a rough lower limit for normal function, then it must be determined what physiological processes are incompatible with the solidification. The implication that changes in activation energy of enzymes or growth are the result of lipid phase changes (13, 16) has been questioned (2). Thermal habitat preference in *Passiflora* correlated with the loss of membrane integrity as measured by ion leakage (11) and with lipid phase-change temperature (12). However, no comparable relationship existed between chilling sensitivity or resistance of these species and the point of changes in temperature dependence of the *in vitro* Hill reaction (5). Other examples of this lack of correlation are also cited (5). This assay did correlate well with chilling behavior in several crop plants (21). Aside from seasonality, we have no physiological data on the plants studied. It may be that there is not one single process that sets the lower limit of growth for all species. A plant's temperature preference, which integrates the thermal properties of its various metabolic systems, may be a useful parameter for comparison to lipid properties.

ACKNOWLEDGMENTS

We thank R. D. Simoni for the parinaric acid, and M. Nobs and J. Ehleinger for seeds used in these studies.

VI. REFERENCES

1. Armond, P A., and Staehelin, L. A. *Proc. Nat. Acad. Sci. USA 76*, 1901-1905 (1979).
2. Bagnall, D. J , and Wolfe, J. A. *J. Exp. Bot. 112*, 1231-1242 (1978).
3. Björkman, O., Mahall, B., Nobs M., Ward, W., Nicholson, F., and Mooney, H. *Carnegie Inst. Wash. Year Book 73*, 757-767 (1974).
4. Björkman, O., Badger, M , and Armond P. A. *Carnegie Inst. Wash. Year Book 77*, 262-276 (1978).
5. Critchley, C., Smillie, R. M., and Patterson, B. D. *Aust. J. Plant Physiol. 5*, 443-448 (1978).
6   Folch, J., Lees, M., and Sloane-Stanley, G. H. *J. Biol. Chem. 226*, 497-509 (1957).
7. Grant, C. W. M., Wu, S. H., and McConnell, H M. *Biochim. Biophys. Acta 363*, 151-158 (1974).
8. Keith, A. D., and Snipes W. *In* "Magnetic Resonance in Colloid and Interface Science" (H. A. Resing and C. G. Wade, eds.), American Chemical Society, Columbus, Ohio (1976).
9. Lyons, J. M. *Annu. Rev. Plant Physiol. 24*, 445-466 (1973).
10. Mulroy, T. W., and Rundel, P. W. *Bio Science 27*, 109-114 (1977).

11. Patterson, B. D., Murata, T., and Graham, D. Aust. J. Plant Physiol. 3, 435-442 (1976).
12. Patterson, B. D., Kenrick, J. R., and Raison, J. K. *Phytochemistry 17*, 1089-1092 (1978).
13. Raison, J. K., *R. Soc New Zeal. Bulletin 12*, 487-497 (1974).
14. Raison J. K., In "Biochemistry of Plants" Vol. 7 (P. K. Stumpf, ed.), Academic Press, New York (1979).
15. Raison J. K., Berry, J. A., Armond, P. A., and Pike, C. S In "Adaptations of Plants to Water and High Temperature Stress" (P. Kramer and N. Turner eds.), Wiley-Interscience, New York (1980).
16. Raison, J. K., and Chapman, E. A. *Aust. J. Plant Physiol. 3*, 291-299 (1976).
17. Schroeder, F., Holland, J. F., and Vagelos, P. R. *J. Biol. Chem. 251*, 6739-6746 (1976).
18. Schreiber, U., and Berry, J. *Planta 136*, 233-238 (1977).
19. Shimshick E. J., and McConnell, H. M. *Proc. Nat. Acad. Sci. USA 71*, 4653 4657
20. Shinitzky, M., and Barenholz, Y *Biochim. Biophys. Acta 515*, 367-394 (1978).
21. Shneyour, A., Raison, J. K., and Smillie, R M. *Biochim. Biophys. Acta 292*, 152-161 (1973).
22. Shreve, F., and Wiggins, I. L. "Vegetation and Flora of the Sonoran Desert." Stanford University, Stanford, CA (1964).
23. Sklar, L. A., Hudson, B. S., and Simoni, R. D. *Proc. Nat. Acad. Sci. USA 72*, 1649-1653 (1975).
24. Sklar, L. A., Hudson, B. S., and Simoni, R D *Biochemistry 16*, 819-829 (1977).
25 Sklar, L. A., Miljanich, G. P., and Dratz E. A. *Biochemistry 18*, 1707-1716 (1979).
26 Tecoma, E. S , Sklar, L. A., Simoni, R. D., and Hudson, B. S. *Biochemistry, 16*, 829-835 (1977).
27. Wolfe, J. *Plant, Cell, and Environment 1*, 241-247 (1978).
28. Wu, S. H , and McConnell, H. M. *Biochemistry 14*, 847-854 (1975).

DIFFERENTIAL THERMAL ANALYSIS
OF TOMATO MITOCHONDRIAL LIPIDS

Adam W. Dalziel and R. W. Breidenbach

Plant Growth Laboratory
University of California
Davis, California

I. INTRODUCTION

Considerable evidence now supports the theory that the physical properties of biological membranes of chilling sensitive species are altered at the critical temperature that causes physiological damage (Lyons, 6).   Studies have also indicated that extracted membrane lipids closely resemble intact membranes in physical characteristics (13, 12, 1, 2, 9, 8).   This study concerns the thermotropic properties of mitochondrial lipids from two ecotypes of the tomato *Lycopersicon hirsutum.*

II. MATERIALS AND METHODS

The two ecotypes used originated from South America as part of a collection made by C. E. Vallejos. One was collected in Ecuador at an altitude of 50 meters (Ecotype I), and the other (Ecotype II) in Peru at 3000 meters.   The two ecotypes differ considerably in their chilling sensitivity , Vallejos, this volume (15).

We grew the tomato plants hydroponically in a fifty percent modified Johnson nutrient solution (5) in a greenhouse at Davis. Average maximum and minimum temperatures were $21.5^{\circ}C$ and $18^{\circ}C$.   Daily growth temperatures ranged between $30^{\circ}C$ and $16^{\circ}C$. After about 100 days the roots were harvested and washed with distilled water.

The roots were chopped into 1-cm segments and homogenized in a Waring blender (two 10-second bursts).   The homogenizing buffer contained 0.25 M sucrose, 50 mM potassium phosphate, 5 mM EDTA (ethylenediamenetetraacetic acid), 0.2% soluble PVP (polyvinylpyroolidone), 0.1% fatty acid poor bovine serum albumin, and 5mM β- mercaptoethanol. The crude homogenate was filtered through four layers of cheesecloth and centrifuged at 750 x g for 10 minutes. The supernatant

319

was centrifuged at 12,000 x g for 15 minutes, and the crude mitochondrial pellet was resuspended in the grinding medium (minus EDTA and PVP). This suspension was centrifuged at 12,000 x g to wash the mitochondria, and the pellet was suspended in a minimal volume of buffer. The purity of mitochondrial preparations was checked by electron microscopy which showed the presence of a small amount of contamination from endoplasmic reticulum.

The mitochondrial suspension was transferred to 50 ml of boiling isopropanol in order to inactivate any residual phospholipase activity. As with all the steps involved in lipid extraction, the sample was kept under a nitrogen atmosphere to prevent oxidative alteration. After boiling for 5 minutes the sample was allowed to cool before adding 400 ml chloroform:methanol (2:1). The rest of the extraction was by the method of Folch et al (1957). The extract was reduced to a small volume of chloroform *in vacuo* ($35^{o}C$), and a sample was transferred to preweighed aluminum crucibles. The remaining chloroform was removed, and the sample was dried for 12 hr over P_2O_5 *in vacuo*. The sample was weighed and 20 µl of 50% ethylene glycol buffer at pH 7.2 was added (50% 2 mM HEPES, 10 mM NaCl, 0.1 mM EDTA; 50% ethylene glycol). When a pH 9 buffer was used it contained 25 mM sodium borate instead of 2 mM HEPES. The sample was sealed hermetically in a nitrogen atmosphere and allowed to hydrate overnight at room temperature.

A Mettler TA 2000 was used to carry out differential thermal analysis (DTA) on these samples, with 50% ethylene glycol as a reference. The samples were first preconditioned by heating and cooling three times over the temperature range of interest. Each scan at a rate of 5°/min was repeated at least two times.

Calorimetric differential thermal analysis differs from conventional DTA in that the signal observed at the differential thermocouples is no longer interpreted primarily as a temperature difference, but rather, as a consequence of varying heat flows to the sample and reference sides.

$$\frac{dH}{dt} = \frac{Ts - Tr}{R} = \frac{\Delta V/S}{R} = \frac{\Delta V}{E}$$

Ts   Sample temperature
Tr   Reference temperature
ΔV Differential thermovoltage
S   Sensitivity of measuring sensor ($S = 100$ µ$V/^{o}C$)
R   Heat resistance between the furnace wall and the crucible
E   Calorimetric sensitivity ($E = R.S$)

III. RESULTS AND DISCUSSION

Figure 1 shows a series of thermograms obtained on mitochondrial lipids extracted from Ecotype I. The baseline thermogram was obtained by scanning over the range of interest without any sample pans in the sample holder. The control sample was rapidly cooled from $90°$ to $-50°$, where it was allowed to equilibrate for 15 minutes. The thermograms begin at $-33.5°C$ and show the sample being heated to $90°C$. A thermovoltage of 2 μV corresponds to a heat flow of 0.15 mW. Comparing the results with the baseline reveals a broad endothermic transition beginning around $-32°C$. The transition may well begin below $-32°C$, but with the buffer used it is not possible to investigate this region. The end point of this transition cannot be accurately determined from a heating scan, because of overlapping contributions from the endothermic process and the signal time constant for the instrument. Even so, it can be estimated to lie between 35 and 45$°C$. The shape of the transition indicated that it might be biphasic, with two broad overlapping transitions.

We found that the shape of this endothermic transition depends on the thermal history of the sample. Figure 1 shows the effect of incubating this sample at $10°C$ for 12 hr. In this case, the sample was rapidly cooled from $90°C$ to $10°C$, incubated isothermally for 12 hr, and then rapidly cooled to $-50°C$, where it was allowed to equilibrate for 15 min. One can now see two partially resolved endothermic transitions. The lower transition, again, begins around $-32°C$, and the upper one appears to begin at $16°C$. After heating the sample to $90°$, the control procedure was repeated and a thermogram similar to the control was obtained. This indicates that the effect of incubating the sample at $10°C$ is completely reversible. We then decided to test the effect of incubating the sample at other temperatures. This is demonstrated by the third thermogram (Fig. 1), which shows the effect of incubating the sample at $20°C$ for 12 hr. The onset of this second transition had now shifted to $22.5°C$. The dependence of this transition on the incubation temperature suggested that a phase separation may be occurring. It is possible that lipids having a melting point above the incubation temperature may segregate into a separate pool and thus melt independently from the remainder to the lipids present.

We then attempted to characterize the kinetics of this process. The samples were incubated at $10°C$ for various periods before being rapidly cooled to $-50°C$. Figure 2 shows results obtained with 5-minute, 30-minute, and 12-hr incubations. It is clear that this effect occurs on a relatively slow time scale, taking several hours to complete. Compared with the types of phase separations detected with synthetic phospholipids, this process appears extremely slow (4, 8, 14).

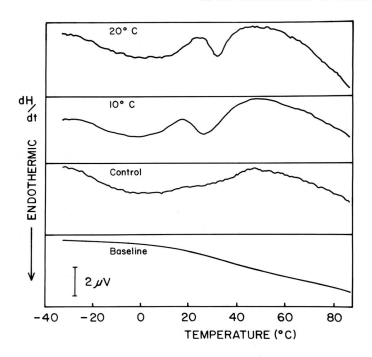

FIGURE 1.   The effect of incubation temperature on the thermotropic properties of mitochondrial lipids isolated from *Lycopersicon hirsutum f. glabratum (Ecotype I).* The sample contained 11.7 mg of lipid and 20 μl 50% ethylene glycol buffer. The sample was either cooled rapidly to -50°C (control) or incubated at 10°C or 20°C for 12 hrs prior to cooling. The 2 μV scale corresponds to a heat flow of 0.15 mW.

Interactions among phospholipid polar head groups offers one possible explanation for this behaviour. Plant lipids usually contain substantial amounts of phosphatidylethanolamine.   This phospholipid resembles phosphatidylserine in that interactions are possible between adjacent polar head groups when the amino group is protonated ($-CH_2CH_2N^+H_3$). Also like phosphatidylserine, it is less hydrated than phosphatidylcholine, so that any aggregates formed would probably produce anhydrous stacks. Since such interactions are possible only when the amino group is protonated, we decided to test the effect of pH on this system.

Samples of mitochondrial lipids from a second ecotype of *Lycopersicon hirsutum* (Ecotype II) were prepared at two different pH values. A detailed comparison of the thermotropic properties of these two

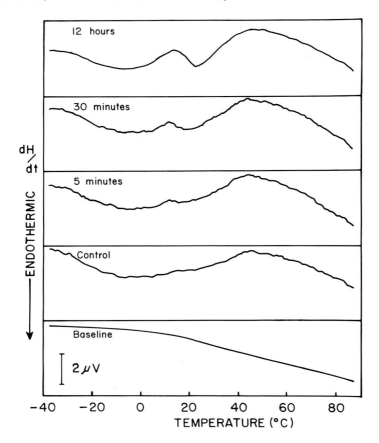

FIGURE 2.   The effect of incubation time on the thermotropic
properties of mitochondrial lipids isolated from Lycopersicon hirsutum f.
glabratum (Ecotype I).   The sample contained 11.7 mg of lipid and 20
μ 1 50% ethylene glycol buffer. The sample was incubated at 10°C for 0
min (control) 5 min; 30 min; or 12 hrs before rapidly cooling to -50°C. 2
μV ≡   0.15 mW.

ecotypes will be published elsewhere. The differences between Ecotype I
and II shown in Figure 1 and Figure 3 controls are largely accounted for by
differences in sample size. We have made no attempt to estimate the
enthalpy of the broad transitions detected, because of the difficulty of
drawing an accurate baseline for such broad transitions.

By incubating a pH 7.2 control sample at 10°C for 10.5 hr, two
partially resolved endothermic transitions were again detected. The
second of these had an onset temperature of about 15°C. Both the pH 9.0
control and the pH 9.0 sample incubated at 10°C for 10.5 hr had thermal
characteristics similar to those of the corresponding samples at pH 7.2.

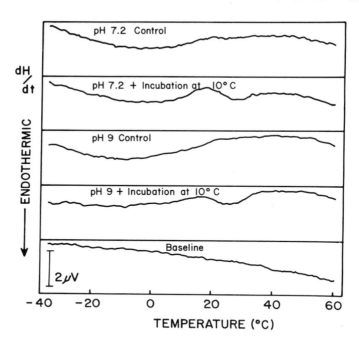

FIGURE 3. The effect of pH on the thermotropic properties of mitochondrial lipids isolated from Lycopersicon hirsutum (Ecotype II). The sample contained 5.0 mg of lipid and 20 μl 50% ethylene glycol buffer:pH 7.2 or pH 9 (see text). The samples were either cooled rapidly to -50°C (control) or incubated at 10°C for 10.5 hrs prior to cooling. 2 μV ≡ 0.15 mW.

This leads us to believe that if a phase separation is in fact taking place, it is not due primarily to the phosphatidylethanolamine polar head group.

Since plant phospholipids are highly unsaturated, it is somewhat surprising that a portion of the lipids discussed in this study melt at relatively high temperatures. It is possible that disaturated phospholipids are present in these preparations, similar to results of Miljanich (11). Alternatively, the phospholipids may be interacting with other lipid components, altering the phase transition temperature (10). Further studies will be required to determine which lipid components are contributing to the observed transitions.

In conclusion, we have used differential thermal analysis to detect a broad endothermic transition in mitochondrial lipids from *Lycopersicon hirsutum*. We have found that the thermotropic properties of these lipids are affected by incubating the sample at temperatures within the range of the endothermic transition. Finally, we were unable to obtain evidence that the phosphatidylethanolamine polar head group plays a significant role in influencing the effect of incubating the sample at $10^{\circ}C$.

ACKNOWLEDGMENT

We would like to thank Mettler Instrument Corporation for the use of the TA 2000 thermal analysis system. We would also like to thank Dr. R. Jones and E. Crumm for their help with the electron microscopy.

IV. REFERENCES

1. Ashe, G. B., Steim, J. M. *Biochim. Biophys. Acta 233*, 810-814 (1971).
2. Blazyk, J. F., Steim, J. M. *Biochim. Biophys. Acta 266*, 737-741 (1972).
3. Folch, J. H., Lees, M., Sloane-Stanley, G. *J. Biol. Chem. 226*, 497-509 (1957).
4. Jacobson, K., Papahadjopoulos, D. *Biochemistry 14*, 152-161 (1975).
5. Epstein, E. Mineral Nutrition of Plants: Principles and Perspectives. J. Wiley & Son, Inc., Chapter 3., 38-39 (1972).
6. Lyons, J. M. *Ann. Rev. Plant Physiol. 24*, 445 466 (1973).
7. Mabrey, S., Powis, G., Schenkman, J. B., Tritton, T. R *J. Biol. Chem. 252*, 2929-2933 (1977).
8. Mabrey, S., Sturtevant, J. M. *Proc. Natl. Acad. Sci. USA 73*, 3862-3866 (1976).
9. Melchior, D. L., Steim, J. M. *Ann. Rev. Biophys. Bioeng. 5*, 205-238 (1976).
10. McKersie, B. D., Thompson, J. E. *Biochim. Biophys. Acta 550*, 48-58 (1979).
11. Miljanich, G P., Sklar, L. A., White, D. L., Dratz, E. A. *Biochim. Biophys. Acta 552*, 294-306 (1979).
12. Reinert, J. C., Steim, J. M. *Science 168*, 1580-1582 (1970).
13. Steim, J. M., Tourtellotte, M. E., Reinert, J. C., McElhaney, R. N., Rader, R. L. *Biochemistry 63*, 104-109 (1969).
14. Van Dijck, P W. M., Kaper, A. J., Oonk, H. A. J., De Gier, J. *Biochim. Biophys. Acta 470*, 58-69 (1977).
15. Vallejos, C. E. This volume.

SOME PHYSICAL PROPERTIES OF MEMBRANES IN THE PHASE SEPARATION REGION AND THEIR RELATION TO CHILLING DAMAGE IN PLANTS

Joe Wolfe[a]

Department of Applied Mathematics
Institute of Advanced Studies
Australian National University
Canberra, ACT 2601 Australia

The finite temperature range of biological activity cannot easily be explained by simple chemical kinetics. The influence of the fluidity of membrane lipids has been implicated in enzyme activity, and this paper compares some models for this influence. The activity of living things, and the rates of many biological reactions, have a remarkable temperature dependence: they are nearly all totally inactive below about 260-270 K, and above about 310-320 K (8). A variation of only 10-15% in the temperature spans a range in which the rates of almost all biological processes increase from zero to a finite value, remain constant within a few orders of magnitude and then revert to zero. This is remarkable because of the statistical distribution of energy among potential reactant molecules: just below the observed minimum temperature some of the substrate molecules should, one might expect, have enough energy to react. Clearly something much more complicated than the reaction kinetics of inorganic chemistry is involved (see Fig. 1).

How can one explain such an enormous variation in reaction rate over such a small temperature? Let's consider some possibilities:

Enzyme cooperativity cannot be the answer; such a system is much more sensitive to substrate concentration than to temperature, and observed temperature dependences are not nearly large enough.

The differential temperature dependence of competing reactions could produce a strong temperature dependence in some complicated process such as a growth rate. Chlorophyll production in etiolated plants,

[a] *Present address: Department of Agronomy, Cornell University, Ithaca, NY 14853.*

for example, increases faster with temperature than its photo-destruction, and so there is some threshhold temperature above which a plant can produce chlorophyll, "green up" and grow, and below which it dies. This threshhold temperature will, of course, depend on irradiance (2, 17). However, the chlorophyll production rate itself varies very quickly with temperature and very large irradiance changes are needed to alter the critical survival temperature by 5° or more. So the "competing reactions" approach doesn't solve the basic problem, since individual reactions exhibit the same "on or off" response to temperature.

Sharpe and De Michelle (21) addressed the problem by extending the model of Johnson et al. (9). They assign to the enzyme low temperature inactive, medium temperature active, and high temperature inactive states into which enzyme molecules partition according to Boltzmann statistics. This is pictured in Figure 2, where the n and nm equivalent states at each energy level account for the "entropy of activation" necessary to produce the large temperature dependence. Unfortunately to produce the observed behaviour both the numbers of equivalent states, and the reaction time constants, must be (literally) astronomically high; in fact transitions between these states would never happen.

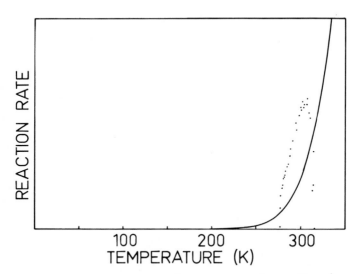

FIGURE 1. A comparison of the rates of reaction for a typical chemical reaction in the low density limit (solid line) and for the Hill reaction in mung bean. [Data from (18).]

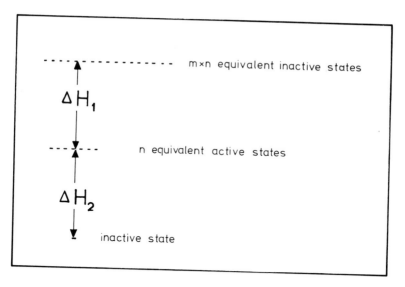

FIGURE 2. *The model of Sharpe and deMichelle (21) requires the enzyme to partion in energy states as shown.*

Membrane Lipid Fluidity

The transitions between different physical phases of substances are often abrupt and this suggested to Lyons and Raison (14) the possibility that the physical state (solid or fluid) of the lipids of cell and organelle membranes determined the activity of the enzymes embedded therein, just as the activity of a fish would depend on the physical phase of the surrounding water. Many enzymes are rendered inactive, or less active, by the removal of their lipid environment [e.g. Jacobs *et al.* (1)], or by its replacement by stronger surfactants and this made the hypothesis even more reasonable. They and other researchers have discovered that many species have membrane lipids of such a composition that they freeze over a range of temperatures corresponding roughly to the temperatures in which the species lives (13). There is also evidence that the adaption to low or high temperatures by an organism involves increasing or decreasing respectively the fraction of low melting point lipids (6, 11, 19, 20, 23), though other workers find that this doesn't happen in some species (4, 22). On the whole, the coincidence of the range of temperatures in which the membrane lipids of an organism might be expected to phase separate and the range of temperature in which it lives strongly suggests that the physical properties of the lipids influence biological activity. Several mechanisms which could mediate such an influence have been described by Wolfe (24).

   Crozier (3) had earlier proposed that the changing slope in Arrhenius plots of biological reactions was due to the transition between ranges of temperature in which different reactions were rate-limiting, but this theory had difficulty in producing sufficiently rapid changes from one regime to another [see e.g. discussion in Johnson *et al.* (8)]. Lyons and Raison (14) were able to overcome this problem by proposing that a certain rate-limiting reaction had one activation energy when its enzyme was in a fluid membrane and another (higher) activation energy in the solid membrane. Thus below some temperature (T_s) an Arrhenius plot exhibits some constant slope, above another temperature (T_f) another constant slope, and between the two (when the lipids phase separate) the reaction rate is the sum of the rates of the reactions occurring in the different phases (see Fig. 3). This approach, however, is necessarily limited to describing only the central part of a typical Arrhenius plot. Since Arrhenius' Law cannot reasonably be applied when the slope is positive (corresponding to a negative activation energy) or negative and very large (if the activation energy were $100\ k_B T$ then the timescale of the reaction would be astronomical), some other approach must be employed to explain the inactivity at high and low temperatures.

   An alternative, simpler form of the general lipid fluidity-enzyme activity theory postulates an enzyme which can only form an activated complex in a fluid lipid environment (See Fig. 4). Above some temperature T_f all the enzyme molecules are active and the reaction follows Arrhenius' Law. Below this temperature progressively more enzyme molecules find themselves in solid domains, and so the reaction rate decreases more rapidly with temperature. A parameter of such a model would be the energy required to move an enzyme molecule from a fluid to a solid domain [i.e. the difference in solubility in the two phases, as observed by Duppel and Dahl (5)]. Such a model explains the low temperature behaviour (and so is perhaps of interest to this conference) but not the reversible or irreversible high temperature inactivation, and so we must look elsewhere.

Bilayer Lateral Compressibility

   Linden *et al.* (12) suggested that enzyme activity may require a certain minimum lateral compressibility of the bilayer since the configuration of the activated state may require displacing the adjacent lipids a little. Now the compressibility of a lipid bilayer will be small for a solid, a few times larger for a liquid, but orders of magnitude larger in the phase separation region (16). Thus the finite temperature range of activity of an enzyme in a bilayer may be that range over which the bilayer may easily be compressed in the plane. Such a model produces qualitatively similar results to the hypothesis that the boundary layer of an active enzyme must be of a limited range of mobility (25).

   It is observed that transport across bilayers is also much greater in the phase separation region than in either fluid or solid region, and there is a striking similarity between the form of the temperature dependence

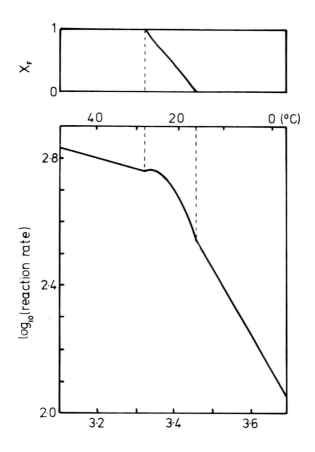

FIGURE 3.   An Arrhenius plot predicted from the theory of Lyons and Raison (14) and the lipid fluid fraction, X_F, on the same scale. Note that one of the constraints on the middle curved section is that it pass through the intersection of the two straight lines.

of transport found by Wu and McConnell (26) and the theoretical compressibility of a mixed lipid bilayer found by Marelja and Wolfe (16). It has also been argued that substantial transient fluctuations in the lipid areas (and thus high lateral compressibility) are required for transport across the membrane (12).

Thus it can reasonably be argued that reactions requiring either transport across a membrane, or reconfiguring of an enzyme in a

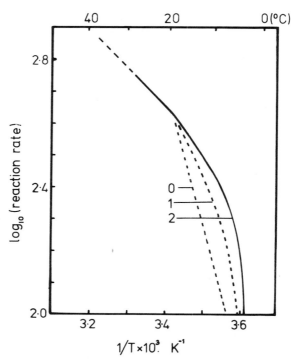

FIGURE 4. An Arrhenius plot predicted from the modified form of the lipid fluidity-enzyme activity theory (see text). The difference in energy of solution is 0x, 1x and 2x the thermal energy as indicated.

membrane will be greatly facilitated by (and perhaps only possible with) the increased compressibility in the phase separation range of temperatures. Now for a binary mixture of lipids whose "tails" differ only by two carbon atoms, the difference between temperatures at the extremes of the phase separation is small (~ 5°C), and the range of biological activity is larger (~ 50°C). A biological membrane, however, can be expected to phase separate over a much wider range of temperatures than such a mixture. First it has several component lipids with a wide range of melting temperatures and second, it has "boundary layer" lipids affected by the presence of intrinsic proteins (10, 15). The hydrophobic region of an enzyme is certain to be less ordered than a frozen lipid tail, and more ordered than a fluid lipid tail, and thus will preferentially adsorb more fluid or less fluid lipids than the bulk well below and above (respectively) the expected phase separation range.

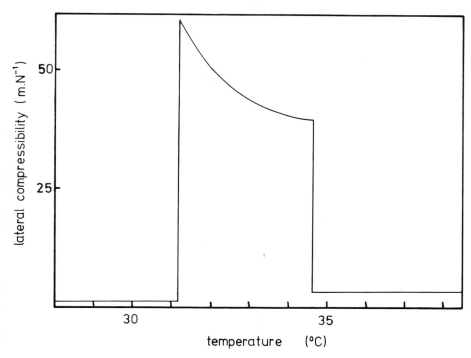

FIGURE 5. Lateral compressibility of a bilayer equally composed of DMPC and DPPC as a function of temperature, using the theory of Marelja and Wolfe (16) and the data of Albrecht et al. (1).

A quantitative analysis of such a system is currently being prepared (Bates, Marelja and Wolfe, unpublished data). Since only a small free energy change is associated with the absorption of boundary lipids over a wide range of temperatures, even in a one lipid component system (15), the compressibility will be high over a temperature range much wider than that over which an unperturbed two lipid component bilayer phase separates.

From the theory of Marelja and Wolfe (16) it is possible to calculate that the compressibility of a bilayer formed from an equal mixture of dipalmitoyl-and dimiristoyl-phosphatidylcholine [for which lipids the required parameters are well known (1)] the compressibility is about 6 m.N^{-1} in the fluid region, 1.5 m.N^{-1} in the solid, and about 50 mN^{-1} in the phase separation (Fig. 5). Suppose that the activation energy of some reaction is $E_o + E_K$ where E_o is the chemical energy involved in attaching the substrate to the enzyme, and E_K is the work done compressing the surrounding lipids to allow the formation of the new configuration. If E_K were ~ 1.7 k_BT (1 k cal mole^{-1}) in the middle of the separation range, where it would have little effect on the reaction kinetics, then it would quickly increase to ~ 50 k_BT in the solid range, where it would effectively "turn off" the reaction, since the chance of thermal excitation providing that much energy is negligible. Of course E_K cannot be simply deduced from the slope of an Arrhenius plot since it is a function of temperature.

I know of no independent measurements of E_K so it would have to be an adjustable parameter of the model. Other parameters such as protein-lipid interaction energy and/or lipid compressibilities are also not known with sufficient accuracy, and therefore, as an exercise in curve fitting, such a model would be doomed to success. However, it is elsewhere suggested how these other parameters could be independently obtained from experiment (15, 16, 24), and so the theory may in principle be tested.

A corrolary of the theory is that the low temperature limit of a biological reaction depends not only on the proportion of short-chained lipids in the membrane, but also on the features of the hydrophobic regions of the enzyme, and how easily they adsorb a boundary layer of lipids (24)..

ACKNOWLEDGMENT

During the writing of this paper I have enjoyed useful discussions with Stjepan Marelja.

REFERENCES

1.   Albrecht, O., Gruler, H. and Sackmann, E. *J. de Physique 39,* 301-313 (1978).
2.   Bagnall, D. J. This volume (1979).
3.   Crozier *J. Gen. Physiol. 7,* 189-216 (1924).
4.   de la Roche, I. A. and Andrews, C. J. *Pl. Physiol. 51,* 468-473 (1973).
5.   Duppel, W. and Dahl, G. *Biochim. et Biophys. Acta 426,* 408-417 (1976).
6.   Gerloff, E. D., Richardson, T. and Stahmann, M. A. *Pl. Physiol. 41,* 1280-1284 (1966).
7.   Jacobs, E. E., Andrews, E. C., Wohlrab and Cunningham, W. *In* "Structure and Function of Cytochromes", Univ. of Tokyo Press, Tokyo (1967).
8.   Johnson, F. H., Eyring, H. and Polissar, M. J. "The Kinetic Basis of Molecular Biology", Wiley, New York (1954).

9.   Johnson, F. H., Eyring, H. and Williams, R. W. *J. Cell Comp. Physiol. 20*, 247–268 (1942).
10. Jost, P. C., Griffiths, O. H., Capaldi, R. A. and Vanderkooi, G. *Proc. Nat. Acad. Sci. USA 70*, 480–484 (1973).
11. Kuiper, P. J. C. *Pl. Physiol. 45*, 684–686 (1970).
12. Linden, D., Wright, K. L., McConnell, H. M. and Fox, C. F. *Proc. Nat. Acad. Sci. USA 70*, 2271–2275 (1973).
13. Lyons, J. M. and Asmundsen, C. M. *J. Amer. Oil. Chim. Soc. 42*, 1056–1058 (1965).
14. Lyons, J. M. and Raison, J. K. *Plant Physiol. 45*, 386–389 (1970).
15. Marelja, S. *Biochim. et Biophys. Acta 455*, 1–7 (1976).
16. Marelja, S. and Wolfe, J. *Biochim. et Biophys. Acta* (In press) (1979).
17. McWilliam, J. R. and Naylor, A. W. *Pl. Physiol. 39*, 262–268 (1967).
18. Nolan, W. G. and Smillie, R. M. *Pl. Physiol. 59*, 1141–1145 (1977).
19. Paton, J. C., McMurchie, B. K., May, B. K. and Elliott, W. H. *J. Bact. 135*, 754–759 (1978).
20. Raison, J. K., Berry, J. A., Armond, P. A. and Pike, C. S. *In* "Adaption of Plants to Water and High Temperature Stress", Wiley, New York. (In preparation) (1980).
21. Sharpe, P. J. H. and DeMichelle, D. W. *J. Theor. Biol. 64*, 649–670 (1977).
22. Siminovitch, D., Singh, J. and de la Roche, I. A. *Cryobiology 12*, 144–153 (1975).
23. Willemot, C. *Pl. Physiol. 55*, 356–359 (1975).
24. Wolfe, J. *Plant, Cell and Environment 1*, 241–247 (1978).
25. Wolfe, J., Ph.D. Thesis, Australian National University, Canberra, Australia (1979).
26. Wu. S. H. and McConnell, H. M. *Biochem. and Biophys. Res. Com. 55*, 484–491 (1973).

LIPID PHASE OF MEMBRANE AND CHILLING INJURY IN THE BLUE-GREEN ALGA, *ANACYSTIS NIDULANS*

Norio Murata, Taka-Aki Ono and Naoki Sato

Department of Biology
College of General Education
University of Tokyo
Komaba, Meguro-ku
Tokyo 153 Japan

I.   *ANACYSTIS NIDULANS*

The blue-green algal cells are composed of two kinds of membranes, the plasma membrane and the thylakoid membrane, but are devoid of a differentiated organelle of photosynthesis, the chloroplast.   The thylakoid membrane is the site of the primary processes of photosynthesis including the absorption of light, and the photosynthetic electron transport and phosphorylation reactions.   The plasma membrane is the barrier between the cytoplasm and the outer medium and also the site of active transport.

Although most blue-green algae are resistant to chilling treatment, *Anacystis nidulans* is injured when exposed to chilling temperatures (2, 6).   Moreover, the chilling-sensitivity depends on the growth temperature (13).   This alga is also characterized by the fact that it contains only saturated and monounsaturated fatty acids (4, 14), while most other blue-green algae contain polyunsaturated fatty acids (1).

We have been studying the effects of temperature on photosynthesis in *A. nidulans,* and have revealed the correlation between the physical phase of thylakoid membrane lipids and the temperature dependence of the photosynthetic processes.   Table I summarizes the temperature dependence of the membrane lipid phase, the photosynthetic processes and chilling injury in cells grown at 38° and 28^oC.

TABLE I.   *Temperatures for Characteristic Points in Cells Grown at 38° and 28°C*

Measurement	Material	Temperatures for characteristic points (°C)	Ref.
Lipid phase transition of thylakoid membranes			
Spin label	M	13 <u>24</u>	(9)
Chlorophyll a fluorescence	C	13 <u>24</u>	(8)
Thermal analysis	M	14	*
Photosynthetic processes			
Photosynthetic O_2 evolution	C	13 <u>24</u>	(9)
H_2O→ Dichlorophenolindophenol	M	12 <u>21</u>	(12)
H_2O→ Ferricyanide	M	7 13 <u>12</u>	(12)
Photophosphorylation	M	21 <u>22</u>	(12)
Delayed fluorescence	C	14 <u>22</u>	(10)
Pigment state 1 and 2 shift	C	13 <u>22</u>	(9)
Chilling injury			
Potosynthesis	C	5 10	Fig. 2 **
H_2O→ Benzoquinone	C	3 <u>11</u>	**

*Ono, Murata and Fujita (unpublished) and **Ono and Murata (unpublished). C and M correspond to "intact cells" and "thylakoid membranes", respectively. The numbers with and without underlines were obtained in cells grown at 38° and 28°, respectively.

II.   LIPID PHASE OF THYLAKOID MEMBRANE.

Thylakoid membranes were prepared by disrupting the lysozyme-treated cells by sonic oscillation or French pressure treatment (11). Chemical analysis indicated that the lipids amounted to about 30% of the dry weight of the thylakoid membranes. The physical phase of the membrane lipids was studied by EPR spectroscopy of a spin label (9), the fluorescence yield of chlorophyll a (8) and calorimetric differential thermal analysis (Ono, Murata and Fujita, unpublished data). These studies indicated that the phase transition between the liquid crystalline and the phase separation states occurred at about 24° or $13^{\circ}C$ in cells grown at 38° or $28^{\circ}C$, respectively (Table I). The results obtained by Verwer et al. (15) by freeze-fracture electronmicroscopy of the thylakoid membrane is compatible with our findings.

III. TEMPERATURE DEPENDENCE OF PHOTOSYNTHESIS AND RELATED PROCESSES

The Arrhenius plot of photosynthetic oxygen evolution in intact cells approximated two straight lines with a break point at 24° or $13^{\circ}C$ in cells grown at 38° or $28^{\circ}C$, respectively (9).

Temperature dependence of the photosynthetic electron transport and phosphorylation reactions was studied in the thylakoid membrane preparation (12). An Arrhenius plot of the Hill reaction with dichlorophenolindophenol revealed a break point at 21° or $12^{\circ}C$ in the membrane from cells grown at 38° or $28^{\circ}C$, respectively. The Arrhenius plot of photophosphorylation mediated by N-methylphenazonium methylsulfate revealed a break point at 21° or $12^{\circ}C$ in the membrane from cells grown at 38° or $28^{\circ}C$, respectively.

It is noted that the abrupt change in the apparent activation energy of these reactions appeared at the temperature region of the lipid phase transition (see Table I). These findings suggest that the lipid phase of the thylakoid membrane has great influence on the photosynthetic electron transport and phosphorylation reactions.

The site of plastoquinone in electron transport is most likely affected by the lipid phase, since the rate-determining step of electron transport from H_2O to system I is located at the oxidation reaction of reduced plastoquinone. Moreover, plastoquinone is considered to be dissolved in the lipid layer of the thylakoid membrane and thus to be most sensitive to fluidity change in membrane lipids.

The effect of the lipid phase on photophosphorylation may be best explained by the fact that the membrane leaks ions when the membrane lipids are in the phase separation state (5). In the chemiosmotic mechanism of phosphorylation, the gradient of ions across the thylakoid membrane is the motive force to produce ATP (7).

In this respect it is worthwhile noticing the temperature dependence of delayed fluorescence of chlorophyll a (10). Figure 1 shows the temperature dependence of the yield of delayed fluorescence in intact cells. The change from the high to the low level of delayed fluorescence occurred at the temperature region centering at 22° or $14^{\circ}C$ in cells grown at 38° or $28^{\circ}C$, respectively. It is known that the delayed fluorescence is markedly influenced by the concentration of H^{+} accumulated inside the thylakoid under illumination (3). Thus, these results suggest that the thylakoid membrane leaks H^{+} from inside the thylakoid when the membrane lipids are in the phase separation state.

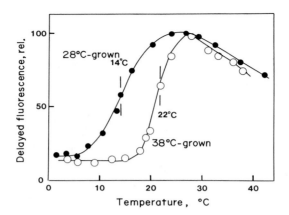

FIGURE 1. *Temperature dependence of delayed fluorescence of chlorophyll a in intact cells of Anacystis nidulans grown at 38° and 28°C (10). The delayed fluorescence emitted at 0.05-1.75 msec after repetitive excitation flashes was measured with a Becquerel-type phosphoroscope. The steady state level of delayed fluorescence after 10 minutes of repetitive illumination was plotted against temperature.*

On the other hand, the Hill reaction with ferricyanide in the thylakoid membrane responded to temperature in a different manner. The break point in the Arrhenius plot appeared at 13° or 7°C in the membranes from cells grown at 38° or 28°C, respectively. These temperatures were much lower than the temperatures of lipid phase transition of the corresponding thylakoid membranes. It seems reasonable, however, to assume that these points are related to the lipid fluidity, because they depended on the growth temperature. So far, we do not have any clear explanation for these characteristic points in the Hill reaction with ferricyanide.

IV. CHILLING INJURY

When the cells of *A. nidulans* are exposed to chilling temperatures, they lose photosynthetic activities. Figure 2 shows that 10° and 5°C were critical temperatures for the cells grown at 38° and 28°C, respectively. The electron transport reaction measured by the Hill reaction with 1,4-benzoquinone also suffered chilling injury in the same temperature region. Another experimental result indicated that K^+ and Mg^{2+} leaked from the cytoplasm when the cells were treated at chilling

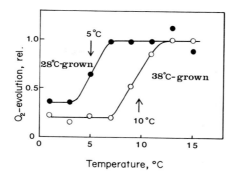

FIGURE 2. *Effect of chilling treatment on activity of the photosynthetic oxygen evolution in intact cells grown at 38° and 28°C. The cells were treated in culture medium at the designated temperatures for 1 hour in the dark and then the photosynthetic activity was measured at 30°C. The rate of oxygen evolution at the maximum level was 200 or 300 µ moles O_2/mg chl/hour in cells grown at 38° or 28°C, respectively.*

temperatures. It is noted that temperatures critical for chilling injury and ion leakage were much lower than temperatures for the lipid phase transition of the thylakoid membrane (Table I). This may suggest that the lipid phase of the thylakoid membrane is not involved in chilling injury of intact cells. It is plausible to assume that the chilling injury is related to the lipid phase of the plasma membrane, and that chilling injury is induced by the leakage of ions from the cytoplasm when the plasma membrane is in the phase separation state. However, the temperature dependence of lipid phase in the plasma membrane of *A. nidulans* has not been studied.

V. EFFECT OF GROWTH TEMPERATURE ON LIPID AND FATTY ACID COMPOSITIONS

The growth temperature-dependence of the phase transition of thylakoid membrane lipids led us to study the effect of growth temperature on the lipid and fatty acid composition. Table II shows that the growth temperature did not significantly affect the relative composition of various lipid classes. Although Sato *et al.* (14) reported a higher content of DGDG (digalactosyldiglyceride) in cells grown at 38°C, their result might have been produced by using the cells at the late culture stages in which DGDG content was found to be abnormally high.

TABLE II.   Lipid Composition in Cells Grown at 38° and 28°C

Lipid	Growth temperature	
	38°C	28°C
MGDG (monogalactosyldiglyceride)	54%	57%
DGDG (digalactosyldiglyceride)	14%	11%
SQDG (sulfoquinovosyldiglyceride)	11%	11%
PG (phosphatidylglycerol)	21%	21%

Cells were cultivated at constant temperatures for more than 7 days. Relative amounts of lipid classes were determined based on fatty acid contents. It is noted that no phosphatidylcholine was present in the blue-green alga.

TABLE III.   Positional Distribution of Fatty Acids in the Four Lipid Classes in Cells Grown at 38° and 28°C.

Lipid	MGDG				DGDG				SQDG				PG			
Growth temperature (°C)	38		28		38		28		38		28		38		28	
Position	1	2	1	2	1	2	1	2	1	2	1	2	1	2	1	2
14:0	1	1	1	2	1	1	1	1	1	0	1	0	1	0	0	0
14:1	3	0	7	0	4	0	6	0	1	0	3	0	1	0	2	0
16:0	3	45	1	43	3	45	2	41	13	50	5	49	4	49	3	45
16:1	41	2	40	5	40	3	40	7	34	0	41	1	39	1	44	3
18:0	0	2	0	0	0	1	0	0	0	0	0	0	0	0	0	0
18:1	2	0	1	0	2	0	1	1	1	0	0	0	5	0	1	2

Fatty acids bound to 1-position were hydrolyzed by lipase of *Rhizopus delemar*. The monoacylglyceride produced were purified by thin layer chromatography with silica gel and fatty acids bound to 2-position were analyzed. It is noted that there were no polyunsaturated fatty acids in this alga.

Table III shows the effect of growth temperature on the positional distribution of fatty acids in the four lipid classes. Although the variation in fatty acid composition was complex, the following conclusions could be drawn when the growth temperature was lowered from 38° to 28°C; at 1-position, 16:0 $\rightarrowtail$ 14:1 (replacement of 16:0 by 14:1) in MGDG (monogalactosyldiglyceride) and DGDG, 16:0 $\rightarrowtail$ 16:1 and 16:0 $\rightarrowtail$ 14:1 in SQDG (sulfoquinovosyldiglyceride) and 18:1 $\rightarrowtail$ 16:1 in PG (phosphatidylglycerol); and at 2-position, 16:0 $\rightarrowtail$ 16:1 and 18:0 $\rightarrowtail$ 16:1 in MGDG, 16:0 $\rightarrowtail$ 16:1 in DGDG and 16:0 $\rightarrowtail$ 16:1 and 16:0 $\rightarrowtail$ 18:1 in PG. The main changes in fatty acids with growth temperature can be summarized as follows: the desaturation and shortening of chain length at the 1-position and the desaturation at the 2-position. These variations in fatty acid composition, though relatively small, are compatible with the growth temperature dependence of the phase transition temperature of membrane lipids, since the desaturation and the shortening of chain length in fatty acids are known to decrease the phase transition temperature of membrane lipids.

Figure 3 shows how the fatty acid composition responded after the shift in growth temperature from 38° to 28°C. A most remarkable change was seen at the 1-position of SQDG where 16:0 was converted to 16:1 within 10 hours. Similar but smaller changes were found at the 1-position of MGDG and PG. In addition to this fast change in desaturation at the 1-position, two slow changes were found to occur in several tens of hours: the shortening of chain length at the 1-position in all of the lipid classes and the desaturation (mostly 16:0 $\rightarrowtail$ 16:1) at the 2-position in MGDG, DGDG and PG.

Experimental results (not shown here) indicated that fatty acid synthesis was remarkably repressed within 10 hours after the shift in growth temperature. This suggests that 16:0 bound to the 1-position of SQDG, MGDG and PG is desaturated without *de novo* synthesis of fatty acids.

IV. SUMMARY

The relationship of the lipid phase of membranes to the temperature dependence of photosynthetic processes and chilling injury was studied in the blue-green alga *Anacystis nidulans* that was grown at different temperatures. Experimental results of spin label, chlorophyll *a* fluorescence and calorimetric differential thermal analysis indicated that the transition of lipid phase of thylakoid membrane between the liquid crystalline and the phase separation states occurred at about 24° or 13°C in cells grown at 38° or 28°C, respectively. Arrhenius plots of electron transport and phosphorylation reactions revealed that a break point appeared in the temperature region of the lipid phase transition of the thylakoid membrane. On the other hand, chilling injury of photosynthesis occurred below 10° or 5°C in cells grown at 38° or 28°C, respectively. The lipid phase of plasma membranes might be involved in the chilling

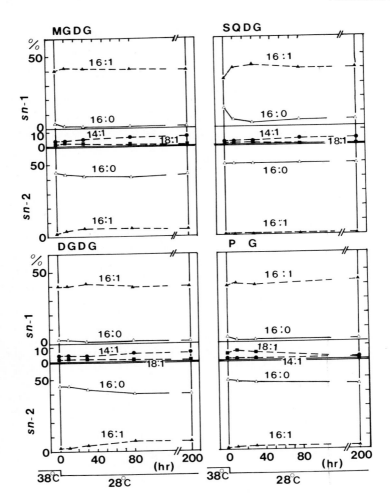

FIGURE 3.   Time course of fatty acid contents at two positions in lipid classes after the shift in growth temperature from 38° to 28°C. Before the shift of temperature, the cells were cultivated at 38°C for more than 7 days.

injury. It was found that K^+ and Mg^{2+} leaked out from the cytoplasm at the temperature region of chilling injury.

The algal cells responded to the growth temperature by varying the fatty acid composition. When the cells grown at 28°C were compared with those grown at 38°C, the desaturation was higher at the 1- and 2- positions and the chain length was shorter at the 1-position of the glycerol moiety of lipids. The variation in fatty acid composition after a rapid shift of growth temperature was composed of a fast change (desaturation at the 1-position) and two slow changes (desaturation at the 2-position and shortening of chain length at the 1-position). These alternations in fatty acid composition are compatible with the growth temperature dependence of the transition temperature of lipid phase in the thylakoid membrane, since the desaturation and the shortening of chain length in fatty acids are known to increase the fluidity and thus decrease the transition temperature of lipid phase of membrane.

V. REFERENCES

1. Erwin, J. A. *In* "Lipids and Biomembranes of Eukaryotic Microorganisms." (J. A. Erwin, ed.), pp. 41-143. Academic Press, New York (1973).
2. Forrest, H. S., van Baalen, C. and Myers, J. *Science 125*, 699-700 (1957).
3. Haveman, J. and Lavorel, J. *Biochim. Biophys. Acta 408*, 269-283 (1975).
4. Hirayama, O. *J. Biochem. 61*, 179-185 (1967).
5. Inoue, K. *Biochim. Biophys. Acta 339*, 390-402 (1974).
6. Jansz, E. R. and Maclean, F. I. *Can. J. Microbiol. 19*, 381-387 (1972).
7. Mitchell, P. *Ann. Rev. Biochem. 46*, 996-1005 (1977).
8. Murata, N. and Fork, D. C. *Plant Physiol. 56*, 791-796 (1975).
9. Murata, N., Troughton, J. H. and Fork, D. C. *Plant Physiol. 56*, 508-517 (1975).
10. Ono, T. and Murata, N. *Biochim. Biophys. Acta 460*, 220-229 (1977).
11. Ono. T. and Murata, N. *Biochim. Biophys. Acta 502*, 477-485 (1978).
12. Ono, T. and Murata, N. *Biochim. Biophys. Acta 545*, 69-76 (1979).
13. Rao, V. S. K., Brand, J. and Myers, J. *Plant Physiol. 59*, 965-969 (1977).
14. Sato, N., Murata, N., Miura, Y. and Ueta, N. *Biochim. Biophys. Acta 572*, 19-28 (1979).
15. Verwer, W., Ververgaert, P. H. J. T., Leunissen-Bijvelt, J. and Verkleij, A. J. *Biochim. Biophys. Acta 504*, 231-234 (1978).

MOLECULAR CONTROL OF MEMBRANE FLUIDITY

Guy A. Thompson, Jr.

Department of Botany
The University of Texas
Austin, Texas

It is very clear from the casual observations of laymen as well as the work of scientists that chilling temperatures have a deleterious effect on many plant species. Yet despite vast losses of crop plants due to chill damage, we still have little understanding of why one species can easily survive low temperatures while another cannot.

One property of the plant cell known to be affected by low temperature is the physical state of its membranes. Earlier papers in this symposium have discussed the concept of "membrane fluidity" and described the lipid-lipid and lipid-protein interactions that affect this parameter. It is known from a variety of fundamental studies that the activities of membrane-embedded enzymes proceed most expeditiously when the fluidity of their membrane environment is within an optimal range. Lowering the cell's temperature can reduce membrane fluidity (and most enzymatic activities) to undesirable levels.

It is remarkable that many organisms are capable of overcoming the rigidifying effect of low temperature on membranes by enzymatically altering membrane lipid composition so as to restore a functional degree of fluidity. This paper describes some of the molecular mechanisms by which this "temperature acclimation" is achieved.

I. REQUIREMENTS FOR ACCLIMATION OF CELL MEMBRANES TO LOW TEMPERATURE

Two important criteria must be met if an organism is to adapt its membranes to chilling temperatures without cell damage. In the first place, it must possess the biochemical capacity to effect lipid compositional changes that will increase fluidity. Most commonly this involves increasing the number of double bonds in lipid fatty acids. Secondly, it must be capable of distributing the altered lipids from the sites of their enzymatic modification to all other membranes of the cell.

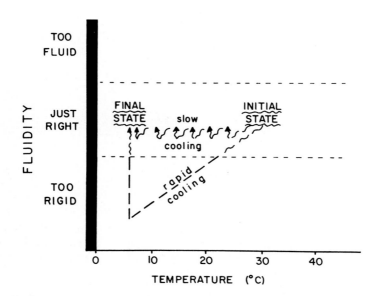

FIGURE 1. *Representation of an organism's ability to maintain its membrane fluidity within an optimum range during a slow temperature reduction but not a fast one.*

Accomplishing these two steps, particularly the latter one, requires a certain amount of time, especially in a structurally complex cell. If the temperature change is gradual, the response of the cell is usually quick enough to prevent membrane fluidity from ever decreasing to a suboptimal level (Fig. 1). On the other hand, sudden chilling can decrease membrane fluidity so rapidly that lipid modification and intracellular dissemination are barely capable (and sometimes incapable) of restoring optimal conditions before irreversible damage is done.

In the ensuing discussion I shall try to emphasize not only the nature of the biochemical modifications that take place but also the time involved in effecting them.

II. A COMPARISON OF TEMPERATURE ACCLIMATION MECHANISMS IN NATURE

I shall restrict my comments to the strategies employed by aerobic organisms, since anaerobes, such as the bacterium *Escherichia coli*, employ an enzymatic pathway not found in higher plants (18). But

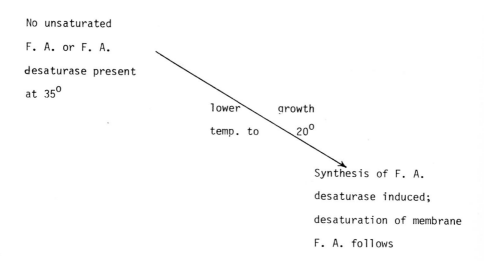

FIGURE 2. Membrane response to low temperature in the aerobic bacterium Bacillus megaterium.

current understanding suggests that it is pertinent to consider the mechanism that some aerobic bacteria have for coping with low temperature. The laboratory of Fulco has investigated temperature acclimation in *Bacillus megaterium* (4). When grown at 35°, these cells contain no unsaturated fatty acids and no fatty acid desaturase activity (Fig. 2). A variety of experiments has shown that a temperature drop leads to the induction of fatty acid desaturase synthesis in amounts adequate to modify membrane lipids within 1.5 hours.

Much less is known regarding the molecular mechanism for temperature acclimation in higher plants, in spite of the great economic importance of the process. In 1969 a remarkably simple scheme for temperature control of fatty acid desaturase activity was proposed by Harris and James (6). It was based on the known participation of molecular oxygen as a cosubstrate in fatty acid desaturation (Fig. 3). Evidence was produced to show that the increased solubility of O_2 in plant cell cytoplasm at low temperature (O_2 is 1.7 times more soluble in water at $10^{o}C$ than at $40^{o}C$) is sufficient in itself to increase the rate of reaction significantly during a 5 hr incubation. There still remains considerable doubt, however, as to whether under normal environmental conditions O_2 availability ever becomes rate-limiting in plants. If this occurs, it would be expected only in non-photosynthetizing tissues, which would be utilizing but not generating O_2.

I shall devote most of the remaining discussion to the process of low temperature acclimation in the ciliated protozoan *Tetrahymena pyriformis*. Considerable data on the molecular control of acclimation in

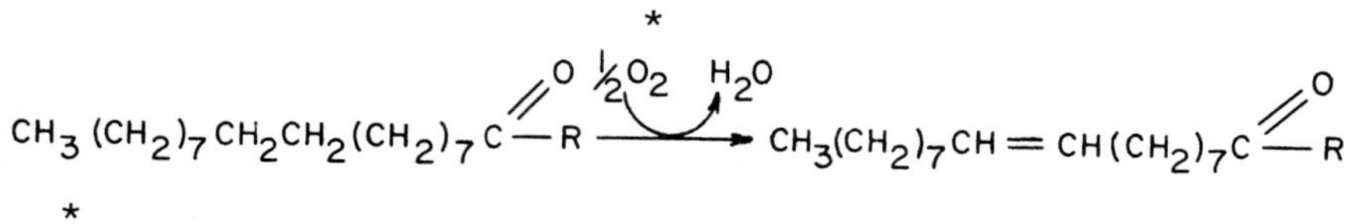

$$CH_3(CH_2)_7CH_2CH_2(CH_2)_7C\overset{O}{\underset{}{\diagdown}}R \xrightarrow{\overset{\frac{1}{2}O_2 \quad H_2O}{}} CH_3(CH_2)_7CH=CH(CH_2)_7C\overset{O}{\underset{}{\diagdown}}R$$

higher levels of the cosubstrate, O_2, drive the reaction towards the right.   (O_2 is 1.7 X more soluble in water at 10° than at 40°).

FIGURE 3. *Membrane response to low temperature in higher plants. Because higher levels of the cosubstrate, O_2, are found at a low temperatures (O_2 is 1.7 times more soluble in water at 10° than at 40°), the fatty acid desaturation reaction is driven to the right.*

Tetrahymena have recently been accumulated in my laboratory and several others, and there is good reason to believe that this cell might serve as a useful model system for understanding temperature acclimation in more advanced organisms.

II.   LOW TEMPERATURE ACCLIMATION IN *TETRAHYMENA*

A.   *Spatial and Temporal Membrane Relationships*

In many respects, *Tetrahymena* is a typical eukaryotic cell.   It contains the usual assortment of intracellular organelles (Fig. 4), and its metabolism is also similar to that of higher organisms (3), particularly higher plants (7) (except that it is non-photosynthetic).

Each functionally different membrane in *Tetrahymena* contains its own characteristic proportions of structural lipids (19). These consist mainly of the sterol-like isoprenoid tetrahymanol and three principal phospholipids (Fig. 5). Because each membrane has different proportions of these lipids and differing degrees of unsaturation in its component fatty acids, the physical properties, i.e., fluidity, of any one membrane type might also be expected to differ from those of the others. This has indeed been shown to be the case at least for the 5 - 6 functionally distinct membranes that have been tested (19).   Once again, there is every reason to believe that in this respect, *Tetrahymena* resembles higher plant cells. We have examined the acclimation of *Tetrahymena* to low temperature by measuring several membrane properties after shifting cells over a period of 30 min. from a growth temperature of 39° to a final temperature of 15° (11).   This reduction of 24° in cell

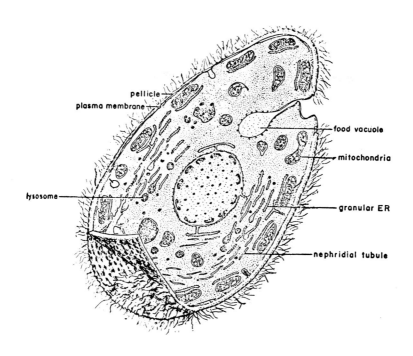

pellicle
plasma membrane
food vacuole
mitochondria
lysosome
granular ER
nephridial tubule

FIGURE   4.   Diagrammatic   cross-section   of   Tetrahymena
pyriformis.

temperature completely stops cell growth and causes a very abnormal
physical state in the membranes.   Fig. 6 illustrates the lateral
aggregation of intramembranous proteins that can be observed by freeze
fracture electron microscopy of one particular membrane.
     By analyzing various functionally-different membranes, we were
able to establish that the initial lipid response to low temperaure occurs
in microsomal membranes. Fig. 7 records some of these changes. The
solid line and squares report the temperatures at which intramembranous
particle-free areas could first be detected by freeze fracture electron
microscopy.   This has been correlated with membrane fluidity.   In
addition, membrane fluidity, as judged by fluorescence polarization
measurements of the extracted lipids (solid lines and circles), increased
extremely rapidly in the cells chilled to $15^{o}C$. Finally, the number of
double bonds in the phospholipid fatty acids (dashed lines and Δ)   increased from about 108/100 molecules just after the shift to about
128/100 molecules within 30 min.   We have plotted these three
parameters together to show that the fluidity of the microsomal lipids
increases at a rate that is closely correlated with the increase in fatty
acid unsaturation.

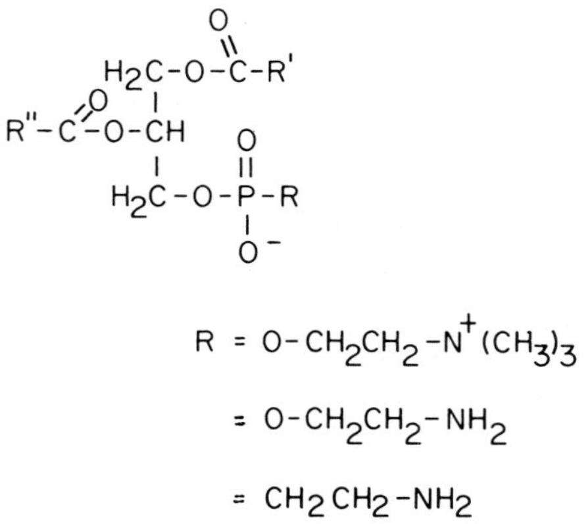

$$
\begin{array}{c}
\quad\quad\quad\quad O \\
\quad\quad\quad\quad \parallel \\
H_2C-O-C-R' \\
\quad O \quad | \\
R''-C-O-CH \quad\quad O \\
\quad\quad | \quad\quad\quad \parallel \\
H_2C-O-P-R \\
\quad\quad\quad\quad | \\
\quad\quad\quad\quad O^-
\end{array}
$$

$$ R = O-CH_2CH_2-\overset{+}{N}(CH_3)_3 $$

$$ = O-CH_2CH_2-NH_2 $$

$$ = CH_2CH_2-NH_2 $$

FIGURE 5. The three principal phospholipid classes of Tetrahymena membranes. R represents, from top to bottom, the polar head groups of phosphatidylcholine, phosphatidylethanolamine, and 2-aminoethyl-phosphonolipid. The major fatty acid constituents are myristic (14:0), palmitic (16:0), palmitoleic (16:1), oleic (18:1), linoleic (18:2), and α-linolenic (18:3) acids.

Significantly, other membranes of the cell do not exhibit such a rapid response to temperature change. This is illustrated in Fig. 8. Here the fluorescence polarization data for microsomal lipids is compared with equivalent data for pellicle lipids isolated in the same experiments. Whereas the microsomal lipids show their most dramatic increase in fluidity in the first few minutes during and after chilling, the pellicle lipids exhibit little change during this period and become more fluid only after a long lag. This lag is required to transport the newly desaturated lipids from the site of their desaturation in the microsomes to the outlying membranes. Dissemination of lipids to the least accessible cellular membrane, that enclosing the cilia, requires an even longer time.

It is data such as these that have brought home to us the importance of time in the temperature acclimation process. In Tetrahymena and other structurally complex cells, the lipid dissemination step is likely to be rate limiting.

B.   Molecular Mechanism for Desaturase Activation

While it was easy to establish that low temperature causes a rise in fatty acid desaturase activity, determining the precise cause for this at the molecular level was much more difficult. We have tested various possibilities. Let us begin with the concept, mentioned earlier, that the desaturase cosubstrate, O_2, might be a rate-limiting factor. We have

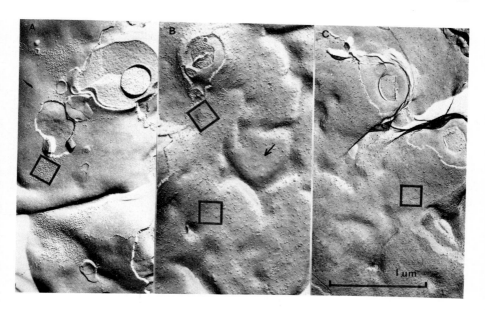

FIGURE 6. The effects of rapidly decreasing temperature on the physical state of the Tetrahymena outer alveolar membrane (part of the pellicle), as visualized by freeze-fracture electron microscopy. A) pronounced lateral movement and aggregation of intramembranous protein particles in cells chilled from a growth temperature of 39.5^0 to 0.5^0C immediately prior to fixation. B) the initial appearance of particle-free regions in membranes chilled from 39.5^0 to 30^0. C) control cells fixed at 39.5^0. The 4×10^4 nm^2 frames may be used for quantifying the degree of aggregation. For details see Martin et al. (10).

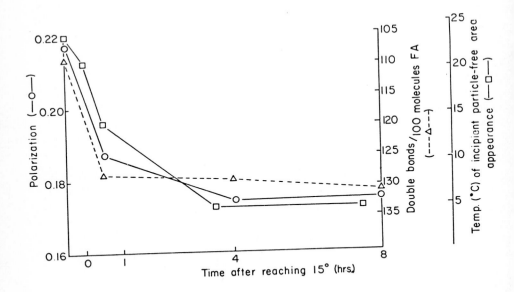

FIGURE 7. Comparison of the changes of three independently measured properties of Tetrahymena microsomal membranes during temperature acclimation by cells chilled from 39.5° over a 30 min. period. Data from freeze-fracture electron microscopic observations of membrane particle redistribution ( ), fluorescence polarization of diphenylhexatriene in membrane lipids (O), and the number of double bonds in phospholipid fatty acids (△).   The change in fatty acid unsaturation was due mainly to a decrease in 14:0 and 16:0 and an increase in 18:3. For details see Martin and Thompson (11).

analyzed the lipids of $39°$-cells and $15°$-cells grown under various O_2 tensions, as illustrated in Fig. 9. In all cases the fatty acid composition was characteristic of the growth temperature, not the O_2 tension (16).

A second possibility was that low temperature induces the synthesis of new fatty acid desaturase molecules in a manner somewhat analogous to the situation in *Bacillus megaterium*. This has been tested in various ways, mainly involving comparisons of temperature-shifted cells incubated with or without inhibitors of protein synthesis (5, 12, 17). It seems clear that the principal changes in fatty acid composition that are triggered by low temperature are observed not only in normal cells but also in cells incapable of synthesizing proteins (17). On the other hand Fukushima *et al.* (5) have recently reported evidence for an increased

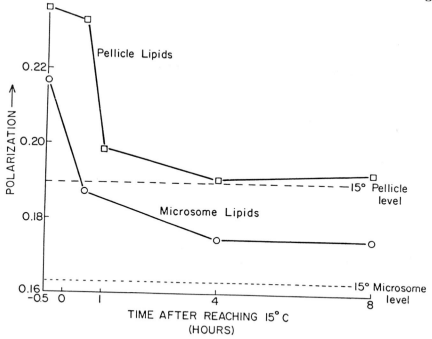

FIGURE 8.   Time course of changes in diphenylhexatriene polarization in membrane lipids of 39.5°-acclimated cells following a shift to 15°C over a 30 min. period as in Fig. 7. At the times indicated, cells were fractionated, and total lipids from purified pellicles and microsomes were utilized for polarization measurements. The dashed lines show polarization values found in the two fractions from cells fully acclimated to 15°. For details see Martin and Thompson (11).

content of palmitoyl-CoA desaturase in cells shifted to 15°. The quantitative contribution of this apparent induction of enzyme synthesis remains to be determined.

The bulk of our work with *Tetrahymena* favors a third mechanism as a major force for controlling fatty acid desaturase activity (19). We believe that any environmental change which markedly reduces membrane fluidity causes an increase in fatty acid desaturase activity, probably not in absolute terms but rather in comparison to the rate of fatty acid synthesis. Such an "activation" is postulated to result from the imposition of greater constraints on the membrane-embedded enzymes by the more rigid lipids surrounding them. This interpretation was first suggested as a possible explanation for the desaturation activity in cells chilled from 39° to 15°. But in those experiments it was hard to rule out a direct effect of temperature on the desaturase enzymes.   In

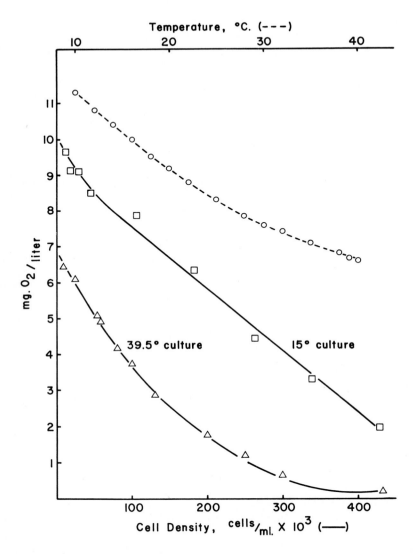

FIGURE 9.   Effect of cell density (lower abscissa) upon the
concentration of dissolved O_2 in Tetrahymena cultures grown at 15 and
$39.5^O C$. Both cultures were in the logarithmic growth phase throughout
the measurement period. The upper curve (broken line) indicates the O_2
concentration in air-saturated water at various temperatures (upper
abscissa).   The fatty acid patterns of phospholipids isolated from 39°
cells at any O_2 tension resembled each other. Likewise, the fatty acid
composition of 15° phospholipids, while different from the 39° pattern,
did not vary over the observed range of O_2 tension (16).

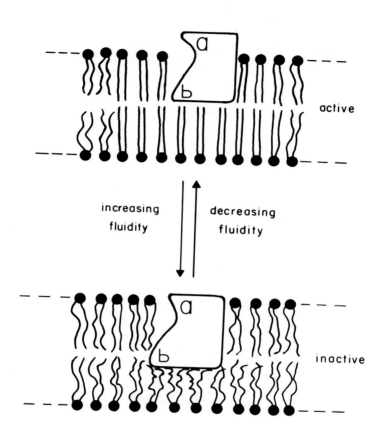

active

increasing   decreasing
fluidity   fluidity

inactive

FIGURE 10. Diagrammatic representation of fatty acid desaturase
movement perpendicular to the plane of the membrane caused by
decreased membrane fluidity.

collaboration with Nozawa and colleagues at Gifu University, we have
studied a variety of conditions in which the fluidity of microsomal
membranes was altered isothermally (19). I shan't give all the details of
these studies, but it is quite apparent from experiments in which Tetrahymena incorporated exogenous polyunsaturated fatty acids (8),
branched chain (methoxy) fatty acids (9), or general anesthetics (13) into
its membranes that an increase in membrane fluidity can be counted on
to reduce fatty acid desaturase activity. We have found only one
apparent exception, ethyl alcohol (14), which seems to fluidize the
membrane in a unique fashion that is not directly pertinent to our present
discussion.

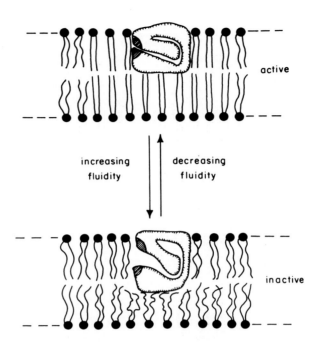

FIGURE 11.   *Diagrammatic representation of the fatty acid desaturase active site (hatched areas) being stabilized in a membrane of suboptimal fluidity.*

Our hypothesis calls for fatty acid desaturase activity to be in a real sense regulated by the fluidity of the membrane environment.   No concrete evidence is available to suggest how this regulation is achieved, but a number of hypothetical models might be constructed.   There is some freeze-fracture evidence that in *Tetrahymena* microsomes, unlike the pellicle membrane, lipid phase separations result in a displacement of proteins perpendicular to the plane of the membrane rather than in a lateral direction (9, 20).   Thus it is conceivable that the fatty acid desaturase molecules are forced into a new position, as shown in Fig. 10, thus orienting the active site more favorably with respect to the substrate.   This is in agreement with the biochemical and ultrastructural observations of Wunderlich *et al.* (20).   Another equally plausible but largely unsupported possibility is that a liquid–crystalline lipid environment would simply not be cohesive enough to maintain the desaturase molecules in their fully active configuration (Fig. 11).   We hope to test some of these models using physical chemical techniques, such as the microcalorimetric approach of Brandts, *et al.* (1).

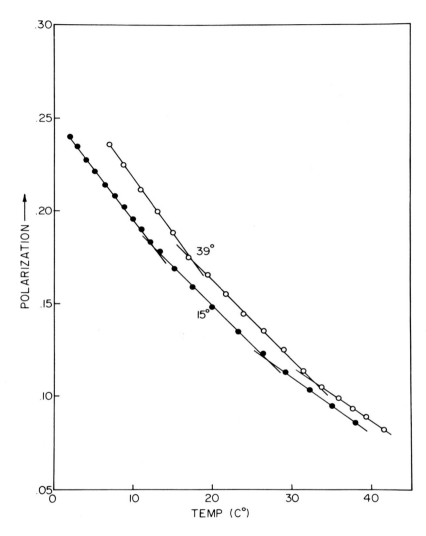

FIGURE 12. The influence of temperature on diphenylhexatriene
polarization in multibilayer vesicles of microsomal phospholipids from
Tetrahymena grown at 39° and 15°C. Data from Dickens, et al. (2).

Although it is far from being proved, we feel that the concept of
fluidity-regulated desaturase activity offers the most viable means of
explaining temperature acclimation in *Tetrahymena*. If it is correct,
then a very exciting generalization may be postulated. Any external or

intracellular factor capable of altering membrane fluidity will affect fatty acid desaturase activity and induce an acclimation response of sorts. It is therefore not surprising that we should observe an "acclimation" of *Tetrahymena* to the presence of the unnatural methoxy-fatty acids (9); anesthetics (13); elevated salt concentrations (Mattox & Thompson, unpublished observations) and a variety of other factors (15). How elegantly simple it would be for a cell to need only one detection and response mechanism for such varied types of environmental stress!

While extrapolation of this hypothesis to higher plants is clearly premature, I shall proceed to do it with no further apology. Imagine the practical benefits of being able to "harden" a commercial vegetable or fruit crop prior to cold exposure by employing a chemical alternative to low temperature or being able to immediately counteract the effect of chilling temperatures with a chemical alternative to high temperature.

Such ideas are not to be taken seriously at the present time. But neither can they be totally discounted. We prefer to let these vague dreams of glory provide motivation for our continued basic investigations of this esoteric creature *Tetrahymena*.

Correlating the average physical properties of membrane lipids (as estimated by fluorescence polarization, electron spin resonance, X-ray diffraction, etc.) with changes in fatty acid desaturase activity still falls far short of explaining precisely how the effect is achieved. We are currently devoting considerable effort towards understanding this relationship in more detail. Much of our work has focused on the microsomal membranes and their lipids. In experimental plots of temperature vs. polarization of the probe diphenylhexatriene in microsomal lipids, we have observed two very reproducible changes of slope, or "break points" (2). Fig. 12 shows that the break points at 17° and 32° in the microsomal phospholipids isolated from 39°-grown cells are replaced by similar break points at 12° and 27° in microsomal phospholipids from 15°-grown cells. Plots made using total lipids (phospholipids and neutral lipids) gave, in each case, break points at slightly lower temperatures. Preliminary experiments with freshly isolated but still intact microsomal membranes indicate that the break points occur at the same characteristic temperatures observed in extracted lipid preparations.

Unfortunately, we have little insight as to what the break points mean. By mixing increasing proportions of 15° lipids with 39° lipids, we were able to show that the higher temperature break point shifted gradually from 32° down to 27°, indicating that the same perturbation triggers the slope change in each curve. The two low temperature break points also seem to have a common origin.

Various related experiments, such as catalytic hydrogenation of the microsomal lipids or mixing the natural lipids in differing proportions with a fully saturated synthetic phospholipid - dipalmitoylphosphatidyl-choline -suggested that the break points may signal the initiation or termination of a phase separation involving certain discreet phospholipid

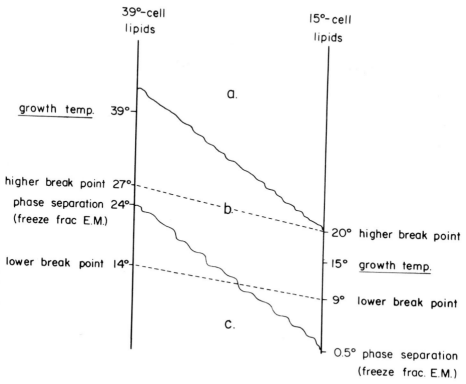

FIGURE 13. A rationalization of the observed physical properties of
39°- and 15°-grown Tetrahymena microsomes and microsome-derived
lipids.   The left vertical axis depicts the physical behavior of
preparations from 39°-grown cells while the right vertical axis
represents the same features of preparations from 15°-grown cells. The
lines connecting equivalent events on the two axes are not meant to
imply that linear responses would necessarily be found in cells grown at
intermediate temperatures. Wavy lines divide the system roughly into
three regions. (a) Region of fully miscible lipids present in the liquid-
crystalline phase; (b) region characterized by the presence of many
relatively small liquid-crystalline and/or gel phase molecular
assemblages that are immiscible in the bulk lipid phase. The dotted line
connects those temperatures at which the first sizable discrete
population of phospholipid molecular species begins undergoing a liquid-
crystalline to gel transition upon cooling; (c) region featuring a rapid
increase in the extent of highly ordered gel domains effecting a
microscopically detectable reorientation of membrane integral proteins.
The dotted line representing the second observed break point connects
temperatures at which another discrete group of phospholipids
commences undergoing a phase transition. Although the break points are
detectable by fluorescence polarization, the impact of the phase
transition on the overall structure of the membrane depends, as with the
higher break point, upon the percentage of available phospholipid
molecules that participate.

molecular species, as proposed by Wunderlich *et al.* (21) from X-ray diffraction data. One of several possible interpretations of the data is shown in Fig. 13. According to this scheme, microsomal lipids measured in temperature range "a" exist entirely in a liquid-crystalline state. As the temperature is lowered towards the growth temperature of the cells from which the lipids came, small clusters of highly saturated lipids are transformed into more rigid aggregates at characteristic temperatures (upper wavy line area "a"). Further decreases in temperature cause a steady rise in the size and number of gel phase domains (area "b"). If an especially prevalent phospholipid molecular species reaches its phase transition temperature, then a rather pronounced change in polarization (break point) might be noted (dotted lines). Area "c" represents that temperature range in which phase separation is sufficiently extensive to produce a large, electron microscopically-visible phase separation.

The purpose of this figure is to illustrate that polarization break points might well occur at any stage during the temperature-induced phase transition of the microsomal lipids. We are currently employing time-resolved fluorescence measurements to determine how many distinct micro-environments exist at any certain temperature. Experimentation along this line will hopefully shed additional light on lipid interaction with the fatty acid desaturases and thereby, the molecular mechanism of low temperature acclimation.

ACKNOWLEDGMENTS

Work from the author's laboratory was supported by grants from the National Institute of General Medical Sciences, the National Cancer Institute, The Robert A. Welch Foundation, and the National Science Foundation.

IV. REFERENCES

1.   Brandts, J. F., Taverna, R. D., Sadasivan, E., and Lysko, K. A.
      Biochim. Biophys. Acta 512, 566-578 (1978).
2.   Dickens, B. F., Martin, C. E., King, G. P., Turner, J. S., and
      Thompson, G. A., Jr., submitted for publication.
3.   Elliott, A. M. *In* "Biology of Tetrahymena", pp 508, Dowden,
      Hutchinson, and Ross, Stroudsburg, Pennsylvania (1973).
4.   Fujii, D. K., and Fulco, A. J. *J. Biol. Chem. 252,* 3660-3670
      (1977).
5.   Fukushima, H., Nagao, S., and Nozawa, Y. *Biochim. Biophys. Acta
      572,* 178-182 (1979).
6.   Harris, P., and James, A. T. *Biochem. J. 112,* 325-330 (1969).
7.   Holz, G. G., Jr. *J. Protozool. 13,* 2-4 (1966).

8. Kasai, R., Kitajima, Y., Martin, C. E., Nozawa, Y., Skriver, L., and Thompson, G. A., Jr. *Biochemistry 15,* 5228-5233 (1976).

9. Kitajima, Y., and Thompson, G. A., Jr. *Biochim. Biophys. Acta 468* 73-80 (1977).

10. Martin, C. E., Hiramitsu, K., Kitajima, Y., Nozawa, Y., Skriver, L., and Thompson, G. A., Jr. *Biochemistry 15,* 5218-5227 (1976).

11. Martin, C. E., and Thompson, G A., Jr. *Biochemistry 17,* 3581-3586 (1978).

12. Nagao, S., Fukushima, H., and Nozawa, Y. *Biochim. Biophys. Acta 530,* 165-174 (1978).

13. Nandini-Kishore, S. G., Kitajima, Y., and Thompson, G. A., Jr. *Biochim. Biophys. Acta 471,* 157-161 (1977).

14. Nandini-Kishore, S. G., Mattox, S. M., and Thompson, G. A., Jr. *Biochim. Biophys. Acta,551,* 315-327 (1979).

15. Nozawa, Y., and Thompson, G. A., Jr. In "Biochemistry and Physiology of Protozoa, 2nd Ed., Vol II, (Levandowsky, M., and Hutner, S. H., eds.) Academic Press, New York, (in press) (1979).

16. Skriver, L., and Thompson, G. A., Jr. *Biochim. Biophys. Acta 431,* 180-188 (1976).

17. Skriver, L., and Thompson, G. A., Jr. *Biochim. Biophys. Acta 572,* 376-381 (1979).

18. Thompson, G. A., Jr. "Regulation of Membrane Lipid Metabolism." Chemical Rubber Company, West Palm Beach, Florida (in press) (1979).

19. Thompson, G. A., Jr., and Nozawa, Y. *Biochim. Biophys. Acta, 472,* 55-92 (1977).

20. Wunderlich, F., Ronai, A., Speth, V., Seelig, J., and Blume, A. *Biochemistry 14,* 3730-3735 (1975).

21. Wunderlich, F., Kreutz, W., Mahler, P., Ronai, A., and Heppeler, G. *Biochemistry 17,* 2005-2010 (1978).

IN VITRO MEMBRANE LIPID RECONSTITUTION AND ENZYME FUNCTION

A. Waring and P. Glatz

Johnson Research Foundation
Department of Biochemistry and Biophysics
University of Pennsylvania
Philadelphia, Pennsylvania

I. INTRODUCTION

The relationship between the physical state of membrane phospholipids and changes in membrane enzyme activity have been reported for a wide variety of biological systems. Lipid related influence on the function of the mitochondrial electron transport chain is of special interest since the organelle has a relatively lipid poor membrane system (4, 13) which shows apparent changes in enzyme mechanism as a function of temperature (3,12). The role of phospholipid in the alteration of enzymatic activity of the mitochondrial electron transport chain has been emphasized by the elimination of temperature dependent changes in enzymatic activity using detergent perturbation of the membrane environment (10) or by modification of the membrane lipid composition upon activation of endogenous phospholipase activity (8, 11). However, the correlative studies have not detailed the nature of the temperature related change in membrane physical state nor have they characterized the specific alteration in enzyme mechanism induced by phospholipid.

The focus of the current study has been an attempt to relate the degree of change in enzymatic activity to the type of alteration of the membrane phospholipid physical character. For comparative purposes two types of phospholipid-protein systems have been examined. The first system considered involves a temperature profile of cytochrome oxidase activity in a heterogenous lipoprotein assembly (the intact inner mitochondrial membrane). The second system consists of tomato mitochondria cytochrome oxidase reconstituted with a homogenous synthetic phospholipid composition having a single, well defined, liquid crystal to gel phase transition in the physiological temperature range.

II. METHODS

Etiolated tomato seedlings (*Lycopersicon esculentum* cv. VF 145-21-4, Petoseed Co.) were grown at 25°C for six days before use. Seedlings (100 gm seedling hypocotyls plus cotyledons) were razor chopped into 200 ml of 4°C grinding medium (0.3 M sucrose, 0.2% w/v MOPS pH 7.4, 1 mM EDTA, 1 mg/ml BSA fraction V, 0.1 mg/ml mercaptobenzothiazole). The slurry was then homogenized for 20 sec using an Ultraturax homogenizer, and mitochondria isolated by differential centrifugation (1), and osmotic swelling to rupture the outer membrane. Mitochondria used in fluorescent probe studies and for lipid extraction, were further subjected to sucrose step density gradient centrifugation to remove microsomal and mitochondrial membrane contaminents (1).

Cytochrome c oxidase activity in mitochondria was measured polarographically with a YSI electrode in a 1.3 ml Gilson chamber and the temperature regulated using a Lauda K2/R refrigerated circulating bath. The assay medium contained 20 mM ascorbate pH 7.4, 10 uM TMPD, 10 uM cytochrome c (sigma type VI) and 10 mM phosphate buffer pH 7.4 The reaction was initiated by the addition of enzyme and the measurements corrected for oxygen consumption without enzyme. Cytochrome oxidase activity in reconstituted phospholipid vesicles was also measured polarographically with 20 mM ascorbate, pH 7.4, 10 uM cytochrome c, and 10 mM phosphate buffer pH 7.4.

Mitochondrial inner membrane phospholipids were extracted with chloroform/methanol 2:1 containing 5% of 28% ammonium hydroxide (8), partitioned according to Folch *et al.*, (5), and flash evaporated to dryness at 20°C.

Mitochrondria were depleted of lipid by detergent-ammonium sulfate fractionation. Twenty volumes of phosphate buffered (0.1 M, pH 7.4) 2% w/v cholate was mixed with one volume of mitochondria. The mixture was homogenized and brought to 50% w/v with solid ammonium sulfate. The pH of the mixture was adjusted to pH 7.4 with ammonium hydroxide and stirred one hour at 4°C. The mixture was then centrifuged at 10,000 x g for 10 min and the precipitate subjected to an additional cycle of fractionation. Lipid depleted mitochondrial protein (3-5 ug lipid P/mg protein) in 2% cholate was reconstituted with dimyristoyl phosphatidylcholine (2 mg phospholipid/mg mitoprotein) by sonication (Branson Sonifier-micro tip – 20 sec), followed by dialysis against 10 mM phosphate buffer, pH 7.4, 4°C for 24 hours to remove cholate (16). Phospholipid liposome vesicles were prepared from extracted mitochondrial phospholipid (0.1 mg PL/ml) or dimyristoyl phosphatidylcholine (0.1 mg PL/ml) in 10 mM phosphate buffer pH 7.4 by sonication (Branson Sonifier-micro tip 60 sec) under argon. Trans-parinaric acid (TPnA) (Molecular Probes, Inc.) as added to mitochondrial membranes or phospholipid liposomes from an ethanol stock solution (0.5 mg/ml). All solutions were deoxygenated with argon and the fluorescent probe to phospholipid ratio was approximately 1:100.

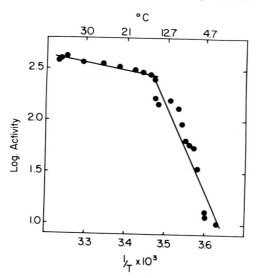

FIGURE 1. *Arrhenius plot of cytochrome oxidase activity in mitochondria.*

Fluorescence lifetime measurements of TPnA in liposomes and membranes was measured by single photon counting. A gated nanosecond flashlamp (Photochem. Res. Assoc.) was used for the excitation of the samples. The lamp flash was optically detected (RCA 1P28), and digitally processed using fast electronic modules (Ortec 474, 473A). The fluorescence was detected with a blue-sensitive phototube (RCA 1P28) and processed with Ortec modules 454, 463, 425 and 425A. The coincidence of the lamp flash and the fluorescence was analyzed using a Ortec 457 time-to-pulse-height converter and a LeCroy 3001 multichannel analyzer. The sample temperature was controlled using a low temperature cuvette holder (Photochem. Res. Assoc.) connected to a refrigerated bath. The chamber was purged with dry nitrogen gas which had been cooled by a heat exchanger immersed in liquid nitrogen which prevented condensation on the optical components and quartz cell.

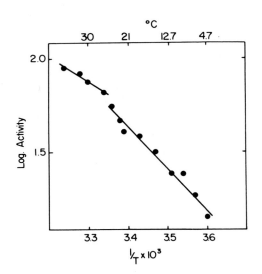

FIGURE 2. *Arrhenius plot of cytochrome oxidase activity (nmoles 0_2/mg protein/min) in DMPC-cytochrome oxidase vesicles (2 mg DMPC/mg protein).*

III. RESULTS

A. *Temperature Dependence of Cytochrome Oxidase in Mitochondria and Phospholipid Vesicles*

Cytochrome c oxidase activity in intact mitochondrial membrane showed a discontinuity in the Arrhenius plot in the temperature range of 16° to 14°C (Fig. 1). At higher temperatures (+32°C) mitochondrial cytochrome c oxidase activity also showed a slight decline in activity.

In contrast to cytochrome oxidase activity in the mitochondrial membrane, activity and oxidase-reconstituted with dimyristoyl phosphatidylcholine had no higher temperature decline (+32°C). However, there was a distinct change in oxidase activity in the DMPC-oxidase vesicles about 24°C. The discontinuity observed for the intact membrane system around 16° to 14°C was not apparent in the reconstituted system (Fig. 2).

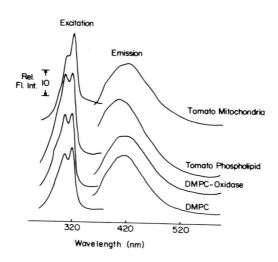

FIGURE 3. *Uncorrected steady state excitation and emission spectra of TPnA in the mitochondrial membrane and phospholipid vesicle systems. Excitation wavelength 320 nm; emission wavelength 420 nm. Half-maximal band pass, 5 nm.*

B.   *Fluorescent Probe Steady State Excitation and Emission Spectra*

Excitation spectra (Fig. 3) of TPnA containing membrane and vesicle systems revealed a shoulder at 294 nm, and two maxima at 306 nm and 320 nm with a valley at 312 nm. The intensity of the 320 nm excitation peak was greater than the 306 nm peak in the intact membrane system. In all other systems the 306 nm - 320 nm peaks were equivalent.

Emission spectra of TPnA were broad (Fig. 3), and showed a slight shift to the longer wavelengths (420 nm - 430 nm) in the intact membrane system. There were no appreciable shifts in excitation-emission wavelengths when the sample temperature was lowered to $+2^{\circ}$C. Mitochondrial membranes and vesicles of extracted mitochondrial lipids had a small amount of endogenous fluorescence (10% of the total fluorescence with TPnA) around 380 nm to 450 nm which was not altered by lowering the temperature to 2°C.

C.   *Fluorescence Lifetimes of TPnA in Vesicles and Membranes*

Examination of the fluorescence decay of TPnA in DMPC-oxidase vesicles, (Fig. 4), showed the presence of two components. Above the transition temperature of DMPC (23°C) there was a fast component and

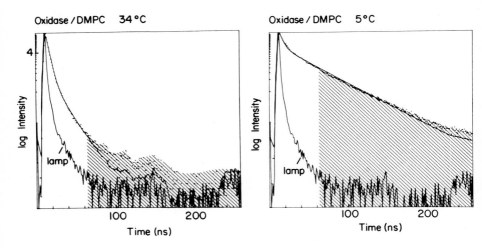

FIGURE 4. Fluorescence decay of TPnA in DMPC-oxidase (2 mg PL/mg protein). The shaded area under the curve shows the time domain of intensity integration (50 to 250 nsec) used for temperature scanning measurements. Excitation filter Perkin-Elmer UV2; emission filter Schott KV399.

a second slower decay (Table I). Below the transition temperature of DMPC there was also short and long components evident (Fig. 4), however the slow component had an enhanced lifetime (Table I) and a higher relative intensity.

In the mitochondrial membrane system TPnA fluorescence decay also approximated a double exponential (Fig. 5). However, TPnA fluorescence lifetime in the mitochondrial system was unchanged for the slower component over the entire experimental temperature range (Table I).

TABLE I. Summary of TPnA Fluorescent Lifetimes in DMPC-Oxidase and Mitochondria

Sample	Temperature	$1/k_1$(ns)	%	$1/k_2$(ns)	%
DMPC-oxidase	33^oC	$4.1+0.2$	83	$14.0+0.3$	17
DMPC-oxidase	5^oC	$4.7\overline{+}0.2$	20	$40.0\overline{+}0.4$	80
Mitochondria	34^oC	$2.5\overline{+}0.2$	86	$15.2\overline{+}0.3$	14
Mitochondria	5^oC	$3.0\overline{+}0.2$	80	$14.9\overline{+}0.3$	20

model curve: $M(T) = A_1 \exp(-k_1 t) + A_2 \exp(-k_2 t)$

normalization $= \int_0^\infty M(t) \, dt = 100\pm 2\%$

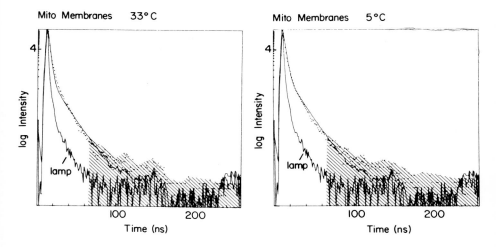

FIGURE 5.   Fluorescence decay of TPnA in mitochondrial
membranes (1 mg/ml). The shaded area under the curve shows the time
domain of intensity integration (50 to 250 nsec) used for temperature
scanning measurements.   Excitation filter Perkin-Elmer UV2; emission
filter Schott KV399.

In the mitochondrial membrane system TPnA fluorescence decay
also approximated a double exponential (Fig. 5).   However, TPnA
fluorescence lifetime in the mitochondrial system was unchanged for the
slower component over the entire experimental temperature range
(Table I).

D.   Fluorescence Intensity Measurements

Trans-parinaric acid fluorescence in DMPC vesicles showed a sharp
increase at 23°C (Fig. 6), similar to that described by Sklar   et al.,
(14) and coincided with heat capacity changes reported by Ladbrooke and
Chapman (9), for DMPC liquid crystal to gel phase transitions using
differential scanning calorimetry.   The enhanced fluorescence paralled
changes in fluorescence lifetime associated with the phospholipid in a
"gel-like" state (14).   Examination of TPnA fluorescence in the DMPC-
reconstituted oxidase vesicle system revealed a temperature dependent
change in fluorescence at 23°C similar to the DMPC vesicles (Fig. 6).
When the fluorescence of TPnA in mitochondrial membranes or in
lipid vesicles prepared from extracted tomato mitochondrial lipids was
measured as a function of temperature there was no significant change
over the temperature range 40° to 2°C (Fig. 6).

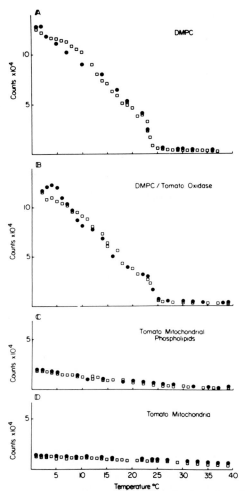

FIGURE 6. Temperature profile of TPnA time domain of fluorescence intensity integration (50 to 250 nsec) in membrane and vesicle systems. Points represent the number of detected fluorescence flashes per 900,000 excitation pulses. Cooling scan O, heating scan ●, at rates of 0.3 °/min. Excitation filter, Perkin-Elmer UV2; emission filter Schott KV399.

IV.   DISCUSSION

The fluorescence probe trans-parinaric acid has been shown to monitor liquid crystal to gel phase transitions *per se* in both artificial (14) and biological membrane systems (15, 18). The occurrence of enhanced lifetimes at and below the temperature of the phase transition in DMPC and DMPC-oxidase vesicles indicates that TPnA monitors the phospholipid phase transition and is not appreciably influenced by the protein component. Probe behavior in tomato mitochondrial membranes and vesicles of extracted mitochondrial phospholipids show no enhanced fluorescent lifetime implying that liquid crystal to gel phase transitions do not occur in the 40° to $2^{\circ}C$ range for these systems. Recent investigations of animal mitochondria using differential scanning calorimetry (6) and TPnA fluorescence (17, 18) indicate that a lipid phase transition occurs at subzero temperature (-5° to $-20^{\circ}C$). In contrast, experiments using electron spin resonance nitoxide probe analysis have revealed a change in probe motional parameters for both extracted phospholipids and intact animal mitochondrial membranes around $20^{\circ}C$ (12) and around $12^{\circ}C$ for tomato mitochondria (8) suggesting that the EPR observations above zero centigrade may represent "viscotropic-fluidity" effects rather than phase transitions in the membrane phospholipids.

The alteration of oxidase enzymatic activity by changes in phospholipid physical state as a function of temperature in both intact membranes and vesicle systems is also of interest. Under the conditions of this study, changes in oxidase activity (i.e. "breaks," discontinuities, or other nonlinear behavior in Arrhenius plots) due to phase transitions cannot be distinguished from those due to fluidity alterations. Enoch *et al.* (2) have observed that in a system which has a close ratio of enzymatic protein to phospholipid, the rate limiting parameter is not related to phospholipid dependent translational diffusion of reacting membrane components. The cytochrome oxidase complex is an analogous assembly in which the rate limiting step or steps may involve the phospholipid dependent intramolecular transport of electrons between heme a components (19, 20).

ACKNOWLEDGMENTS

The authors wish to thank Dr. J. Vanderkooi for the use of the time-resolved fluorescence measurement apparatus and her helpful suggestions. We are grateful to Dr. B. Chance for his enthusiastic discussions on rate limiting steps in intrinsic membrane catalytic proteins. The useful exchanges on fluorescent probe analysis with Dr. L. Bashford and Dr. J. Smith of the Johnson Foundation probe group are gratefully acknowledged.

V. REFERENCES

1. Douce, R., Mannella, C. A., and Bonner, W. D. *Biochim. Biophys. Acta 292,* 105-116 (1973).
2. Enoch, H. G., Catala, A., and Strittmatter, P. *J. Biol. Chem. 251,* 5095-5103 (1976).
3. Erecinska, M. and Chance, B. *Arch. Biochem. Biophys. 151,* 304-315 (1972).

4. Fleisher, S., Klouwen, H., and Brierley, G. *J. Biol. Chem. 236,* 2936-2941 (1972).
5. Folch, J., Lees, M., and Sloane-Stanley, G. H. *J. Biol. Chem. 226,* 497-509 (1957).
6. Hackenbrock, C. R., Hochli, M., and Chau, R. M. *Biochim. Biophys. Acta 445,* 466-484 (1976).
7. Keith, A. D., Aloia, R. C., Lyons, J. M., Snipes, W., and Pengelly, E. T. *Biochim. Biophys. Acta 394,* 204-210 (1975).
8. Keith, A. D., Breidenbach, R. W., Lyons, J. M., and Waring, A. (unpublished data) (1976).
9. Ladbrooke, B. D., and Chapman, D. *Chem. Phys. Lipids 3,* 304-367 (1969).
10. Raison, J. K., Lyons, J. M., and Thomson, W. W. *Arch. Biochem. Biophys. 142,* 83-90 (1971).
11. Raison, J. K., and Lyons, J. M. *Proc. Natl. Acad. Sci. U.S. 68,* 2092-2094 (1971).
12. Raison, J. K., Lyons, J. M., Mehlhorn, R. J., and Keith, A. D. *J. Biol. Chem. 246,* 4036-4040 (1971).
13. Schwertner, H. A., and Biale, J. B. *J. Lipid Res. 14,* 235-242 (1973).
14. Sklar, L. A., Hudson, B. S., and Simoni, R. D. *Biochemistry 16,* 819-812 (1977).
15. Tecoma, E. S., Sklar, L. A., Simoni, R. D., and Hudson, B. S. *Biochemistry 16,* 829-835 (1977).
16. Vik, S. B., and Capaldi, R. A. *Biochemistry 16,* 5755-5759 (1977).
17. Waring, A., Chance, B., and Glatz, P. *XIth International Cong. of Biochem. Abstract* (1979).
18. Waring, A., Glatz, P., and Vanderkooi, J. M. *Biochim. Biophys. Acta* (submitted) (1979).
19. Yoshida, S., Orii, Y., Kawato, S. and Ikegami, A. *In* "Cytochrome Oxidase" (T. E. King et al., eds.) pp. 231-240. Elsevier/North Holland Biomedical Press, Amsterdam, New York, Oxford (1979).
20. Yu, C. A., Yu, L., and King, T. S. *J. Biol. Chem. 250,* 1383-1392 (1975).

THE INFLUENCE OF FATTY ACID UNSATURATION
ON FLUIDITY AND MOLECULAR PACKING
OF CHLOROPLAST MEMBRANE LIPIDS

David G. Bishop, Janette R. Kenrick

Plant Physiology Unit, CSIRO Division of Food
Research and School of Biological Sciences
Macquarie University
North Ryde, Sydney 2113, Australia

James H. Bayston, Athol S. Macpherson

CSIRO Division of Chemical Technology
South Melbourne, 3205, Australia

Stanley R. Johns and Richard I. Willing

CSIRO Division of Applied Organic Chemistry
Melbourne, 3207, Australia

I.   INTRODUCTION

The role of membrane lipids in the susceptibility of certain plants to chilling injury at temperatures between about $10^{\circ}C$ and $15^{\circ}C$ has been largely deduced from discontinuities in Arrhenius plots of biological parameters, and of spin label motion in membranes and membrane lipids (14, 15, 21). The use of discontinuities in Arrhenius plots to determine critical temperatures in biological systems has been the subject of controversy for over fifty years. In the present context its use has been questioned on theoretical grounds (1, 9, 30) and in a recent review, Schreier et al., (25) have drawn attention to the possibility of artefactual discontinuities arising in Arrhenius plots of spin-label motion, due to the incompatibility of the experimental measurements and the formal equations used to derive motion parameters.

The fatty acid composition of animal and plant membrane lipids is such that phase separations and transitions might not be expected to

375

occur above 0°. This is particularly true in chloroplasts of algae (except certain Cyanophyta) and higher plants, whose fatty acid composition is highly unsaturated (2). Chloroplasts also contain only very low levels of sterols (3) and the great bulk of chlorophyll molecules are bound to specific proteins and cannot be considered as part of the bulk lipid (28). However, phase transitions in membrane lipids including those of the chloroplasts have been invoked as being responsible for the chilling sensitivity of such plants as maize, tomato (22) and mung bean (23). Much of this data has been obtained by measurements of spin label motion in chloroplast membranes and their lipids, using the motion parameters whose validity has been questioned by Schreier *et al.* (25).

This paper describes an examination of the temperature characteristics of the chloroplast polar lipid from three species of higher plants, both chilling sensitive and resistant, and compares the properties of glycolipids with widely varying fatty acid composition, by monolayer and ^{13}C nuclear magnetic resonance techniques.

II. EXPERIMENTAL

Chloroplasts were isolated from greenhouse-grown tomato *(Lycopersicon esculentum var. Floridade)*, maize *(Zea mays var. GH390)*, and peas *(Pisum sativum var. Massey Gem)* by methods previously described (3). The chloroplast polar lipid fraction was prepared by chromatography on columns of Kieselgel 60 (Merck) (17), except that the polar lipid fraction was eluted with $CHCl_3:CH_3OH$, 1:1. Fatty acid analysis was performed by gas liquid chromatography of methyl esters (5).

Spin label motion was measured in a Varian E4 spectrometer with controlled temperature regulation as described by Raison and Chapman (23) at a constant temperature increase of $0.5^{\circ}C/min$. Spin label (to give a final concentration of $10^{-4}M$) was added to a solution of lipid (2 mg) in $CHCl_3$ and the solvent removed under N_2. Then 0.2 ml of 20 mM Tris-acetate, 1 mM EDTA, pH 8.0 was added and the sample mixed for 2 min on a vortex mixer in the presence of glass beads. Three parameters of spin label motion were measured, correlation time (τ_0) as described by Melhorn *et al.* (16), order parameter (S_0) as described by Gaffney (8) and partition as described by Shimshick and McConnell (26). The spin labels used were 5NS-A (3-oxazolidenyloxy-2-(3 - carboxypropyl)-2-tridecyl-4,4-dimethyl), 12NL (1-octadecanoyl-2-[3-oxazolidenyloxy-2-(10-carboxydecyl)-2-hexyl-4,4-dimethyl] -3sn-glycerylphosphorylcholine), ---a gift from Dr. C. C. Curtain--, 12NS-Me (3-oxazolidenyloxy-2-(10-carbmethoxydecyl)-2-hexyl-4,4-dimethyl), and 5N10 (3-oxazolidenyloxy-2-pentyl-2-butyl-4,4-dimethyl).

Differential scanning calorimetry was performed in gold pans in a Perkin Elmer DSC-2 calorimeter, using samples of lipid (6-8 mg) hydrated with 10 µl of 20 mM Tris-acetate, 1 mM EDTA, pH 8.0

containing 30% v/v ethylene glycol, as described by van Dijck *et al.*
(7). Both heating and cooling scans were run at 5°/min.

The isolation of highly purified chloroplast lipids for monolayer
studies and the technique used for the measurement of force-area curves
will be described in a forthcoming publication (Bishop *et al.*in preparation). ^{13}C-NMR measurements were performed on a Varian CFT-20
spectrometer as previously described (12, 13).

III.   RESULTS

A   *Fatty Acid Composition of Chloroplast Lipids*

Chloroplast membrane lipids are composed predominantly of
glycolipids, unlike most other biological membranes, whose lipid
components are phospholipids. The major chloroplast lipids are MGG
(monogalactosyldiacylglycerol) and DGG (digalactosyldiacylglycerol),
with smaller amounts of SL (sulfoquinovosyldiacylglycerol) and PG
(phosphatidylglycerol) (4, 6). The fatty acid composition of the total
polar lipid from chloroplasts of three species is shown in Table I. The
composition is similar in all species although the chilling resistant
species, pea, contains a higher level of saturated fatty acids and has a
lower double bond index (DBI) than the two chilling sensitive species,
tomato and maize. The fatty acid compositions give no indication that
the bulk membrane lipids would undergo a phase transition above 0°C.
The presence of significant amounts of molecular species which might
undergo phase separations from the bulk lipid, such as those containing
two saturated acids, is also eliminated.   Although higher plant
chloroplast SL may contain up to 50% of saturated fatty acids, molecular
species containing two saturated acids are only very minor components
(29).

B.   *Spin Label Motion in Chloroplast Lipids*

The most common parameters of spin label motion which have been
used to detect phase transitions and separations in membrane lipids are
the order parameter, correlation time and partition parameters. The
limitations of these parameters and the fact that they may give rise to
artefactual discontinuities in Arrhenius plots has been pointed out by
Schreier *et al.* (25).   Figure 1 shows the Arrhenius plots of the
parameters S_g and τ_o measured with the spin label 5NS-A, and that of
τ_o measured with 12NS-Me, in tomato chloroplast polar lipid. While the
plot of S_g for 5NS-A is approximately a straight line, those of τ_o for
both 5NS-A and 12NS-Me obviously do not fit a single straight line, and it
is difficult to distinguish whether the data fit a smooth curve, or two or
three intersecting straight lines. It is also difficult to reconcile the

TABLE I.   Fatty Acid Composition of Chloroplast Polar Lipids

Acid	Tomato	Maize	Pea
	%	%	%
16:0	9.2	7.2	19.2
16:1	0.8	2.8	5.1
16:2	-	-	0.4
16:3	-	-	8.7
18:0	1.5	0.8	1.9
18:1	1.9	0.3	1.4
18:2	12.8	3.0	12.0
18:3	73.7	85.9	50.9
Double bond index[a]	2.48	2.67	2.10

[a]*The double bond index (DBI) was calculated as Σ (fatty acid %) • (number of double bonds in fatty acid) • 10^{-2} for all unsaturated fatty acids in the molecule.*

straight line plot obtained for the measurement of S_g with 5NS-A with that of the τ_o plot obtained from the same spectral data. Although the numerical change in log S_g over the range 3°-36^oC shown is only small, plots of S_g or 2T against temperature (data not shown) gave no indication of discontinuities. The values obtained in partition experiments using 5N10 gave Arrhenius plots which appeared to be curves, but these measurements, which were based on heights of peaks, rather than areas, may also give rise to artefactual discontinuities in the plots (25).

We have, however, attempted to fit straight lines to the Arrhenius plots of spin label motion, to produce discontinuities and to derive 'apparent transition temperatures' wherever possible. The values obtained are shown in Table II. While the S_g values consistently produced straight line plots over the range, there did not appear to be any consistency in the apparent transition temperatures derived from τ_op plots, either within one species, or between species based on their chilling sensitivity. It is noticeable however that the apparent transition temperature for measurements of τ_o with the various spin labels are always in the order 5NS-A > 12NL ≳ 12NS-Me indicating that the choice of spin-label may exert an effect on the apparent transition temperature. Such an effect has previously been reported by Overath

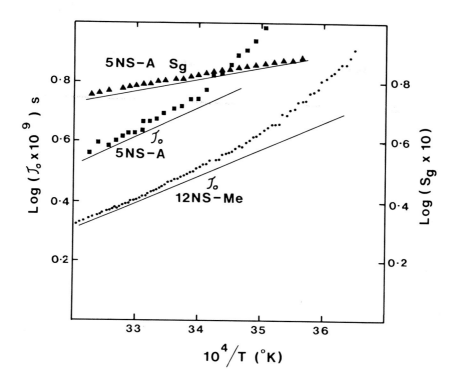

FIGURE 1. *Arrhenius plots of spin-label motion in the polar lipid of tomato chloroplasts. The values for S_g and τ_o with 5NS-A have been derived from the same spectra. The straight lines drawn beside each set of points merely serve to assist in interpreting the shape of the lines created by the data points.*

and Trauble (18) who found variations of $7-8^{\circ}C$ in the transition temperature of the membrane phospholipids of *E. coli*, depending on the spin label used. It should be stressed, however, that accurate fitting of straight lines to data which deviate as little from a straight line as those obtained with 12NS-Me, is a difficult undertaking.

TABLE II.   *Apparent Transition Temperatures Derived from the Motion of Spin Labels in Multibilayers of Chloroplast Polar Lipids*

	Spin label	Parameter measured	Apparent transition temperature °C
Tomato	5NS-A	S_g	None
		τ_o	21
	12NL	τ_o	17
	12NS-Me	τ_o	12
	5N10	Partition	17
Maize	5NS-A	S_g	None
		τ_o	14
	12NL	τ_o	12
	12NS-Me	τ_o	8
Pea	5NS-A	S_g	None
		τ_o	15
	12NL	τ_o	9
	12NS-Me	τ_o	6

C. Differential Scanning Calorimetry

The thermotropic properties of chloroplast polar lipid, measured by differential scanning calorimetry are shown in Figure 2. In each case, a small broad transition could be detected over the range from -20 to +15°C. Enthalpy values (ΔH) for these transitions, when comapred to a purified phospholipid such as dioleylphosphatidylcholine (ΔH, 14.5 cal/gm) (7) suggest that less than 5% of the lipid is participating in the transition. The possibility that small amounts of pigment, still remaining in the lipid preparations, could be responsible for the transition has not yet been eliminated. The fact, however, that the transition is observed over the same range for all three samples, seems to eliminate it as a factor in chilling sensitivity.

D. Monolayer Experiments

Monolayer studies on membrane lipids have in general been confined to synthetic phospholipids with a clearly defined fatty acid composition. However, naturally occurring lipids contain a diversity of fatty acids, although MGG and DGG of higher plant chloroplasts are characterized by

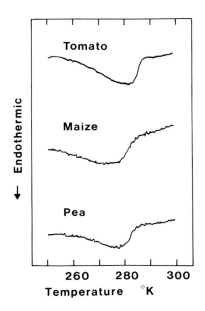

FIGURE 2. Thermotropic properties of chloroplast polar lipid measured by differential scanning calorimetry. Approximate energy contents (ΔH) of the transitions were : tomato, 0.4 cal/gm; maize, 0.5 cal/gm; pea 0.3 cal/gm.

high contents (80-95%) of α-linolenic acids (all *cis*-9,12,15-octadeca-trienoic acid). We have studied the molecular packing properties of chloroplast polar lipids, by isolating individual lipids with widely differing fatty acid composition, from diverse plant and algal chloroplasts, and comparing their force-area curves in monolayer experiments (Bishop *et al.*, unpublished data).

Force-area curves for synthetic phosphatidylcholines containing eighteen-carbon fatty acids are shown in Figure 3. The di-18:0 compound forms a very condensed monolayer. More expanded monolayers are formed as the degree of unsaturation of the fatty acid chains is increased, although the major change occurs with the introduction of one double bond per fatty acid molecule.

The force-area curve of hydrogenated silver beet MGG (Fig. 4) is similar to that of di-18:0 phosphatidylcholine, in being very condensed, and approaching the limiting area occupied by two long-chain saturated acids. The MGG from silver beet, which contains 94% trienoic fatty acids (DB1, 2.9) does not however form a monolayer as expanded as di-18:3 phosphatidylcholine, but rather, one which is similar to di-18:1 phosphatidylcholine, indicating that head group interactions in the MGG molecule are stronger than in the phosphatidylcholine molecule. This is

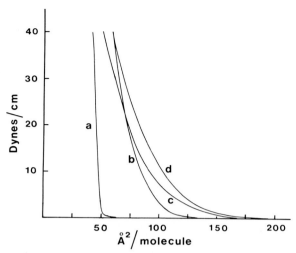

FIGURE 3. Force-area curves of synthetic phosphatidylcholines. (a) di 18:0-PC, (b) di-18:1-PC, (c) di-18:2-PC, (d) di-18:3-PC.

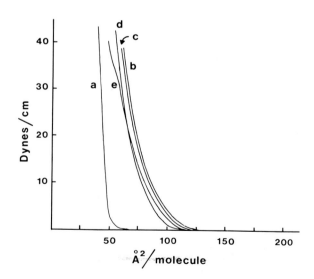

FIGURE 4. Force-area curves of chloroplast monogalactosyldiacylglycerols. (a) hydrogenated silver beet MGG, DBI, 0; (b) silver beet MGG, DBI, 2.9; (c) Ulva MGG, DBI, 3.3; (d) Synechococcus MGG, DBI, 1.0; (e) Anabaena MGG, DBI, 1.5.

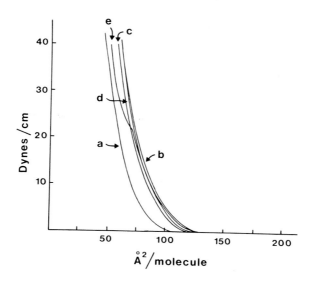

FIGURE 5. *Force-area curves of chloroplast digalactosyldiacylgly-cerols. (a) hydrogenated silver beet DGG, DBI, 0; (b) silver beet DGG, DBI, 2.6; (c) Ulva DGG, DBI, 1.9; (d) Synechococcus DGG, DBI, 1.3; (e) Anabaena DGG, DBI, 1.7.*

confirmed by the fact that the force area curves of three other naturally occurring MGG's, whose DBI's range from 1.0 to 3.3, are similar to those of silver beet MGG. Increases in the degree of fatty acid unsaturation above more than one double bond per molecule have little effect on the packing properties of the MGG molecule.

The force area curves of DGG's whose DBI's range from 0 to 2.6 are shown in Figure 5. The fully saturated compound forms a monolayer which is more expanded than that of the corresponding MGG and which does not compress to the same extent. This no doubt reflects the larger size of the DGG head group, but is also apparent that the naturally occurring DGG's are also dominated by head-group interactions.

The force-area curves for the four major polar lipids of silver beet chloroplasts are shown in Figure 6. The curves for MGG and DGG are almost identical, despite the disparity in size of their headgroups, and the only minor differences in their fatty acid composition. Comparison of the MGG and DGG curves from each of the other species (Figs. 4 & 5) confirms the similarity in the force-area curves of the two galactolipids from any one source. Although both SL and PG form more condensed monolayers than the galactolipids, they do not exert any condensing effect on the galactolipids, as the force-area curve of a mixture of the

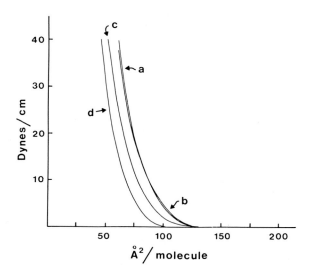

FIGURE 6. Force-area curves of silver beet chloroplast polar lipids (a) MGG, DBI, 2.9; (b) DGG, DBI, 2.6; (c) SL, DBI, 1.5; (d) PG, DBI, 1.3. The force-area curve of a mixture of the four lipids in the molar proportions in which they occur in the chloroplast (MGG:DGG:SL:PG, 50:33:13:4) is superimposable on the curves for MGG and DGG.

four lipids in the proportions in which they occur in the chloroplast, is identical to that of the two galactolipids. (This curve is not shown in Fig. 6 simply for clarity). Thus the packing properties of the chloroplast membrane polar lipid is determined by the MGG and DGG and is remarkably similar for a variety of species with widely differing fatty acid composition.

E. ^{13}*C-Nuclear Magnetic Resonance*

We have extended our previous studies of chloroplast lipids by ^{13}C-NMR (6, 11, 12, 13) by comparing the motional properties of some of the MGG's used in the monolayer experiments. ^{13}C-NMR measurements of silver beet DGG in methanol permit the identification of all three glycerol carbon atoms, all twelve sugar carbon atoams and fifteen of the eighteen carbon atoms in the fatty acyl chain, on the basis of their chemical shifts, and the longitudinal relaxation times (T_1's) of all these carbon atoms have been measured (6, 12).

TABLE III. Longitudinal Relaxation Time (T_1) of Individual Carbon Atoms of Hydrogenated Chloroplast Galactolipids in Methanol

Carbon atom	T_1 (Sec)	
	MGG	DGG
Fatty acyl chain		
2	0.45	0.38
3	0.7	0.38
4	1.1[a]	0.58
5	1.1[a]	1.4[a]
6	1.1[a]	1.4[a]
7-14	0.9[b]	1.2[b]
15	1.1[a]	1.4[a]
16	2.9	3.5
17	4.1	4.6
18	3.3	5.0
Sugar molecules*		
1″	0.45	0.21
2″	0.39	0.27
3″	0.35	0.21
4″	0.27	0.15
5″	0.42	0.21
6″	0.34	0.10
1″	–	0.26
2″	–	0.23
3″	–	0.21
4″	–	0.20
5″	–	0.26
6″	–	0.26
Glycerol molecule*		
1′	0.13	0.14
2′	0.26	0.27
3′	0.17	0.12

*See Fig. 2 in Bishop et al., (6) for identification of carbon atoms.
[a,b] Average value of unresolved peaks.

Table III shows the T_1's of the individual carbon atoms in the sugar head groups, glycerol molecules and fatty acyl chains for hydrogenated silver beet MGG and DGG. The T_1's of the sugar headgroups of hydrogenated DGG are lower than those of MGG showing that the

TABLE IV. *Longitudinal Relaxation Time (T_1) of Individual Carbon Atoms of Linolenic Acid Molecules in Chloroplast Glycolipids*

Carbon atom	T_1 (sec)			
	Silver Beet			Ulva
	MGG	DGG	SL	MGG
2	0.50, 0.47	0.30, 0.25	0.29	0.52
3	0.72	0.53	0.44	0.76
4-6	0.84	0.72	0.67	0.79
7	1.1	0.91	0.86	0.80
8	1.2	1.5	1.0	1.1[b]
9	1.9	2.1	1.2	1.7[b]
10	1.9	2.1	1.2	2.4[b]
11	2.3	2.9[a]	2.6[a]	2.7[b]
12	3.6	3.9	2.8	3.4[b]
13	3.6	3.9	2.8	3.4[b]
14	4.5	2.9[a]	2.6[a]	2.7[b]
15	7.4	7.6	4.6	6.6[b]
16	7.5	7.4	4.6	7.0[b]
17	12.0	11.1	9.7[a]	10.4[b]
18	8.6	8.9	6.5[a]	8.1[b]

[a],[b] *Average value of unresolved peaks.*

headgroups are more tightly packed. Similarly the T_1's of the carbon atoms in the fatty acyl chain, which are near to the headgroups (carbons 2-4) are shorter in hydrogenated DGG than in MGG because of the tighter packing of the headgroup. However, towards the methyl end of the fatty acyl chain, the T_1's of the DGG carbon atoms are longer than those in MGG because the larger size of the DGG headgroup permits more motional space at the methyl end of the chain, even though DGG headgroups are more tightly packed. These results are in excellent agreement with the monolayer measurements of these compounds.

The T_1's of carbon atoms in the 18:3 fatty acids in silver beet MGG, DGG, SL and *Ulva* MGG are shown in Table IV. Silver beet MGG and DGG contain essentially only 18:3 acid (12), while SL contains approximately 50% 18:3 and 50% 16:0 (13) and *Ulva* MGG contains 30%

18:3 with 57% tetraenoic acids (16:4 and 18:4). Comparison of the T_1's of the fatty acyl carbon atoms in silver beet MGG and DGG show that overall motion is similar in both compounds except at the carboxyl end of the chain. This fact agrees well with the interpretation of the force-area curves. The differences in motion at the methyl end of the chains in hydrogenated MGG and hydrogenated DGG is not apparent in the naturally occurring compounds. In contrast, motion of the carbon atoms in the 18:3 chain of SL is considerably less than that of the corresponding atoms in MGG and DGG, because of the presence in the SL molecule of the 16:0 chain (13). However, a further increase in the degree of unsaturation of the companion acids in the acyl chains of MGG, as occurs in *Ulva* MGG, clearly does not have a marked effect on the motion of the 18:3 chain (Table IV), even though exact determination of some T_1's is difficult because some averaging with signals of carbon atoms from the tetraenoic acid chains occurs. Again this result is in excellent agreement with the monolayer experiments (Fig. 4) in which it was found that silver beet and *Ulva* MGG's have almost identical force-area curves.

IV. DISCUSSION

The hypothesis that chilling sensitivity in higher plants is due to phase changes in part or all of the membrane lipids carries an implicit assumption that all membranes in a particular plant show the phase transition at the same temperature (22, 23). The present results indicate that the component lipids of the thylakoid membranes from a variety of photosynthetic cells, have very similar physical properties. On the basis of fatty acid composition, spinlabel motion and differential scanning calorimetry, we have not found it possible to differentiate between the properties of chloroplast membrane polar lipids, with respect to the degree of chilling sensitivity of the plant. The validity of the major physical techniques used to establish the presence of phase changes, spin label motion and fluorescence, has come under serious questioning (20, 25). Apparent discontinuities in Arrhenius plots have been found above $0°$ in a variety of plants, both chilling sensitive and resistant (20). The high degree of unsaturation of chloroplast membrane polar lipids would indicate that significant phase changes would not occur above $0°$. Differential scanning calorimetry measurements of higher plant MGG and DGG have indicated that phase transitions occur at about $-30°$ and $-50°C$ respectively (27). The transition temperatures of dioleyl phosphatidylcholine ($-22°C$) and stearyl-oleyl-phosphatidylcholine ($3°C$) (19), molecules with a lower degree of unsaturation than those found in chloroplast membranes are still below the range at which chilling symptoms develop in sensitive plants. Although head group composition can also effect the temperature of phase transitions, such effects are diminished when unsaturated acids are present in the molecule (7). It does not appear possible that chloroplast membrane lipids contain

significant amounts of molecular species of lipids containing two
saturated fatty acids, as such acids are only minor components of higher
plant MGG and DGG (10, 24, 29), but these two lipids appear to establish
the packing characteristics of the chloroplast membrane lipid.

The monolayer and ^{13}C-NMR studies described herein show that
galactolipids with a diverse fatty acid composition have very similar
physical properties, provided that they have a double bond index of one,
or greater. The available evidence with synthetic phospholipids which
contain at least one monounsaturated chain suggests that such
compounds will have phase transitions around 0°C or lower. Most
naturally occurring lipids have such a degree of unsaturation or higher.

We conclude therefore that on the available evidence phase
transitions or separations would not be expected to occur above 0° in the
bulk lipids in chloroplast membranes of either chilling-sensitive or
chilling-resistant plants and that evidence already obtained from
Arrhenius plots of spin label motion, or measurements using fluorescent
labels must be treated with great caution. Consideration of the
membrane lipid fatty acid composition of other membranes in higher
plants leads us to similar conclusions in those cases also. Whether
chilling sensitivity in higher plants is due to temperature-induced
conformational changes in membrane proteins which initiate changes in
the molecular ordering of membrane lipids remains an open question.

V.   REFERENCES

1. Bagnall, D. G. and Wolfe, J. A. *J. Exptl. Bot. 29,* 1231-1242 (1978).
2. Benson, A. A. *In* "Structure and Functions of Chloroplasts" (M. Gibbs, ed.) pp. 129-148. Springer, Berlin (1971).
3. Bishop, D. G. *Arch. Biochem. Biophys. 154,* 520-526 (1973).
4. Bishop, D. G. *Photochem. Photobiol. 20,* 281-299 (1974).
5. Bishop, D. G. and Smillie, R. M. *Arch. Biochem. Biophys. 137,* 179-189 (1970).
6. Bishop, D. G., Nolan, W. G., Johns, S. R. and Willing, R. I. *In* "Light Transducing Membranes: Structure, Function and Evolution." (D. W. Deamer, ed.) pp. 269-288. Academic Press, New York (1978).
7. van Dijck, P. W. M., de Kruijff, B., van Deenen, L. L. M., de Gier, J. and Demel, R. A. *Biochim. Biophys. Acta 455,* 576-587 (1976).
8. Gaffney, B. J. *In* "Spin Labeling Theory and Applications" (L. J. Berliner, ed.) pp. 567-571. Academic Press, New York (1976).
9. Gray, B. F., Gray, P. and Kirwan, N. A. *Combustion and Flame 18,* 439-459 (1972).
10. Jamieson, G. R. and Reid, E. H. *Phytochem. 10,* 1837-1843 (1971).
11. Johns, S. R., Leslie, D. R., Willing, R. I. and Bishop, D. G. *Aust. J. Chem. 30,* 813-822 (1977a).

12. Johns, S. R., Leslie, D. R., Willing, R. I. and Bishop, D. G. *Aust. J. Chem. 30,* 823-834 (1977b).
13. Johns, S. R., Leslie, D. R., Willing, R. I. and Bishop, D. G. *Aust. J. Chem. 31,* 65-72 (1978).
14. Lyons, J. M. *Ann. Rev. Plant Physiol. 24,* 445-466 (1973).
15. Lyons, J. M. and Raison, J. K. (This volume) (1979).
16. Mehlhorn, R., Snipes, W. and Keith, A. *Biophys. J. 13,* 1223-1231 (1973).
17. Nolan, W. G., and Bishop, D. G. *Arch. Biochem. Biophys. 190,* 473-473-482 (1978).
18. Overath, P. and Trauble, H. *Biochemistry 12,* 2625-2634 (1973).
19. Phillips, M. C., Hauser, H. and Paltauf, F. *Chem. Phys. Lipids 8,* 127-133 (1972).
20. Quinn, P. J. and Williams, W. P. *Prog. Biophys. Molec. Biol.* (in press) (1979).
21. Raison, J. K. *J. Bioenergetics 4,* 285-309 (1973).
22. Raison, J. K. *In* "Mechanisms of Regulation of Plant Growth" (R. L. Bieleski, A. R. Ferguson and M. M. Cresswell, eds.), pp. 487-497. Royal Society of New Zealand, Wellington (1974).
23. Raison, J. K. and Chapman, E. A. *Aust. J. Plant Physiol. 3,* 291-299 (1976).
24. Rullkotter, J., Heinz, E., and Tulloch, A. P. *Z. Pflanzenphysiol. 76,* 163-175 (1975).
25. Schreier, S., Polnaszek, C. F. and Smith, I. C. P. *Biochim. Biophys. Acta 515,* 375-436 (1978).

TEMPERATURE REGULATION
OF PLANT FATTY ACYL DESATURASES

Paul Mazliak

Laboratoire de Physiologie Cellulaire
Universite P. & M. Curie
Paris, France

It has been repeatedly observed that the lipids from plants grown at low temperature are generally enriched in unsaturated fatty acids as compared with the lipids from the same species grown at higher temperatures (4). However, exceptions to this general trend have been signalled (1).

I. INCREASE IN FATTY ACID UNSATURATION AT LOW TEMPERATURE

The general rule implying that the mean unsaturation of plant fatty acids decreases as the growth temperature increases was illustrated recently in our laboratory, in two series of analyses concerning flax stem and rape-seed lipids (Tables I and II). At the lowest temperatures, lipids were enriched in the most unsaturated fatty acids (linolenic acid in the case of flax and linoleic plus linolenic acids in the case of rape-seeds). If one admits the classical series of desaturations postulated for linolenic acid biosynthesis:

$$C_{18:0} \xrightarrow[-2H]{D_1} C_{18:1} \xrightarrow[-2H]{D_2} C_{18:2} \xrightarrow[-2H]{D_3} C_{18:3}$$

where D_1, D_2 and D_3 represent three different acyl desaturases, the preceding results mean that low temperatures favour specifically the functioning of the last desaturases of the chain (D_3, linoleyl desaturase and D_2, oleyl-desaturase) while higher temperatures provoke the relative accumulation of intermediate compounds in this chain (oleic and linoleic acids).

It is generally thought that the increase in unsaturation of plant lipids at low temperature first concerns membrane lipids (16). The fluidity or viscosity of these membranes is primarily determined by the

391

TABLE I.   Influence of Temperature on the Fatty Acid Compositions of
           Flax Stem Lipids
           The percentages given in this table are the means of results
given by three different samples. Each sample contained four stem apex
(ca 300 mg) (from Moutot and Mazliak, unpublished).

Linum usitatissimum L. (flax)

Varieties	% total fatty acid weight						
	C_{14}	C_{16}	$C_{16}\Delta^{3t}$	C_{18}	$C_{18:1}$	$C_{18:2}$	$C_{18:3}$
Ocean							
22°C	3.2	14.6	3.2	1.6	1.7	15.1	61.3
27°C	3.8	19.1	3.4	2.2	2.4	20.6	48.6
Hera							
22°C	3.1	15.4	3.0	2.0	2.2	16.0	58.1
27°C	2.9	14.6	3.3	2.0	3.3	21.9	51.9
Prekuli 665							
22°C	3.9	15.7	2.9	1.7	1.3	12.2	62.2
27°C	2.7	20.2	2.2	0.8	2.4	22.2	46.9

TABLE II. Influence of Temperature on the Fatty Acid Composition of
           Rape Seeds, Eight Weeks after Flowering
           Average data for two separate samples, from Tremolieres, et
al, (18).

Brassica napus L. var. Primor (0erucic) (Rape seed)

Maturation temperature (°C)	% total fatty acid weight					
	Daylength 16 hr				Daylength 9 hr	
	12°	17°	22°	27°	17°	22°
C_{16}	8,2	9	9,5	7	7,6	8,5
$C_{16:1}$	0,6	0 6	0,7	0,9	0,3	0,6
C_{18}	1,6	2	2,5	2	2,1	0,6
$C_{18:1}$	47,5	48,2	57,2	56	50	54
$C_{18:2}$	28,5	26,5	19,5	21,5	26	24,7
$C_{18:3}$	13,5	13,7	10	9	13,7	11,2

degree of unsaturation of the acyl residues present in lipid molecules. This fluidity affects the activity of membrane-bound enzymes and the mobility of membrane proteins. So, it would be first to maintain a correct membrane fluidity, to allow the proper functions of cell membranes, that the increase in fatty acid unsaturation induced by low temperature, would occur. As a matter of fact, the enrichment of membrane lipids in polyunsaturated fatty acids, at low temperature, has been observed in some plant organelles, for example mitochondria (Table III).

TABLE III. *Influence of Growth Temperature on the Fatty Acid Composition of Isolated Wheat Motochondria. From Miller et al. (14).*

Triticum aestivum L. (wheat)

Variety	Growth temperature (^{o}C)	moles percent				
		C_{16}	C_{18}	$C_{18:1}$	$C_{18:2}$	$C_{18:3}$
Marquis	2°	26.1	1.9	2.6	16.9	52.8
Marquis	24°	23.1	1.3	5.8	39.9	29.6
Capelle	2°	23.3	0.7	3.6	16.4	56.0
Capelle	24°	26.7	1.0	4.2	38.4	29.9
Rideau	2°	23.9	0.9	5.0	15.0	55.3
Rideau	24°	23.8	1.0	5.6	37.0	32.8
Kharkov	2°	22.8	1.0	4.9	17.2	54.2
Kharkov	24°	24.4	1.7	6.2	37.6	30.1

However, in one single case in our laboratory, this increase in lipid unsaturation of mitochondrial membranes at low temperature, has not been observed: this was for mango fruits stored at different temperatures. In that case, chilling injury developed over the fruits stored at low temperature and we have been able to show: 1) that no increase in lipid unsaturation occurred in mango mitochondria (Table IV); and 2) that the succinate oxidation capacity of the organelles was markedly reduced, which could probably be related to the development of chilling injury (10).

TABLE IV. *Influence of Storage Temperature on the Fatty Acid Composition of Mango Mitochondria Isolated after 28 Days of Storage, from Kane et al., (10).*

Mangifera indica L., cv. Amelie (mango)

Storage temperature (°C)	moles percent					
	C_{16}	$C_{16:1}$	C_{18}	$C_{18:1}$	$C_{18:2}$	$C_{18:3}$
4	34	24	3	14	15	10
8	34	26.5	3	12.5	16	8
12	26	35.5	4	15	13.5	6
20	26	37	2	15	12	8

II. REGULATION OF UNSATURATED FATTY ACID SYNTHESIS FROM ^{14}C-ACETATE

To understand the presumed adaptive ability of many plants to change their membrane fluidity at low temperature, it was necessary to study the regulation of plant fatty acid desaturases by temperature and other factors.

First we have checked, with labeled precursors, that the biosynthesis of polyunsaturated fatty acids was effectively more active at low temperature. On the one hand, using [14]CO_2 as a precursor photosynthesized by flax stems, we have found that the biosynthesis of linoleic and, overall, linolenic acids was really much more intense in these stems at 22°C, as compared with the biosynthesis of the same acids at 27°C (Fig. 1). On the other hand, using NH_4-^{14}C-oleate as a precursor, we have observed that exposing rape-seeds to low temperatures (12-17°C) during the maturation of the seeds induced higher oleate and linoleate desaturation in the siliqua (Table V).

A simple explanation for the accumulation of polyunsaturated fatty acids in plants, at low temperature, was offered ten years ago by Harris and James (8). Using 2-^{14}C-acetate as a lipid precursor, these authors showed that the relative amount of unsaturated fatty acids synthetized by the slices of various seeds (castor bean, flax, sunflower) increased with a decrease in the temperature or with an increase in the O_2 concentration of cell fluids. As it had been shown that fatty acid dehydrogenations required O_2 and reduced pyridine nucleotide as cofactors, they concluded that the formation of unsaturated fatty acids by plants was indeed controlled by the O_2 concentration in cell sap. Thus as the temperature decreased, the increased concentration of O_2 in solution would activate the desaturation systems. The O_2 content of tissues, however, cannot be taken as the only aspect of temperature

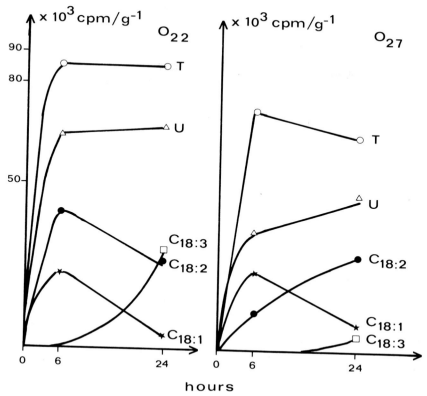

FIGURE 1. Incorporation of $^{14}CO_2$ into the fatty acids of flax stems (variety Ocean). A pulse of $^{14}CO_2$ was given for 10 min to plants grown either at $22°C$ (O_{22}) or $27°C$ (O_{27}). This pulse was followed by various periods of chase, the lengths of which are indicated in abscissa. The radioactivities of the different fatty acids present in one gram of fresh matter were determined by radio-gas chromatography at the end of each chase-period and plotted in ordinates $(cpm.g^{-1})$ -T : total fatty acids - U : total unsaturated fatty acids - (from Moutot Jolivet and Mazliak, 1978, unpublished).

action on fatty acid desaturation. Since all desaturases require O_2, it is to be expected that synthesis of all unsaturated fatty acids will be reduced if O_2 is present in limiting amounts only. In fact, it has been found in many tissues that decreases in polyunsaturated fatty acids were counterbalanced by higher proportions of oleic acid instead of saturated ones (for a review see Beringer (3) (see also the data of Tables I - II). Harris and James (8), using ^{14}C-acetate as a lipid precursor, were not able to prove directly that the activities of the plant fatty acid desaturases proper were dependent on temperature and/or O_2 concentration. We decided to check this hypothesis directly upon true plant fatty acid desaturases. We used for this study several plant desaturation systems that previously we had been able to get function *in vitro*, i.e., 1) the oleate-desaturating systems from various oleaginous seeds (6); and 2) the microsomal oleyl CoA-desaturase from potato tuber (2).

III. REGULATION OF LINOLEIC ACID BIOSYNTHESIS (^{14}C-OLEATE DESATURATION) IN OLEAGINOUS SEEDS

Sunflower seeds, rape seeds or flax seeds were harvested three weeks after flowering, i.e. at the time of the greatest polyunsaturated fatty acid biosynthesis. Thin slices, cut with a razor blade, were incubated in aqueous solutions of NH_4-(1-^{14}C)-oleate at various temperatures or under various atmospheres (N_2, O_2 or air). ^{14}C-linoleic and ^{14}C-linolenic acids formed within the slices after different incubation periods were analyzed by radio-gas liquid chromatography (Fig. 2). Figure 3 shows the kinetics of oleate desaturation at a fixed temperature (20°C). The greater part of the ^{14}C-oleate (given as a substrate) is desaturated into ^{14}C-linoleic acid during the first 24 hours; it is only after that time that ^{14}C-linolenic acid begins to appear in small amounts.

Figure 4 shows the variation of oleate desaturation with temperature in sunflower seeds. Surprisingly, we could observe an optimal temperature for desaturation at about 27°C and it is clear that contrary to what was expected, low temperatures (2°, 10°C) did not activate linoleic acid formation. Similar results were obtained with flax seeds (Fig. 5) and rape seeds (data not shown).

Figure 6 shows the variation of oleate desaturation in sunflower seed slices with O_2 concentration in the atmosphere above the slices (then in cell saps). As expected, no desaturation of oleate was observed under N_2 but the same desaturation intensities were observed under air (21% O_2) and pure oxygen (100% O_2). Therefore, there is no simple relationship between O_2 concentration and oleate desaturation activity within sunflower seed. Exactly similar results were found with flax or rape seeds (data not shown).

IV. REGULATION OF THE MICROSOMAL OLEYL-CoA DESATURASE FROM POTATO TUBER

This is an inducible system which appears during aging of thin potato slices in an aerated humid medium (2, 19). The microsomal system has been thoroughly studied *in vitro*: 1-^{14}C-oleyl-CoA desaturation re-

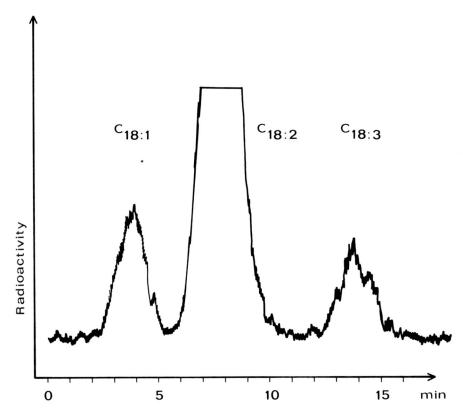

FIGURE 2. ^{14}C-oleate desaturation into ^{14}C-linoleate and ^{14}C-linolenate by slices of rape-seeds (see text for incubation conditions). The radio-gas chromatogram of ^{14}C-fatty acid methyl esters is presented.

quires oxygen, reduced nicotinamide adenine dinucleotide (NADH) and bovine serum albumin. The optimal pH is 7.2; SH-complexing agents, cyanide (9), heavy metals (13) and detergents (12) inhibited oleyl-CoA desaturation. The linoleyl residue formed by desaturation of oleyl-CoA appears always esterified within a phosphatidylcholine molecule and the question has been raised whether oleyl-phosphatidylcholine was not the true substrate of the desaturation system (17). The microsomal system is presently envisioned (7) as comprising three different proteins embedded in the membrane lipid matrix (Fig. 7).

Drs. Cherif and Kader (5), in our laboratory, have studied *in vitro* the oleyl-CoA desaturating system of potato microsomes and the effects of varying temperature and oxygen concentration. Figure 8 presents the kinetics of desaturation of oleyl-coenzyme A *in vitro*, at various tem-

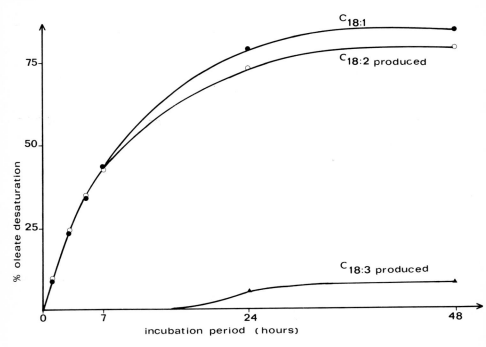

FIGURE 3. Kinetics of ^{14}C-oleate desaturation by sunflower seed
slices, at 20°C (for incubation conditions, see text).

peratures. The highest desaturation activity (34%) was obtained at 30°C;
lower (20°C) or higher (35°C) temperatures decreased the intensity of
desaturation. It is interesting to notice on the graph that for the first 15
min of incubation, the slope of the curves representing the desaturation
activities is nearly similar for these three temperatures. However, the
level of the plateau reached after 30 min varies according to the
temperatures, the highest level being observed at 30°C. At low
temperatures (2 and 10°C), the desaturation was strongly reduced. At
10°C for instance the slopes of the curve representing desaturation
activity against time is nine times lower than the slope at 30°C, when
measured for the first 15 min. The plateau was reached at 10°C after a
longer time of incubation (90 min) than at higher temperatures. At 2°C,
the percentage of desaturation was almost null; a low desaturation
activity (7%) was noticed after 90 min of incubation at this temperature.

    Thus, contrarily to the hypothesis of Harris and James (8), these
experiments show that the microsomal oleyl-CoA desaturase of potato
tuber presents its optimal activity at moderated temperatures and
approximatively follows classical enzymic kinetics. At low temperature,
although the solubility of oxygen in the incubation medium was greatly
increased, oleyl-CoA desaturase activity was very low.

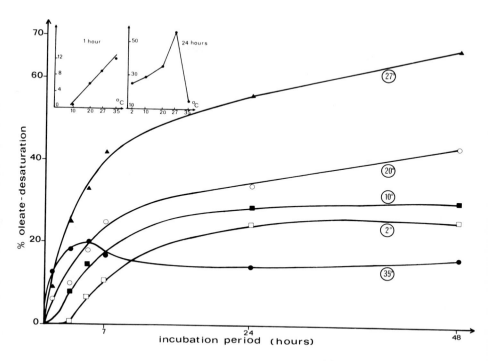

FIGURE 4. Variation with temperature of ^{14}C-oleate desaturation by sunflower seed slices incubated for different periods of time as indicated in abscissa. Insets: variation of ^{14}C-oleate-desaturation with temperature (abscissa) for 1 hour (left) or 24 hours (right) of incubation.

The desaturation kinetics of ^{14}C-oleyl-CoA in vitro at a fixed temperature ($30^{\circ}C$) and under atmospheres with various oxygen and nitrogen concentrations are represented in Figure 9. The results show a clear correlation between the desaturation activity and the oxygen content of the atmosphere only within the range 0 - 20% of oxygen. No variation could be observed in the desataturation rates for the first ten minutes, between 5 and 20% of oxygen in the atmosphere, but after this time a plateau was soon reached, with the highest level of desaturation corresponding to 20% of oxygen. No desaturation at all was observed under pure nitrogen. Contrary again to what was expected, no increase in oleyl-CoA desaturase activity was observed under pure oxygen, but instead, some inhibition occurred as compared with the functioning of the system under normal air. These results confirm that oxygen is a

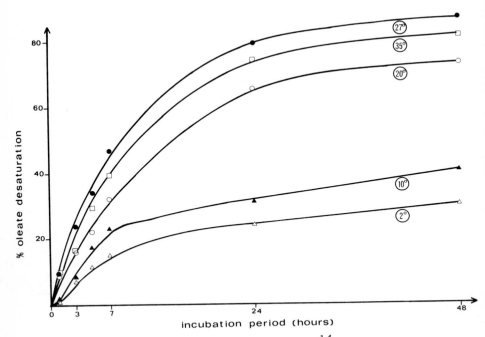

FIGURE 5. *Variation with temperature of ^{14}C-oleate desaturation by flax-seed slices incubated for different periods of time, as indicated in abscissa.*

limiting factor for the desaturation but this gas does not appear to be the sole or the main controlling factor of the desaturation activity.

VI.   CONCLUSION

        Based on the results reported in this paper, the following conclusions may tentatively be proposed:

1.   Different temperatures during plant development may modify considerably the fatty acid composition of plant lipids.   In general, lowering the environmental temperature produces a higher content of polyunsaturated fatty acids in cell membranes or reserve lipids.

2.   Temperature affects the biosynthesis of unsaturated fatty acids directly and not only by increasing O_2 concentration in the cell sap. The formation of the different unsaturated fatty acids is catalyzed by different desaturases which might have different temperature optima.

Thus, low temperatures would favour polyunsaturated fatty acid accumulation while higher temperatures would favour oleic (or palmitoleic) acid accumulation.

3.   The molecular mechanism underlying these temperature effects are not presently understood, despite considerable study. A stimulating working hypothesis has been put forth recently by Nozawa, Thompson and co-workers (11, 15) indicating that the membrane fluidity itself is the regulating factor controlling the activity of the membrane-bound fatty acid desaturases. Following these views, certain temperature changes provoking alterations of membrane fluidity would be decisive for triggering the onset of fatty acid desaturation in cell membranes. As the work of Nozawa and Thompson (11, 15) was conducted on the protozoan *Tetrahymena*, new studies are necessary to check the validity of their hypothesis as far as plant tissues are concerned.

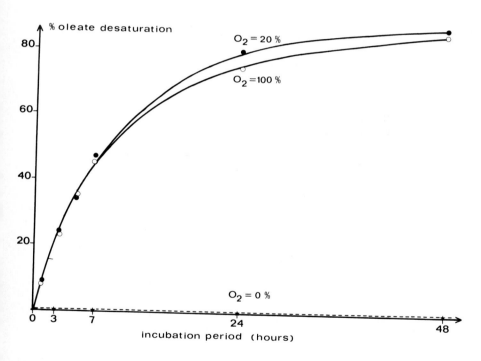

FIGURE 6. Variation, with oxygen concentration, at $27^{o}C$, of ^{14}C-oleate desaturation by sunflower seed slices incubated for different periods of time, as indicated in abscissa.

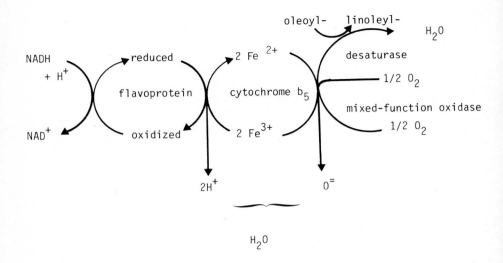

FIGURE 7. The oleyl-desaturation system of plant microsomes.

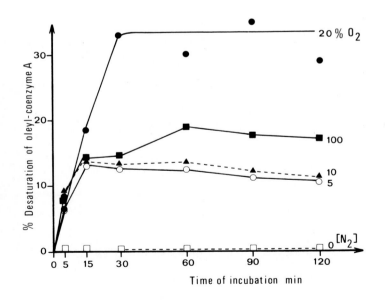

FIGURE 8. Temperature effect on the activity of the microsomal oleyl-coenzyme A desaturase of potato tuber. The incubation temperature, varying from 2° to $35^o C$ is indicated in the graph.

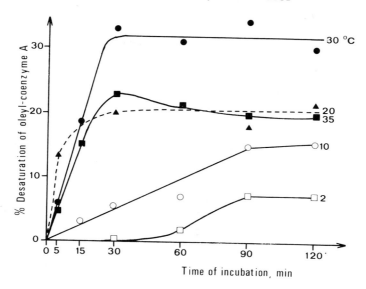

FIGURE 9. Effect of oxygen on the activity of microsomal oleyl-coenzyme A desaturase of potato tubers at 30°C. The oxygen concentration of the atmosphere, varying from zero (N₂ = pure nitrogen) to 100% is indicated in the graph.

TABLE V. ^{14}C-oleate Desaturation Activities at 22°C by Seeds from Rape Seeds (Winter Variety Primor (0 erucic) Grown at Different Temperatures during Four Weeks after Flowering NH₄-^{14}C-oleate was injected in vivo, at 22°C, into different samples of siliqua; 24 hours later, the seed fatty acids from the different samples were analyzed by radio-gas chromatography, from Tremolieres et al., (18).

	% total fatty acid radioactivity					
	Daylength 16 hr				Daylength 9 hr	
Maturation temperature	12°	17°	22°	27°	17°	22°
C 18:1	25.2	58.5	86.3	84	79	93.3
C 18:2	53.2	35.2	13.7	16	15	6.7
C 18:3	21.5	6.2	traces	–	5.5	–

VII. REFERENCES

1.   Appelqvist, L. A. *In* "Recent Advances in the Chemistry and Biochemistry of Plant Lipids" (T. Galliard and E. I. Mercer, editors) pp. 247-286, Acad. Press, London (1975).
2.   Ben Abdelkader, A., Cherif, A., Demandre C. and Mazliak, P. *Europ. J. Biochem., 32,* 155-165 (1973).
3.   Beringer, H. *In* "Proceedings of the 13th IPI - Colloquium of the International Potash Institute" held in York, England, 123-133 (1977).
4.   Canvin, D. T. *Can. J. Bot., 43,* 63-69 (1965).
5.   Cherif, A. and Kader, J. C. *Potato Res., 19,* 21-26 (1976).
6.   Cherif, A. and Mazliak, P. *Rev. F. Corps Gras, 25,* 15-20 (1978).
7.   Fulco, A. J. *Ann. Rev. Biochem., 43,* 215-241 (1974).
8.   Harris, P. and James, A. T. *Biochim. Biophys. Acta, 187,* 13-18 (1969).
9.   Kader, J. C. *Biochim. Biophys Acta, 486,* 429-436 (1977).
10. Kane, O., Marcellin, P. and Mazliak, P. *Plant Physiol., 61,* 634-638 (1978).
11. Martin, C. E., Hiramitsu, K., Kitajima, Y., Nozawa, Y., Skriver, L. and Thompson, G. A. *Biochemistry, 15,* 5218-5233 (1976).
12. Mazliak, P. *C. R. Ac. Sc., 281,* 1471-1474 (1975).
13. Mazliak, P., Decotte, A. M. and Kader, J. C. *Chem. Biol. Int., 16,* 115-119 (1977).
14. Miller, R. W., de La Roche, I. and Pomeroy, M. K. *Plant Physiol., 53,* 426-433 (1974).
15. Nozawa, Y. and Kasai, R. *Biochim. Biophys. Acta, 529,* 54-66 (1978).
16. Raison, J. K. *In* "Rate Control of Biological Processes - Symposia of the Society for Experimental Biology, XXVII", pp. 485-512, Cambridge University Press (1973).
17. Stymme, S. and Appelqvist, L. A. *Europ. J. Biochem., 90,* 223-229 (1978).
18. Tremolieres, H., Tremolieres, A. and Mazliak, P. *Phytochemistry, 17,* 685-687 (1978).
19. Willemot, C and Stumpf, P. K. *Can. J. Botany, 45,* 579-584 (1967).

CHEMICAL MODIFICATION OF LIPIDS IN CHILLING SENSITIVE SPECIES

Judith B. St. John

U.S. Department of Agriculture
Science and Education Administration
Agricultural Research
Agricultural Environmental Quality Institute
Weed Science Laboratory
Beltsville, Maryland

Our basic studies into the mechanism of herbicide action suggested that substituted pyridazinones might be useful tools with which to alter the fatty acid composition of plant membrane lipids. We initially reported on substituted pyridazinones and showed that the effects of Sandoz 6706 [4-chloro-5-(dimethylamino)-2-(α,α,α-trifluoro-m-tolyl)-3 (2H)-pyridazionone] were inhibition of photosynthesis and interference with chloroplast pigment development (2) and inhibition of the formation of the polar lipids (3). The inhibition in the formation of the polar lipids was related to an increase in linoleic acid (18:2) accompanied by a decrease in linolenic acid (18:3) content of the polar lipids. We suggested that Sandoz 6706 might act by preventing the formation of chloroplast membranes, resulting from inhibition of the formation of the polar lipids, chlorophylls, and carotenoids necessary for chloroplast membrane formation. We subsequently demonstrated that the fatty acid composition of the major lipids of the chloroplast membranes, the mono- and digalactosyl diglycerides, can be definably altered with various substituted pyridazinones (6). The galactolipid fatty acid composition of wheat (*Triticum aestivum* L. cv. 'Arthur') can be altered so that there is a decrease in linolenic acid accompanied by an increase in linoleic acid without a shift in the relative ratio of saturated to unsaturated fatty acids. The fatty acid composition can be shifted to a higher proportion of saturated acids, and the fatty acid composition of the monogalactosyl diglycerides can be altered in preference to the digalactosyl diglycerides.

Mention of a trademark or proprietary product does not constitute a guarantee or warranty of the product by the U.S. Department of Agriculture and does not imply its approval to the exclusion of other products that may also be suitable.

Structure-activity comparisons (9) revealed that inhibition of the Hill reaction and photosynthesis by the substituted pyridazinones is related to the structure of the parent compound, pyrazon [5-amino-4-chloro-2-phenyl-3(2H)-pyridazinone]. Trifluoromethyl substitution of the phenyl ring of pyrazon, mono-methyl substitution of the amine, or substitutions at both positions result in inhibition of carotenoid biosynthesis. Both substitutions were required for maximum effect. Dimethyl substitution of the amine of pyrazon is related to a decrease in linolenic acid accompanied by an increase in linoleic acid without a shift in the ratio of saturated to unsaturated fatty acids of the membrane lipids; this substituted pyridazinone (Sandoz 9785 = BASF 13 338, hereinafter referred to as BASF 13 338) reduced levels of membrane bound linolenic acid but did not interfere with chlorophyll or carotenoid biosynthesis. In addition, of 49 analogs evaluated (10), substitution of the phenyl ring in the 2 position, halogen substitution in the 4 position, and various substituted amino groups in the 5 position resulted in compounds showing the highest activity. Variation of substituents at these positions allowed shifting the ratio of membrane-bound linoleic to linolenic acid from the control value of 0.50 to 6.7 for BASF 13 338. The trifluoromethyl substitution of the phenyl ring and mono-methyl substitution of the amine are related to a shift toward a higher proportion of saturated fatty acids of chloroplast membranes.

Studies with higher plants aimed at elucidating the role of membrane lipids and their component fatty acids in plant responses to chilling have relied on approaches which yielded indirect correlations. Numerous studies have compared chilling-resistant and chilling-sensitive plant species and demonstrated that chilling-resistant plant species contain more unsaturated fatty acids, particularly linolenic acid (18:3) (4). Secondly, as the temperature of growth decreases, lipid unsaturation increases with the predominant increase being linolenic acid. When we discovered that pyridazinones could be used to block linolenic acid biosynthesis, we realized that for the first time it would be possible to chemically manipulate the linolenic acid of plant membranes in a given plant species and study plant response to temperature.

We have used BASF 13 338 to relate the level of linolenic acid in the membrane lipids of cotton (*Gossypium hirsutum* L.) root tips to chilling resistance in cotton seedlings under controlled environmental conditions (9). Seeds were germinated at 15, 20, 25, and 30°C. As the temperature of growth decreased from 30 to 15°C, the linolenic acid content of the polar lipid fraction of untreated root tips increased, as evidenced by a shift in the ratio of linoleic to linolenic acid (18:2/18:3) from 2.67 at 30°C to 1.16 at 15°C, Treatment with BASF 13 338 reduced the low-temperature-induced increase in linolenic acid, as evidenced by an 18:2/18:3 ratio of 2.56 compared with the control ratio of 1.16. BASF 13 338 lowered the levels of linolenic acid at all temperatures. Thus, BASF 13 338 also inhibits linolenic acid synthesis in non-photosynthetic tissues in the absence of light and blocks the increased synthesis of linolenic acid during chill-hardening. When hesisnon-hardened control and BASF 13 338 treated seedlings were chilled at 8°C, the BASF treated seedlings wilted within 24 hours. After 4 days at 8°C, the seedlings were returned to 30°C. The control seedlings rehydrated and resumed normal growth, whereas the treated seedlings remained wilted after 7 days and either died or continued to appear abnormal. The hardened control and hardened BASF 13 338 treated seedlings showed fewer latent symptoms of chilling injury, although a greater number of abnormal seedlings occurred in the latter. Thus, decreased levels of linolenic acid in the membranes of root tips correlated with increased sensitivity to chilling in cotton.

Willemot (11) subsequently observed that when wheat plants were treated with BASF 13 338 for 36 hours before frost hardening, both the accumulation of linolenic acid and development of freezing tolerance were inhibited. Willemot's work is discussed in detail in another chapter of this volume.

We have recently extended our studies to the effects of BASF 13 338 on linolenic acid levels and hardiness of cereals under controlled environmental and field conditions. BASF 13 338 was soil-incorporated at rates of 0, 2.8, 5.2, or 11.2 kg/ha. Wheat (*Triticum aestivum* L. cvs. 'Arthur' and 'Potomac') , barley (*Hordeum vulgare* L.), and rye (*Secale cereale* L.) were sown in the fall. Treated plants showed early symptoms of low temperature injury as evidenced by frost banding. Frost banding as a symptom of low temperature injury in cereals was first reported by Richards (5); injury appeared to occur first at the soil surface and the bands became visible after leaf elongation. Incidence and severity of frost banding were markedly increased in BASF 13 338-treated seedlings. Severity also increased as concentration of the chemical increased. Barley appeared to be the most sensitive and rye the least sensitive.

Analysis of fatty acid composition of membrane lipids from shoots of seedlings exposed to hardening temperatures (as low as -1.6°C) revealed that seedlings grown on soil incorporated with BASF 13 338 had significantly lower levels of linolenic acid (Table I). Reduced levels of linolenic acid were related to reduced survival and reduced tillering (Table II, tillering data only). Tables I and II are taken from St. John *et al.*, (8).

In the absence of BASF 13 338, all species accumulated linolenic acid to the same level and yet species and varietal differences in sensitivity to BASF 13 338 were pronounced. Arthur wheat was the most sensitive to the chemical, followed by Potomac wheat, Monroe barley and Abruzzi rye. Abruzzi rye differed from the other species in that the change in ratio of membrane-bound linoleic to linolenic acid leveled off at a treatment rate of 5.6 kg/ha. However, BASF 13 338 continued to reduce tilelring of Abruzzi rye through a rate of 11.2 kg/ha. We believe that our combined data indicate that increased levels of membrane-bound linolenic acid are associated with the adaptation of plants to functionality at low temperatures, but that other factors limit hardening and possibly distinguish levels of sensitivity between cultivars and

J. B. St. John

TABLE I.   Effects of BASF 13 338 Applied Preemergence on the Ratio
           of Membrane-Bound Linoleic (18:2) to Linolenic (18:3) Acid
           Under Field Conditions.

B Species	BASF 13 338 (kg/ha)			
	0	2.8	5.6	11.2
	(18:2/18:3)	(18:2/18:3)	(18:2/18:3)	(18:2/18:3)
Arthur wheat	0.16a[+]	0.68b	1.29c	2.30d
Potomac wheat	0.16a	0.67b	1.00c	1.68d
Monroe barley	0.14a	0.31b	0.59c	0.87d
Abruzzi rye	0.19a	0.28b	0.60c	0.54c

[+]Values across columns with common letters are are not
significantly different at the 5% level by use of Duncan's multiple range
test.

TABLE II.   Effect of BASF 13 338 Applied Preemergence on Tillering of
            Winter Cereals During Winter of 1977-78.

Species	BASF 13 338 (kg/ha)			
	0	2.8	5.6	11.2
	-----mean number of tillers/meter of row-----			
Arthur wheat	244a[+]	174b	124c	11d
Potomac wheat	185a	173a	60b	5c
Monroe barley	164a	138a	60b	3c
Abruzzi rye	199a	133b	99b	9c

[+]Means across columns with common letters are not significantly
different at the 5% level by use of Duncan's multiple range test.

species. The data do show that for a given species, levels of linolenic acid
can indicate susceptibility to chilling injury and might be used as a
screening tool for isolation of more resistant lines.   However, the
correlation between linolenic acid levels and chilling is not clear, and
caution is advised when crossing species.

The critical question dealing with the role of membrane lipids in plant response to chilling were posed in the introductory chapter of this volume. It is clear that molecular ordering of membrane lipids can be altered by variations in fatty acyl composition in microbial systems. Can the fatty acyl composition have any influence on the molecular ordering of the lipids in the more complex membranes of higher plants? What is the relationship between fatty acyl composition of plant membranes and the temperature at which lipid order changes to a critically low level? We have shown that substituted pyridazinones can be used to alter the fatty acyl composition of membranes in higher plants and provide a tool for answering these questions within a given plant species. We have also correlated reduced levels of linolenic acid in membranes to increased sensitivity to chilling. De La Roche (1) has recently suggested that it is the inhibition of photosynthesis by the pyridazinones, which we first reported (2), that is responsible for increased sensitivity to chilling. Regardless of the interplay between photosynthesis and linolenic acid level (i.e., is photosynthesis inhibited by the pyridazinones because of decreased levels of linolenic acid in chloroplast membrane?) the problem of the role of fatty acyl composition in molecular ordering of the lipids of plant membranes is still amenable to attack using substituted pyridazinones.

REFERENCES

1. De La Roche, A. I. *Plant Physiol. 63,* 5-8 (1979).
2. Hilton, J. L., Scharen, A. L., St. John, J. B., Moreland, D. E., and Norris, K. H. *Weed Sci. 17,* 541-547 (1969).
3. Hilton, J. L., St. John, J. B., Christiansen, M. N., and Norris, K. H. *Plant Physiol. 48,* 171-177 (1971).
4. Lyons, J. M. *Ann. Rev. Plant Physiol. 24,* 455-466.
5. Richards, B. L. *Proc. Utah Acad. Sci. Arts Lett. 11,* 3-9 (1934).
6. St. John, Judith B. *Plant Physiol. 57,* 38-40 (1976).
7. St. John, Judith B. and Christiansen, M. N. *Plant Physiol. 57,* 257-259 (1976).
8. St. John, J. B., Christiansen, M. N., Ashworth, E. N., and Gentner, W. A., *Crop Sci. 19,* 65-69 (1979).
9. St. John, J. B. and Hilton, J. L. *Weed Sci. 24,* 579-582 (1976).
10. St. John, J. B., Rittig, F. R., Ashworth, E. N., and Christiansen, M. N. *In* "Advances in Pesticide Science" (H. Geissbuhler, ed.) Part 2, Pergamon Press. Oxford and New York, 272-273 (1979).
11. Willemot, C. *Plant Physiol. 60,* 1-4(1977).

CHEMICAL MODIFICATION OF LIPIDS DURING FROST HARDENING OF HERBACEOUS SPECIES

Claude Willemot

Research Station, Agriculture Canada
Ste-Foy, Quebec, Canada

I. INTRODUCTION

It is generally accepted that the main targets of dehydration damage caused by extracellular ice formation are the cell membranes. It has therefore been an obvious approach to look for changes in membrane lipids and proteins correlated with increased frost tolerance at low hardening temperature. These changes would meet the biophysical requirements for membranes either to resist dehydration stresses at below freezing temperatures or to function at low above freezing temperatures and to allow the further changes which lead to frost hardening. The present discussion will be limited to those changes which occur during low temperature acclimation in membrane lipids of herbaceous plants which acquire intermediate levels of frost tolerance in the range of $-15^{\circ}C$ to $-30^{\circ}C$.

II. CHANGES IN MEMBRANE LIPIDS DURING FROST HARDENING

Which changes likely to influence biophysical characteristics of membranes have been observed during low temperature acclimation? A simple model of a membrane is illustrated in Figure 1. (1) General increase in phospholipids and increase in phospholipid turnover; (2) changes in polar heads of phospholipids; (3) increase in the degree of unsaturation of the fatty acids; (4) changes in sterol concentration and composition.

A. General Increase in Phospholipid Content

A general increase in phospholipids has been observed amongst others in cereals (23), in winter wheat (5, 32), and in alfalfa (11). This

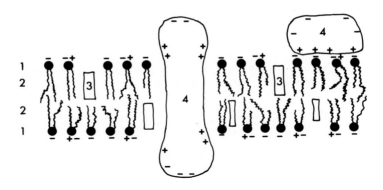

MEMBRANE COMPONENTS:

1 Polar heads of phospholipids
2 Fatty acids
3 Sterols
4 Proteins

FIGURE 1. Simple membrane model showing lipid components likely
to influence biophysical characteristics of the membrane.

response to low temperature is analogous to that of trees (24, 36) (Fig. 2).
Percentage dry weight greatly increases in most hardening tissues. The
increase in phospholipids was however shown to reflect a specific
increase in membrane material per cell, by electron microscopy (22) and
by expressing it on a dry weight and a DNA basis (25) (Table I). It did not
merely reflect cell division at low temperature.

TABLE I.   Augmentation of Phospholipid and Protein in Insoluble
            Homogenates of Summer and Winter Black Locust Bark
            Tissues*

Tissue	Temperature at LD_{50}	$\dfrac{Phospholipid}{DNA}$	$\dfrac{Protein}{DNA}$
Summer	$-10^{\circ}C$	4.7	14.0
Winter	$-196^{\circ}C$	11.1	33.0

*From Singh et al, (25).

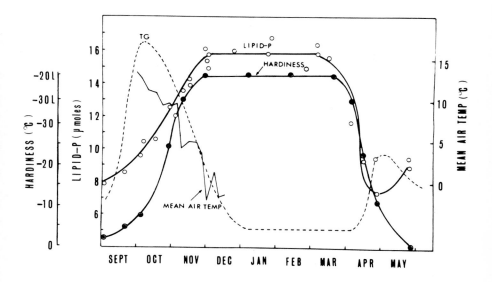

FIGURE 2. Seasonal changes in frost hardiness and total amount of phospholipids in poplar cortex. Frost hardiness was expressed as the minimum temperature at which the cortex survived freezing without injury. -20 L and -30 L on hardiness scale: tissues were slowly prefrozen to -20 or -30°C prior to immersion in liquid nitrogen (-196°C). TG: triglyceride from cortex. From Yoshida (36).

In cereal seeds germinating at low temperature, the increase in phospholipids was shown to occur on a dry weight basis (5, 23). Spring and winter types, which greatly differed in frost hardiness, showed the same extent of increase in membrane material at low temperature (23) (Table II). In twelve day old winter wheat plants, which were beyond the germination stage, phospholipids increased during frost hardening in root and crown on a fresh weight basis, but decreased on a dry weight basis (32) (Fig. 3). There was no difference in lipid P content between hardy Kharkov and less hardy Champlein. The specific increase in membrane material observed by Redshaw and Zalik (23) and de la Roche et al. (5) seems therefore more related to germination at low temperature, than to frost hardening. The general increase in dry weight, including phospholipids, observed by Willemot et al. (32) may be primarily a low temperature response. When young winter wheat plants were fed ^{33}P after various times of hardening, phospholipid biosynthesis was shown to be strongly stimulated in root and crown, and this stimulation was much

TABLE II.   Polar Lipid Phosphorus Content of Cereal Varieties at
            Different Stages of Vernalization*

Variety	Weeks of vernalization	Polar lipid P mg/100 g dry sample
Sangaste fall rye	0	6.5
	6	10.4
Prolific spring rye	0	6.4
	6	10.3
Kharkov winter wheat	0	7.1
	6	10.4
Red Bobs spring wheat	0	7.1
	6	10.8

*From Redshaw and Zalik (23).

TABLE III.   Lipid P Content in Roots of Alfalfa Hardy Cultivar Rambler
             and Less Hardy Caliverde at Various Times of Hardening*

Variety	Days of hardening	μg P/g fresh weight $\pm$ SD
Rambler	0	76.0 +4.01
	12	92.9 +2.25
	24	87.5 +2.36
	36	107.3 +1.74
Caliverde	0	78.2 +2.24
	12	72.9 +2.68
	24	68.3 +2.07
	36	73.1 +2.34

*From Grenier and Willemot (11).

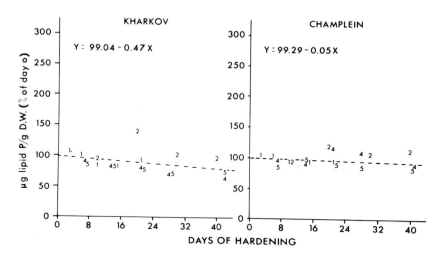

FIGURE 3. Lipid P content (% of the mean on day 0) of root and crown of winter wheat after various times of frost hardening. The numbers refer to separate experiments. From Willemot et al., (32).

greater in hardy Kharkov than in frost sensitive Champlein (29) (Fig. 4). This increase in phospholipid biosynthesis came later than the increase in frost tolerance. Potential for frost hardening correlated with ability to synthesize phospholipids at low temperature, but phospholipid biosynthesis was not a prerequisite for frost hardening. It may be required to maintain a high level of frost tolerance, and was interpreted as a repair mechanism.

Young alfalfa plants of the hardy cultivar Rambler showed an increase in lipid P in the root on a fresh weight basis (11) (Table III). This represented however a decrease on a dry weight basis. The change in lipid P content correlated with the hardening ability of the cultivars, but the increase was not specific to membrane material. When young alfalfa plants were fed ^{33}P, incorporation into the root lipids decreased progressively during frost hardening (11) (Fig. 5). The decrease was greatest in the frost sensitive cultivar Caliverde. Smolenska and Kuiper (26) showed no increase in lipid P content of rape leaves and roots at low temperature.

From these data it is difficult to conclude what significance a general increase in phospholipids and membrane material has with respect to frost hardening at low temperature.

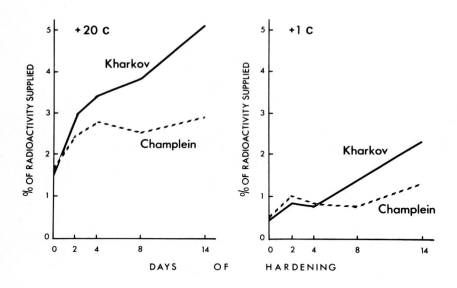

FIGURE 4. Incorporation of ^{33}p into lipids of root and crown of two cultivars of winter wheat, hardy Kharkov and less hardy Champlein, during 12 hours at $20^{o}C$ and 24 hours at $1^{o}C$, after various times of frost hardening. From Willemot (29).

B.   Changes at the Level of Polar Heads of Phospholipids

Few data have been gathered on changes in proportions of phospholipid classes in plants at low temperature. In the bark of the poplar tree a large increase in phosphatidylcholine (PC) and phosphatidylethanolamine (PE) was observed in the fall, with little change in the level of the remaining phospholipids (36) (Fig. 6). These data were confirmed with *Robinia* (24, 25). During frost hardening of alfalfa under controlled conditions PC and PE increased significantly in the roots of the hardy cultivar Rambler, but less in tender Caliverde (10) (Fig. 7). When alfalfa plants were fed ^{14}C-acetate for a constant time after different periods of hardening, synthesis of PC and, to a lesser extent, PE was strongly stimulated in hardy Rambler, but not in Caliverde (13) (Fig. 8). Specific activity of PC was always much greater than that of PE. When alfalfa plants were fed ^{33}P after different times of hardening, only PC showed a relative increase in labeling, and more in Rambler than in Caliverde. No difference in phospholipid composition were found in winter wheat germinating at $24^{o}C$ and $1^{o}C$ (6). When young plants of winter wheat were fed ^{33}P, little change was observed in the

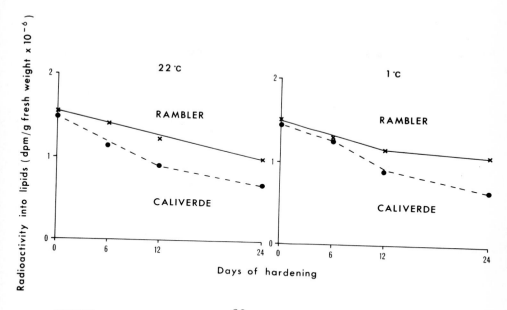

FIGURE 5. Incorporation of ^{33}P into lipids of alfalfa roots during 12 hours at $22°C$ and 24 hours at $1°C$ after various times of frost hardening. From Grenier and Willemot (11).

pattern of phospholipids synthesized in the root and crown (11). In the rape plant, low temperature caused an increase in PE and PC in the hardened leaf tissue, but a decrease in these phospholipids in the unhardened root tissue (26). The available information on changes at the level of the polar heads of the phospholipids at low hardening temperatures is still too fragmentary to draw definite conclusions. Except for wheat, most plants investigated showed an increase in PC and PE during frost hardening. The turnover of PC seems to be higher than that of other phospholipids at low temperature.

C. Increase in Degree of Unsaturation of Fatty Acids

The aspect of lipid metabolism which has been studied the most in connection with frost hardening of herbaceous plants is that of fatty acid desaturation. In 1964 Lyons *et al.* (19) showed a relationship between degree of unsaturation of the fatty acids of mitochondrial membranes, flexibility of these membranes, and chilling resistance of several species. In the following years increased unsaturation of fatty acids at low

418

C. Willemot

TABLE IV. *Fatty Acid Composition of The Polar Lipid Fraction of Cereal Varieties at Different Stages of Vernalization (% of Total Fatty Acids in the Fraction)**

Varieties	Weeks vernalized	16:0	18:1	18:2	18:3
Sangaste	0	20.8	15.3	58.5	5.4
(fall rye)	6	20.7	12.2	40.2	26.9
Prolific	0	22.4	15.2	56.9	5.5
(spring rye)	6	21.4	8.8	39.4	30.4
Kharkov	0	18.0	13.7	63.7	4.6
(winter wheat)	6	20.0	8.1	60.2	11.7
Red Bobs	0	19.2	16.3	61.0	3.5
(spring wheat)	6	22.9	7.8	55.4	13.9

*From Redshaw and Zalik (23).

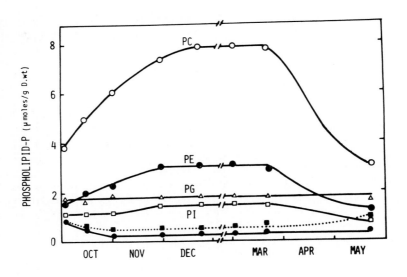

FIGURE 6. *Seasonal changes in individual phospholipids from poplar cortex. The two lowest curves correspond to unidentified acidic phospholipids. From Yoshida (36).*

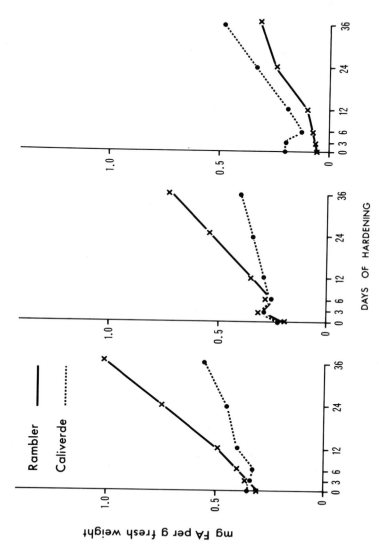

FIGURE 7. Total fatty acids of phosphatidylcholine, phosphatidylethanolamine and triglycerides in roots of alfalfa during frost hardening. From Grenier and Willemot (10).

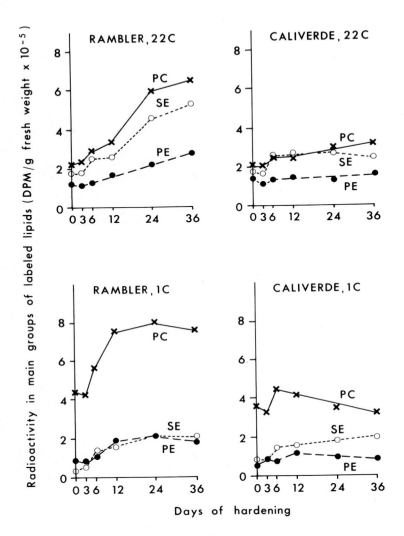

FIGURE 8. ^{14}C-acetate incorporation into phosphatidylcholine, phosphatidylethanolamine and sterol esters of alfalfa roots during six hours at 22°C and during 12 hours at 1°C after various times of frost hardening. From Grenier et al.,(13).

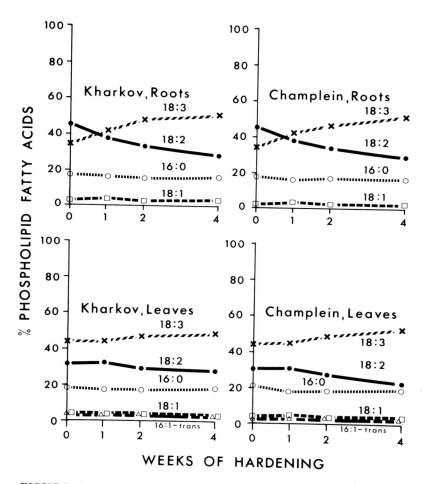

FIGURE 9. *Phospholipid major fatty acids content, expressed as percentage of total fatty acids, in two cultivars of winter wheat during frost hardening. From Willemot et al., (33).*

hardening temperature was shown in almost all species investigated: in alfalfa in the field during the fall (9) and under controlled hardening conditions (10, 12, 13), where the increase, mainly in linoleic acid (18:2), was greatest in the hardy cultivars; in pear trees (16); in wheat and rye seeds germinating at low temperature, where linolenic acid (18:3) increased at the expense of linoleic acid (18:2) (5, 23, 24); in young winter wheat plants beyond the germination stage (32); and in the rape plant (26). Temperature had little influence on fatty acid unsaturation of potato tuber tissue, which has little hardening potential (8). With the publication of two review papers on chilling injury by Lyons (17, 18) it became increasingly probable that unsaturation of fatty acids played an important role in low temperature acclimation. In recent years, however, evidence to the contrary has been accumulating.

Several authors have observed that cultivars varying widely in frost tolerance show the same increase in unsaturation of the fatty acids at low temperature. Redshaw and Zalik (23) showed similar increases in linolenic acid content in fall and spring rye and in winter and spring wheat germinating at low temperature (Table IV). de la Roche *et al.* (7) confirmed their observations with 4 winter wheat cultivars which differed widely in hardening potential (Table V). In 12 day old winter wheat plants, linolenic acid increased to the same extent in the root and crown of hardy Kharkov and tender Champlein (33) (Fig. 9). Poplar (36) and black locust (24, 25) bark tissue acquires extreme frost hardiness in the fall, without increased fatty acid unsaturation. Smolenska and Kuiper (26) showed the same increase in linolenic acid content at low temperature in rape leaves which acquire frost tolerance and in roots which do not harden. The lack of differences in fatty acid unsaturation between hardy and less hardy cultivars did not prove that it was not involved in frost hardening. It could mean that increased unsaturation was a prerequisite to frost hardening, but that less hardy cultivars had their frost tolerance limited by other factors. Two new experimental approaches became recently available to test this possibility: (1) Wilson (35) showed that in the chill sensitive *Phaseolus vulgaris,* in which the degree of unsaturation of fatty acids increased at low temperature, hardening by drought was possible without low temperature treatment and without increased unsaturation of the fatty acids; (2) St. John and Christiansen (27) showed that treatment of germinating cotton seeds with pyridazinones reduced both the tolerance of the seedlings to chilling and the low temperature induced increase in linolenic acid content of the polar lipids in the root tips.

Drought experiments were performed by de la Roche (2) with germinating winter wheat seeds (Table VI) and by Willemot and Pelletier (31) with young winter wheat plants (Table VII). In both cases the results confirmed those of Wilson (35). Significant frost hardening was achieved without low temperature treatment, and with a decrease in the level of linolenic acid.

Treatment of young winter wheat plants with a pyridazinone (BASF 13-338) caused simultaneous inhibition of frost hardening and of synthesis of linolenic acid in the roots at low temperature (30) (Table

TABLE VI. *Fatty Acid Composition of Phospholipids from Epicotyls of Unhardened (24°C), Cold Hardened (2°C) and Desiccation Hardened (24°C) Seedlings of Winter Wheat, cv. Kharkov MC 22**

Treatment	LT_{50}	Fatty acids (mole %)				
		16:0	18:0	18:1	18:2	18:3
24°C	-4°C	23.9	0.4	9.3	27.0	39.4
2°C	-22°C	22.2	0.6	7.2	17.8	52.2
90% R.H., 24°C	-13°C	21.2	1.1	25.2	21.0	31.5

*
From de la Roche (2).

TABLE VIII. *Total (mg per g fresh weight) and Individual Fatty Acids (weight %) of Roots of Frost Hardening Winter Wheat Treated with BASF 13-338**

Treatment	Control		Treated	
Days of hardening	0	14	0	14
Fatty acids				
16:0	17.2 +0.2	17.6 +0.4	16.8 +0.3	18.8 +0.2
18:0	0.6 +0.1	0.5 +0.2	0.6 +0.2	1.1 +0.5
18:1	3.5 +0.5	4.4 +0.8	3.9 +0.5	4.8 +1.9
18:2	41.0 +0.3	32.1 +0.9	50.6 +0.5	49.4 +1.7
18:3	37.8 +0.6	45.3 +1.1	28.1 +0.8	25.9 +0.9
Total fatty acids	0.88 +0.22	1.28 +0.24	0.89 +0.14	0.72 +0.14
Ratio 18:3/18:2	0.92	1.41	0.56	0.52
LT_{50} (^{o}C)	-5.0	-17.7	-5.0	-6.0

*
From Willemot (30).

TABLE VII. *Frost Resistance (LT_{50}) and Fatty Acid Content (mg per g fresh weight) of 12 Day Old Winter Wheat Plants Grown Subsequently at 20°C or Hardened at 1°C, at Two Soil Mosture Levels, 40% (normal) or 10% (drought stressed)*

Days of growing and harden.	Temp of growing and harden.	Moisture level^v	LT_{50} °C	% total fatty acids					Total fatty acids	Ratio 18:3/18:2
				16:0	18:0	18:1	18:2	18:3		
0	20°C	40	-7.2+0.1	18.5	0.3	3.4	42.5	35.3	1.72+0.12	0.83
7	20°C	40	-5.9+0.1	19.0	0.4	3.5	39.8	37.3	1.46+0.06	0.94
	20°C	10	-10.9+0.2	19.8	0.5	4.8	40.2	34.7	2.10+0.25	0.86
14	20°C	40	-6.8+0.1	19.1	0.5	3.8	41.0	35.8	1.56+0.27	0.87
	20°C	10	-9.6+0.1	20.2	0.6	5.5	39.9	33.8	2.15+0.15	0.85
	1°C	40	-19.5+0.2	17.5	0.2	3.5	33.9	44.9	1.83+0.11	1.32
	1°C	10	-23.0+0.1	18.8	0.3	3.8	35.8	41.3	2.47+0.18	1.15
28	1°C	40	-23.9+0.2	17.7	0.2	3.1	28.2	50.7	1.90+0.10	1.80
	1°C	10	-25.0+0.3	19.4	0.4	4.0	30.1	46.2	2.58+0.11	1.54

*From Willemot and Pelletier (31).

^{v}% of soil water holding capacity.

424

TABLE IX. *The Effect of BASF 13-338 and Sandoz 6706 on the Fatty Acid Composition (weight %) and the Freezing Tolerance (LD_{50}, °C) of Kharkov Wheat Seedlings*

Treatment	LD_{50}	Lipid component	Fatty acid composition				
			16:0	18:0	18:1	18:2	18:3
24°C, control	-4	total lipids	17.6	0.5	8.3	23.7	49.9
		phospholipids	23.1	0.7	9.6	31.1	35.5
2°C, control	-23	total lipids	15.7	0.7	5.9	14.1	63.6
		phospholipids	21.9	0.5	7.9	17.6	52.1
2°C, BASF 13-338	-23	total lipids	15.2	0.7	8.9	51.8	23.3
		phospholipids	19.3	0.2	11.1	47.1	22.3
2°C, Sandoz 6706	-23	total lipids	14.7	0.4	10.4	38.4	36.1
		phospholipids	18.5	0.6	12.6	35.3	33.0

*From de la Roche (3).

425

VIII). Subsequent experiments showed that photosynthesis, which is essential for frost hardening of winter wheat (21) was completely inhibited (34). de la Roche (3) repeated this experiment with wheat seeds germinating in the dark at 2°C. These seedlings draw their carbohydrates from the endosperm and do not depend on photosynthesis for frost hardening. His results show unequivocally that an increase in linolenic acid is not a prerequisite to frost hardening in winter wheat (Table IX).

In summary, except possibly for alfalfa, where increased unsaturation at low temperature correlates with hardening ability of the cultivars, the evidence for major involvement of increased fatty acid unsaturation in frost hardening at low temperature is much less strong than it appeared only a few years ago. Increased unsaturation is not a prerequisite to frost hardening in winter wheat and in extremely hardy trees such as the black locust and poplar, and it is apparently not required for chill hardening of *Phaseolus vulgaris*.

D.   Changes in Sterol Content

The least studied aspect of lipid involvement in frost hardening is the fate of sterols at low temperature. Davis and Finkner (1) showed little effect of temperature on total sterol profile of one month old wheat plants. Temperature had no influence on the proportions of sterol lipid classes in potato tuber (8). With a wheat embryo total membrane fraction, de la Roche (4) observed little change in the sterol to phospholipid ratio, but a significant decrease in the ratio of campesterol to sitosterol at low temperature (Table X). This ratio decreased to the same extent in 4 cultivars differing in frost hardening potential. In the root and crown of young winter wheat plants of the hardy cultivar Kharkov, little change was shown during hardening at low temperature in total sterol profile, proportions of sterol containing lipid classes, sterol profile within each class, and lipid P to sterol ratio (Willemot, unpublished). The few data available give little evidence for a significant modification of sterol composition under low temperature conditions.

III.   DISCUSSION

A.   Involvement of Lipids in Frost Hardening

The evidence for involvement of lipids in frost hardening of herbaceous species is much less strong than it appeared initially. Only in alfalfa are the observed facts (increase in lipid P, in PC and PE, and in degree of unsaturation of the fatty acids) correlated with hardening potential of cultivars. Greater ability to synthesize phospholipids at low temperature was shown in the hardy cultivars of wheat and alfalfa. This

TABLE V.   Major Fatty Acid Composition of the Total Membranes from Excised Embryos of Seedlings Germinated and Grown at 2 and 24°C for 5 Weeks and 50 Hours, Respectively*

Cultivar	Growth temperature (°C)	LT_{50} (°C)	Fatty acid composition (mole %)			
			16:0	18:1	18:2	18:3
Kharkov	2	-18	21.4	4.9	30.9	42.0
	24	-2	24.6	7.4	47.5	19.8
Rideau	2	-13	21.5	4.7	33.3	39.2
	24	-2	24.6	6.8	49.8	18.3
Capelle-Desprez	2	-6	22.8	4.6	32.1	39.8
	24	-1	24.7	5.5	48.8	20.2
Marquis	2	-5	22.0	4.5	32.3	40.5
	24	-2	22.8	7.2	52.0	17.4

*From de la Roche et al. (7).

lack of firm evidence does not necessarily mean that lipids are not involved. But further evidence will have to include, amongst others, comparisons between hardy and tender cultivars, the demonstration that specific inhibitors of lipid transformations at low temperature inhibit simultaneously frost hardening, and the biophysical demonstration that lipids fulfil their proposed role in hardening.

B.   Role of Increased Fatty Acid Unsaturation at Low Temperature

It is difficult to accept that the generally observed increase in fatty acid unsaturation at low temperature would not be beneficial to the plants. Several authors have suggested that even if lipid changes were not involved in frost hardening, they would help the plant to function at low temperature. These two events appear however undissociable: frost hardening at low temperature presupposes that membranes function satisfactorily at low temperature, and factors which favor membrane function at low temperature should therefore improve frost hardening.

Increased unsaturation of fatty acids at low temperature is not a prerequisite to frost hardening in winter wheat. Is it then without a role? Is it simply the result of increased O_2 solubility at low temperature (14)? Is the excellent correlation obtained with a cultivar of winter wheat between levels of frost tolerance and of linolenic acid the result of dependence on common factors, low temperature and light, without causal relationship (Willemot, unpublished)?

TABLE X.   *Sterol Composition of Total Membrane from Four Cultivars of Wheat Differing in Freezing Resistance*

Cultivar	Growth temperature (^{o}C)	Sterol/ phospholipids	Sterol composition (mole %) Campesterol	Sitosterol
Kharkov	2	.14	21.0	79.0
	24	.14	32.2	67.8
Rideau	2	.17	23.6	76.4
	24	.14	29.9	70.1
Cappelle- Desprez	2	.16	23.4	76.6
	24	.16	33.5	64.5
Marquis	2	.16	23.8	76.2
	24	.15	32.4	67.6

*From de la Roche (4).

Another interpretation of the data, which would recognize a role for the increased unsaturation of fatty acids at low temperature, without making it a prerequisite to frost hardening, is that when it does not occur at low temperature, the plant uses alternate means to maintain a functional level of membrane fluidity. This could be achieved by a negative feedback control, as shown for *Tetrahymena pyriformis* by Martin *et al.* (20). If this were the case, one should be able to demonstrate better control of membrane fluidity in frost tolerant than in tender cultivars at low temperature. The existence of alternate mechanisms would complicate the demonstration of their involvement in frost hardening because none of these mechanisms would be a prerequisite to hardening.

Interpretation of the data is further complicated by the fact that the biomembrane continuum may be broken up into a number of laterally separated liquid crystalline domains at equilibrium (15). This equilibrium would shift with temperature. Analysis of individual membrane domains may be required for the biophysical interpretation of changes in composition.

C. Conclusion

In conclusion, the available data do not show a simple relationship between lipid composition of membranes and frost tolerance of herbaceous plants. They should not be interpreted as evidence that membrane lipids are not involved, but rather that the relationship is a complex one, involving possibly alternate mechanisms and differentiated membrane domains, and where non-lipid, e.g. protein, involvement may be more important than anticipated.

IV. REFERENCES

1. Davis, D. L., and Finkner, V. C. *Plant Physiol. 52,* 324-326 (1972).
2. de la Roche, I. A. *Acta Hort. 81,* 85-89 (1978).
3. de la Roche, I. A. *Plant Physiol. 63,* 5-8 (1979a).
4. de la Roche, I. A. *In* "Comparative Mechanisms of Cold Adaption in the Arctic." (Symposium, 28th Annual Meeting, A.I.B.S.) (L. S. Underwood, Ed.) Acad. Press, N.Y. (In press) (1979b).
5. de la Roche, I. A., Andrews, C. J., and Pomeroy, M. K. *Can. J. Bot. 50,* 2401-2409 (1972).
6. de la Roche, I. A., Andrews, C. J., and Kates, M. *Plant Physiol. 51,* 468-473 (1973).
7. de la Roche, I. A., Pomeroy, M. K., and Andrews, C. J. *Cryobiol. 12,* 506-512 (1975).
8. Galliard, T. , Berkeley, H. D., and Matthew, J. A. *J. Sci. Fd. Agric. 26,* 1163-1170 (1975).
9. Gerloff, E. D., Richardson, T., and Stahmann, M. A. *Plant Physiol. 41,* 1280-1284 (1966).
10. Grenier, G., and Willemot, C. *Cryobiol. 11,* 324-331 (1974).
11. Grenier, G., and Willemot, C. *Can. J. Bot. 53,* 1473-1477 (1975).
12. Grenier, G., Tremolieres, A., Therrien, H. P., and Willemot, C. *Can. J. Bot. 50,* 1681-1689 (1972).
13. Grenier, G., Hope, H. J., Willemot, C., and Therrien, H. P. *Plant Physiol. 55,* 906-912 (1975).
14. Harris, P., and James, A. T. *Biochem. J. 112,* 325-330 (1969).
15. Jain, M. K., and White, H. B. *In* "Advances in Lipid Research" (R. Paoletti and D. Kritchevsky, eds.), Vol. XV, pp. 1-60. Acad. Press, N.Y. (1977).
16. Ketchie, D. O. *Proc. Am. Soc. Hort. Sci. 88,* 204-207 (1966).
17. Lyons, J. M. *Cryobiol. 9,* 341-350 (1972).
18. Lyons, J. M. *Ann. Rev. Plant Physiol. 24,* 445-466 (1973).
19. Lyons, J. M., Wheaton, T. A., and Pratt, H. K. *Plant Physiol. 39,* 262-268 (1964).
20. Martin, C. E., Hiramitsu, K., Kitajima, Y., Nozawa, Y., Skriver, L., and Thompson, G. A. Jr. *Biochem. 15,* 5218-5227 (1976).
21. Paulsen, G. M. *Crop Science 8,* 29-32 (1968).

22. Pomeroy, M. K., and Siminovitch, D. *Can. J. Bot. 49,* 787-795 (1971).
23. Redshaw, E. S., and Zalik, S. *Can. J. Biochem. 46,* 1093-1097 (1968).
24. Siminovitch, D., Singh, J., and de la Roche, I. A. *Cryobiol. 12,* 144-153 (1975).
25. Singh, J., de la Roche, I. A., and Siminovitch, D. *Nature 257,* 669-670 (1975).
26. Smolenska, G. S., and Kuiper, P. J. C. *Physiol. Plant. 41,* 29-35 (1977).
27. St. John, J. B., and Christiansen, M. N. *Plant Physiol. 57,* 257-259 (1976).
28. Thomson, L. W., and Zalik, S. *Plant Physiol. 52,* 268-273 (1973).
29. Willemot, C. *Plant Physiol. 55,* 356-359 (1975).
30. Willemot, C. *Plant Physiol. 60,* 1-4 (1977).
31. Willemot, C., and Pelletier, L. *Can. J. Plant Sci.* (In Press) (1979).
32. Willemot, C., Hope, H. J., Pelletier, L., Langlois, J., and Michaud, R. *Can. J. Plant Sci. 57,* 555-561 (1977a).
33. Willemot, C., Hope, H. J., Williams, R. J., and Michaud, R. *Cryobiol. 14* 87-93 (1977b).
34. Willemot, C., Hope, H. J., and St-Pierre, J. C. *Can. J. Plant Sci. 59,* 249-251 (1979).
35. Wilson, J. M. *New Phytol. 76,* 257-270 (1976).
36. Yoshida, S. *Contr. Inst. Low. Temp. Sci. Ser. B11,* 1-40 (1974).

CHILL-INDUCED CHANGES
IN ORGANELLE MEMBRANE FATTY ACIDS

E. J. Bartkowski

Celpril Industries
Manteca, California

T. R. Peoples and F. R. H. Katterman

Department of Plant Sciences
University of Arizona
Tucson, Arizona

Symptoms of chilling injury vary among plant and tissue types when they are exposed to various periods of low nonfreezing temperatures between 0 and 15C. Cultivated cotton *(Gossypium hirsutum* and *G. barbadense)* is one such thermally sensitive plant especially during seed germination and seedling emergence. Christiansen (3) found that chilling at the initiation of cottonseed hydration resulted in meristematic radicle abortion and a growth lag. Chilling temperatures for a short time after the onset of germination also caused a reduction in growth rate followed by an incactivation of the radicle cortex due to cellular collapse or disintegration. Thus from the work of Christiansen (3, 4) on short staple cotton *(G. hirsutum)* and results from recent work (2) on long staple cotton *(G. barbadense)* two periods of chilling sensitivity were described for cultivated cotton. The first period was estimated to be at the beginning of germination (1 to 6 hrs) whilst the second period was from 28 to 32 hours later.

Since the radicle tip appears to be the most vulnerable site of chilling damage, our next area of investigation was to study the effect of chilling on subcellular organelle membranes with respect to a change in unsaturated fatty acid composition.

I. MATERIALS AND METHODS

Cottonseed *(G. barbadense)* were germinated for 42 hrs at 34C and served as controls. A second lot was germinated at the same temperature for 30 hrs then 6 hrs at 5C and for a concluding 6 hrs at 34C

431

to give a total germination time of 42 hrs. Ten mm sections of radicle tips from both control and chill treated plants were excised and ground in a standard tris-sucrose buffer at 4C in a chilled mortar and pestle. Nuclear, mitochondrial, and microsomal enriched fractions were obtained by differential centrifugation techniques as described elsewhere (12). After extraction of the organelles with a 2:1 chloroform-methanol mixture, the organic extract was further purified by the Folch wash procedure (9) and then fractionated into polar and neutral classes by silicic acid chromatography (5). The fatty acid components of the polar lipid fraction were methylated and assayed quantitatively by means of gas liquid chromatography (1). The total double bond index (D.B.I.) for both control and chill treated organelles was determined by Eq. (1), summed over all 16 and 18 carbon fatty acids.

$$\text{Eq. (1)} \quad \frac{\Sigma \ (\# \text{ double bonds x mole \%})}{100} = \text{D.B.I.}$$

II   RESULTS AND DISCUSSION

Results from Table I show that the major increase in DBI due to chilling temperature occurs in the microsomal membranes. The chilled mitochondrial membranes, however, show a general decrease in unsaturated fatty acid content mainly oleic and linoleic acids, as shown by a lower DBI. The nuclear membranes exhibit essentially no change in fatty acid composition when exposed to chilling temperatures. Additionally, the data from Table I indicates that the greatest change in unsaturated fatty acid level during radicle tip chilling occurs with the linoleic and linolenic acids of the microsomal membranes. Note that there is a decrease in linoleic acid along with a corresponding increase in linolenic acid. The latter is considered to be formed by stepwise desaturation of stearic acid (10). It has been suggested that the higher levels of linolenic acid observed at lower temperatures is attributable to increased amounts of dissolved oxygen, a required cofactor, at a fixed desaturase activity (10). Assuming that oxygen is not limiting in this study, increased linolenic acid levels may be explained by differences in desaturase activity as a function of temperature. The increased levels of linolenic acid in this study are also compatible with previous work on total lipids isolated from chilled wheat and rye radicles (8). In our study, however, the effect of chilling was localized in the microsomal membranes.

The significant decrease in chilled mitochondrial membrane DBI puts forth (Table I) the possibility that phospholipid transfer from the microsomal to the mitochondrial membranes by the phospholipid exchange proteins (PEP) (15) is diminished by the lower temperature. If indeed the PEP were not affected by exposure to chilling temperature, we would expect to find the microsomal and mitochondrial membranes to have nearly similar membrane DBI's due to their mutual interaction (15).

TABLE I. The Effect of Chilling Temperatures on Mole Percent Fatty Acid Composition of Microsomal, Mitochondrial and Nuclear Membranes Isolated from Gossypium barbadense Radicle Tips

Fatty Acid	Microsomal		Mitochondrial		Nuclear	
	Nonchill	Chill	Nonchill	Chill	Nonchill	Chill
	----mole %----		----mole %----		----mole %----	
16:0	15.05	14.95	30.40	33.10	25.24	26.00
16:1	2.47	4.73	4.75	8.26	0.77	1.27
16:2	7.27	1.29	1.61	4.20	2.23	1.79
18:0	18.56	16.04	19.14	12.72	8.80	10.25
18:1	15.51	22.36	16.79	11.52	16.86	15.98
18:2	19.09	9.89	20.17	13.52	40.02	36.64
18:3	5.54	18.27	2.46	1.79	1.53	2.40
Double Bond Index	0.88	1.04	0.73	0.61	1.06	1.02

With a decreased transport of polar lipids to normally replace degraded mitochondrial membranes, it is possible that the normal process of β-oxidation of the unsaturated fatty acids from the degraded membrane fragments would then become more evident as noted by the reduction of oleic and linoleic acids in chilled mitochondrial membranes (Table I).

TABLE II. The Double Bond Index (DBI) and Specific Activity of DNA Isolated from Nuclei of a Chill-Resistant (#189) and Chill-Sensitive (#200) Seed Lot of Pima S-4 Cotton (Gossypium barbadense L.)

Pima S-4	DNA Specific Activity	DBI
	--------cpm/mg--------	
Lot #189	582b*	1.01 n.s.
Lot #200	417a	1.05

*Means followed by different letters are significantly different at the 5% level according to the Student Newman Keuls Test.

While fatty acid analysis and DBI's are useful for interpreting possible mechanism of chilling injury, they are not reliable for accurately predicting susceptibility or resistance to chilling temperatures (14, 16). This fact is emphasized by the data shown in Table II wherein a comparison is made between the DNA polymerase activity and DBI of nuclei from *G. barbadense* Pima S-4 lots that exhibit relative degrees of chilling sensitivity (1). As can be noted, there is no difference in DBI between the chill sensitive lot 200 and the chill resistant lot 189. However, there is a significant difference between the DNA polymerase activities of the two lots. Previous work from our laboratory (6, 7) has shown that DNA replication in *G. barbadense* radicle tips is dependent upon attachment to various sites of the nuclear membranes and that the level of DNA replication is affected by chilling temperatures. Use of the latter observation enabled us to predict a relative hierarchy of chilling resistance in given lots of *G. barbadense* (7). As previously suggested by Patterson, *et al.*, (14) chilling sensitivity may be related more efficiently to a physical change in the membrane structure during the period of chilling. The latter event can then be noted in a proportionate change of a characteristic biochemical reaction (i.e. respiration) as affected by a change in mitochondrial membranes (11), a difference in DNA replication rates as noted above, and a change in the Hill activity of chloroplasts (13).

III. SUMMARY

Although further experimentation is required, we feel that these data strongly suggest that the microsomal membrane is the focal point of control of chilling resistance in plants. The ability of the microsome to desaturate fatty acids and efficiently transfer these unsaturated fatty acids to other membranes may be the rate limiting step in chilling resistance in higher plants and some lower organisms such as *Tetrahymena* as reported by Thompson (this volume).

IV. REFERENCES

1. Bartkowski, E. J., Buxton, D. R., Katterman, F. R. H., and Kircher, H. W. *Agron. J. 69,* 37 (1977).
2. Buxton, D. R., Sprenger, P. J., and Peglow, E. J., Jr. *Crop Sci. 16,* 471 (1976).
3. Christiansen, M. N. *Plant Physiol. 38,* 520 (1963).
4. Christiansen, M. N. *Plant Physiol. 42,* 431 (1967).
5. Christie, W. W., Noble, R. C., and Moore, J. H. *Analyst. 95,* 940 (1970).
6. Clay, W. F., Katterman, F. R. H., and Bartels, P. G. *Proc.   Nat. Acad. Sci. 74,* 3134 (1975).

7. Clay, W. F., Buxton, D. R., and Katterman, F. R. H. *Crop Sci.* *17,* 342 (1977).
8. Farkas, T., Deri-Hadlaczky, and Bela, A. *Lipids 10,* 331 (1975).
9. Folch, J., Lees, M., and Sloane-Stanley, G. H. *J. Biol. Chem. 226* 497 (1957).
10. Harris, P., and James, A. T. *Biochem. Biophys. Acta 187,* 12 (1969).
11. Lyons, J. M. and Raison, J. K. *Plant Physiol. 45,* 386 (1970).
12. Mascarenhas, J. P., Laties, G. G., and Cherry, J. H. *Meth.   Enzymol., 31:* 588–589 (S. Fleischer & L. Packer Eds.) (1974).
13. Nolan, W. G. and Smillie, R. M. *Plant Physiol. 59,* 1141 (1977).
14. Patterson, B. D., Kenrick, J. R., and Raison, J. K. *Phytochem. 17,* 1089 (1978).
15. Wirtz, K. W. A. *Biochem. Biophys. Acta 344,* 95 (1974).
16. Yamaki, S. and Uritani, I. *Agric. Biol. Chem, 36* 47 (1972).

DIFFUSION RELATIONSHIPS BETWEEN CELLULAR PLASMA MEMBRANE AND CYTOPLASM

Alec D. Keith, Andrea Mastro and Wallace Snipes

Department of Biochemistry & Biophysics
The Pennsylvania State University
University Park, Pennsylvania

The spin label approach has been successfully used by several researchers during the past decade to characterize structural and dynamic properties of model and biological membranes (3, 4, 6, 7). On a comparative scale knowledge of the cytoplasm relating to structural and dynamic properties have developed more slowly. The present treatment concentrates on the approach and early phases of using spin labels to characterize the diffusional environment provided by cellular cytoplasmic spaces. This approach may be very useful in comparing chilling sensitive and chilling resistant strains of agricultural crop plants. It has been previously reported that cytoplasmic streaming may be differentially modified in the two types of strains as a consequence of being treated with low temperatures.

Typical spin label experiments allow four measurements from the spin label signal. These four are, line height (h), line width (W), hyperfine coupling constant (A or T for the isotropic component and tensor terms) and the g-value (g for the isotropic or tensor terms). Sometimes partitioning between hydrocarbon rich zones and water rich zones yields a two component signal. Figure 1 shows most of the direct measurements that can be made on spin label signals including the spectral types that occur from ^{15}N.

I. ROTATIONAL MOTION RANGE

Figure 2 shows the line shapes of spin label signals resulting from solvent viscosity changes. In the fast tumbling range, in an isotropic medium these measurements can be taken relatively accurately. As the viscosity increases and spin label rotational motion slows, the individual Nitrogen hyperfine lines change in shape. In the fast tumbling range (the motion range where the three Nitrogen hyperfine lines have a Lorentzian

A. D. Keith *et al.*

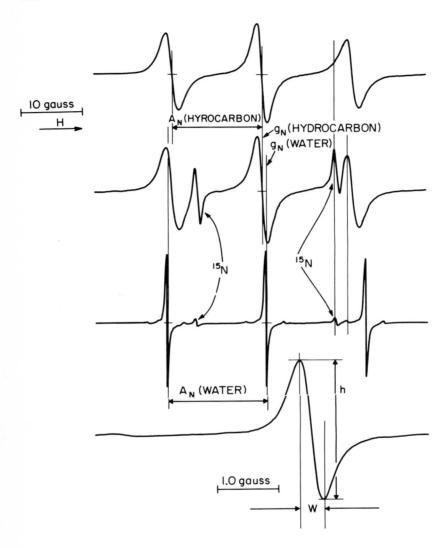

FIGURE 1.   Spin label spectra.   The top spectrum is that of
TEMPONE dissolved in octadecane.   The second spectrum is that of
TEMPONE dissolved in octadecane with a small capillary of ^{15}N
TEMPONE dissolved in water inserted into the octadecane sample. The
third spectrum is that of ^{2}H TEMPONE dissolved in water containing a
small capillary of ^{15}N TEMPONE dissolved in water inserted into the
^{2}H TEMPONE sample. The fourth signal is the midline of the ^{2}H-
TEMPONE signal amplified 10 times on the X axis to illustrate the
method of measurement of linewidths (W) and lineheights (h).

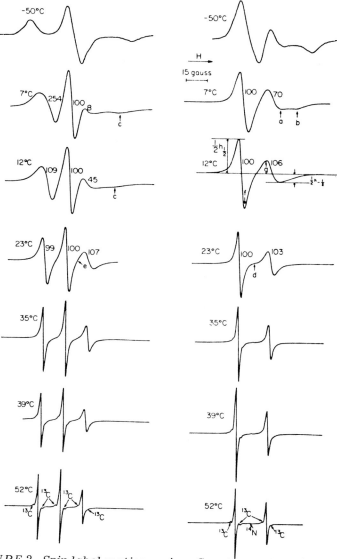

FIGURE 2. Spin label motion series. Spectra on the left are for ^{14}N-TEMPONE in glycerol and on the right are for ^{15}N-TEMPONE in glycerol. The numbers beside the spectral lines at 7°, 12° and $23^{\circ}C$ represent normalized values of the integrated intensities of different spectral lines. Values are normalized to 100. The small letters (a), (b), and (c) show asymmetry and non-averaged tensor contributions due to loss of motional averaging. The small letters (d) and (e) show comparative separation between two spectral lines of ^{14}N and ^{15}N-TEMPONE at the same state of motion. Small letters (f) and (g) point out how half lines may be used to measure the degree of line symmetry.

line shape between inflection points and the integrated intensity (I) of the three lines are equal, e.g. $I_1 = I_0 = I_{-1}$). The integrated intensity can be expressed as $I = kW^2h$ for a first derivative spectral line. As solvent viscosity increases the three Nitrogen hyperfine lines broaden differentially. At x-band microwave frequency, the order of line broadening is $W_{-1} > W_1 > W_0$. Because of the relation expressed above these lines shape changes are most readily seen as line height changes. One of the conventional rotational correlation time (τ_c) calculations,

$$\tau_c = KW_0 \left[\frac{ho}{h_{-1}} \right]^{1/2} - 1, \qquad (1)$$

uses both W and h measurements. Line height changes are more readily seen because,

$$h = \frac{I}{KW^2}, \qquad (2)$$

or simply that h is proportionate to $1/W^2$. As shown in Figure 2, as the viscosity increases a limiting condition arises. In spectra reflecting the intermediate motion range, the high field line (-1) loses integrated intensity before either the low field (1) or mid field lines (0). The numbers shown in Figure 2 show examples of integrated intensity of different resonant lines. The loss of integrated intensity by the high field line results because of incomplete time-dependent averaging of the hyperfine and g-value tensor elements. Measurements taken for the τ_c equation shown above or for other τ_c equations which depend on h and/or W measurements, whether regarded as numerically valid or empirical, result in inaccurate calculated values. For the condition where $I_{-1} > I_0$ the resulting τ values are abnormally increased. The exact motional state where this condition arises has some variability depending on the spin label used and the degree of heterogenerity in the local environments provided by the sample. An Arrhenius plot generated from the spectra shown in Figure 2 is shown in Figure 3. Glycerol under the conditions reported here has no known or reported phase change in the temperature range where the Arrhenius plot shows a "break." The ^{15}N-TEMPONE spectra and τ_c values are included to illustrate that the apparent "break" point in Figure 2 is artificial. This break point occurs because the high field line of ^{14}N-TEMPONE loses integrated intensity before the mid-field line does.

II.   TRANSLATIONAL DIFFUSION MOTION RANGE

Several excellent treatments for various aspects of Heisenberg electron-electron spin exchange have been published (1, 2). The phenomenon of electron spin exchange (ω_{ex}) causes uniform broadening

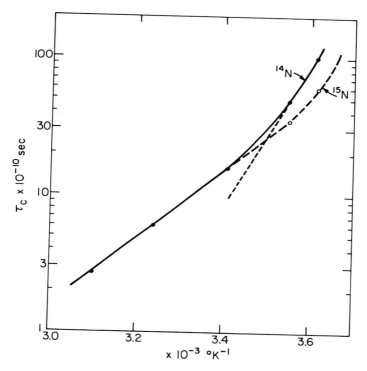

FIGURE 3. Arrhenius plot for ^{14}N-TEMPONE data from Figure 2. Closed line is for ^{14}N-TEMPONE and the dotted line is for ^{15}N-TEMPONE. The curve or break comes about because of loss of integrated intensity of the high field line in both cases.

of all three Nitrogen hyperfine lines as the collision frequency between spin labels or a spin label and some other spin-bearing species increases. Figure 4 shows a spectral series where the spin label concentration in aqueous medium is increased over a concentration range. The concentration range used shows the onset of line broadening, the gradual increase in line width as the concentration dependent line broadening increases, and finally the end result of the three Nitrogen hyperfine lines merging into a single electron spin exchanged narrowed line. The same concentration is also shown with ^{15}N-TEMPONE to illustrate that line shape changes smoothly with no central line appearing until the ^{15}N-hyperfine lines merge together. Figure 5 shows a plot of concentration-dependent line width (ΔH) against the spin label concentration (M). This plot shows the limits of accurate measurement. At the point in the plot

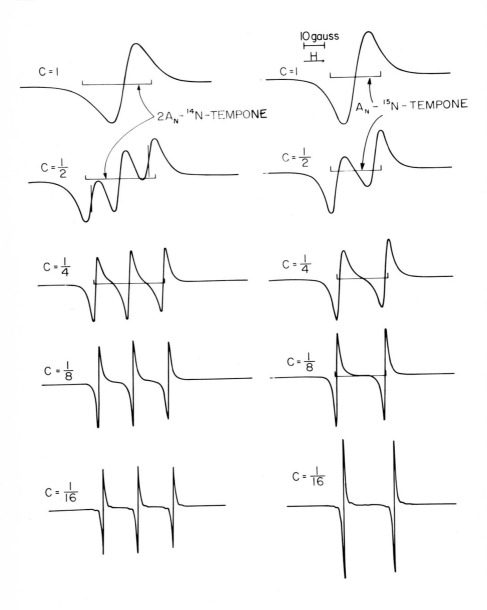

FIGURE 4.   Concentration dependent changes of spin label linewidths. ^{14}N on the left and ^{15}N-TEMPONE on the right. The general pattern of electron-electron spin exchange line broadening and finally spectral narrowing is shown.

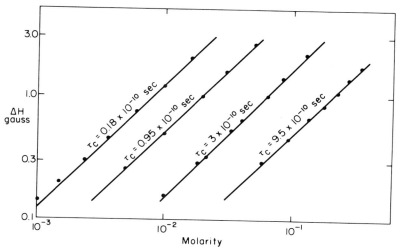

FIGURE 5A. Concentration dependence of line broadening at four
different viscosities. The four lines reading from left to right are:
water, 65% aqueous glycerol, 75% aqueous glycerol, and 85% aqueous
glycerol. τ_c values for each line are shown. The ΔH values for spectra
used for τ_c measurements were less than 0.1G.

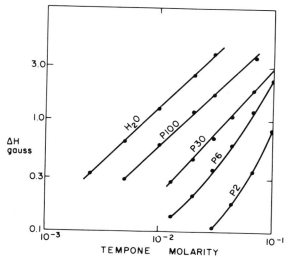

FIGURE 5B. Concentration dependence for line broadening in
confining volume. The P-notation refers to the minimum molecular
weight in thousands excluded from that bead size. All beads were 100
mesh.

where the ratio $\Delta H/M$, falls below 0.007 the values can no longer be accurately measured. The value

$$\frac{\Delta H}{A_N} = \text{approximately } 0.2, \tag{3}$$

is a numerical approximation of the upper limit of accurate line width measurements. Values greater than about 0.2 have too much overlap between adjacent ^{14}N-hyperfine lines.

The general relation between the diffusion constant (D), concentration dependent line broadening (ΔH) and electron-electron spin exchange (ω_{ex}) have been established and are shown below (5).

$$\omega_{ex} = 2 \ \Delta H \tag{4}$$

$$D = \frac{K\omega_{ex}}{M} \tag{5}$$

These relations hold for isotropic media, low viscosity solvents, and isotropic spin labels.

Other paramagnetic species such as the 3d transition series of metals and their chelates can be used to effectively broaden the Nitrogen hyperfine lines of spin labels. These techniques have been developed for use on biological systems using non-toxic paramagnetic spin broadening agents (5).

A. The Effect of Confining Space on Diffusion

Considerations based on the Einstein-Stokes' relation describing diffusional events hold for isotropic media. Considering a small spherical spin label solute, rotational motion and translational diffusion can be described in terms of either. From

$$\tau_c = \frac{4\pi r^3 \eta}{3kT} , \tag{6}$$

and

$$D = \frac{kt}{6\pi r\eta} , \tag{7}$$

diffusion can be expressed in terms of rotation motion,

$$D = \frac{0.22r^2}{\tau_c} \cdot \qquad (8)$$

This relation holds for molecular environments which are isotropic and do not contain barriers to diffusion. For cytoplasm, or polymer formed gels where "rigid" barriers are dispersed in an aqueous medium small molecular solutes may experience greater restriction to diffusion than to molecular rotation. For such cases the effective diffusion constant measured with spin labels (D_{SL}) is distance dependent. The distance-dependency for D_{SL} depends on the spacing, size, shape, and concentration of the polymer field. For example, diffusion within the spaces provided by the polymer network is more rapid than diffusion between different contained-spaces. Therefore, diffusion values arrived at by measuring the collision frequency (ω_{ex}) of a spin label probe will be different depending on whether most of the collisional events occur between molecules residing in the same confined space or between molecules residing in adjacent confined spaces. The mean free path (ρ) between molecular collisions is given by

$$\rho = \frac{\rho_{sol} \; M_{sol} \; MW_{sol}}{\rho_{SL} \; M_{SL} \; MW_{SL}} \; S_{SL}^{2/3} \qquad (9)$$

where MW is molecular weight of the solvent (sol), and spin label (SL), M is molarity, ρ is buoyant density and S is the cubic lattice spacing. The cubic lattice spacing (S) varies with ($M^{1/3}$) by the expression,

$$S = \frac{(10^{27} \overset{o}{A}{}^3/N)^{1/3}}{M^{1/3}} \; , \qquad (10)$$

Where N is Avagadro's number. A characterization of spin label collision frequency in isotropic media led to the equation for diffusion (5).

$$D_{SL} = K\frac{\omega_{ex}}{M_{SL}} \; , \qquad (11)$$

Where K is a proportionality constant arrived at by using the measured value of D obtained from a ^{3}H-spin label employing a capillary method

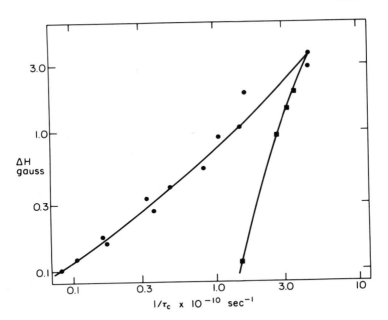

FIGURE 6. *The relation between translational diffusion (ΔH) where $D = K\Delta H/2M$ and rotational motion as a rate process ($1/\tau_c$). Solid circles represent relatively isotropic media (glycerol-water). Solid squares represent ^{14}N-TEMPONE in polyacrylamide beads of different pore sizes.*

(5). The value D_{SL} is equivalent to D in isotropic media; however, in heterogeneous media, D_{SL} becomes smaller with decreasing spin label concentration. Therefore, a plot of D_{SL} vs M_{SL} gives information relating to the structure of the diffusion environment.

The data shown in Figure 6 adds additional clarity to this consideration. Plotting D_{SL} (ΔH is proportionate to D_{SL}) against $1/\tau_c$ reveals, that in isotropic media, the rate of molecular rotation ($1/\tau_c$) varies smoothly with D_{SL}; however, in a confining space environment, that D_{SL} is much more affected than rotational motion. Equations (11) and (12) show the relation between $\Delta H\mu_{ex}$ and D_{SL}.

$$\Delta H = 2\omega_{ex} \qquad (12)$$

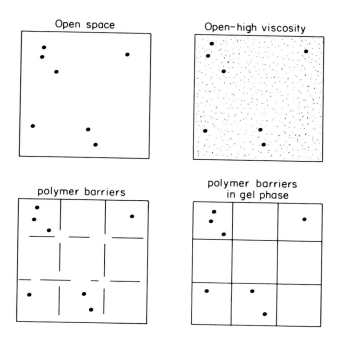

FIGURE 7. Particle distribution in space. The relation between relatively open and confined space is illustrated.

Figure 7 illustrates different degrees of confined space. The upper left of Figure 7 shows one example of particles randomly distributed over a unit area. With no barriers, these particles are free to diffuse about. The upper right illustrates the effect of high viscosity solvent. The product $(D_{SL}) \times (\tau_c)$, should be the same for both cases. The lower left introduces incomplete polymer barriers. For this case the product of $(D_{SL}) \times (\tau_c)$ should be decreased. The lower right illustrates diffusion barriers creating relatively isolated spaces. The product $(D_{SL}) \times (\tau_c)$ is least for this case.

Figure 8 illustrates the differential effect of confining space on three different polymers with respect to rotational motion and translational diffusion. All three polymers interact with the P300 beads to further restrict diffusion. It is of interest to notice that BSA, the most spherical solute of the three, has the greatest cooperativity with confining space, resulting in a large decrease in D_{SL}.

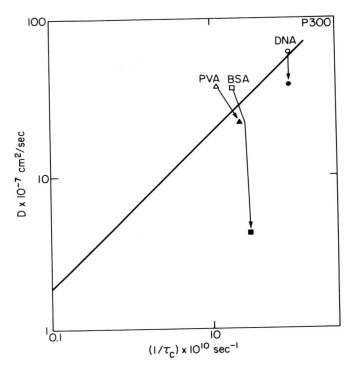

FIGURE 8. *Polymer and confining space effect on diffusion. The solid line shows the ideal relation between D and (1/τ$_c$). The open symbols show values for ^{14}N-TEMPONE in preparations of the polymer shown. PVA, is 115,000 molecular weight polyvinyl alcohol at 10% in water. BSA is 10% Bovine serum albumin in water. DNA, is 4 mg/ml of high molecular weight calf thymus DNA in phosphate buffered saline (standard PBS solution). The closed symbols represent the same preparations with P300 polyacrylamide beads added to the aqueous medium and centrifuged in a capillary to insure that all spin label is inside the beads.*

Figure 9 illustrates the effect of different concentrations of sucrose on molecular motion in free and confined space. The straight reference line shows the ideal relationship between $1/τ_c$ and D for perfectly isotropic media. The open circles illustrate that increasing concentrations of sucrose increases the product (D_{SL}) x $(τ_c)$. The closed circles illustrate that confining space further limits D_{SL} while essentially having no effect on $τ_c$.

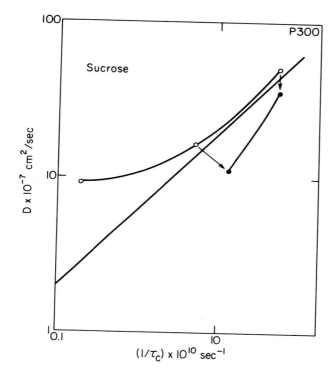

FIGURE 9. Sucrose effect on diffusion. Three concentrations of sucrose are shown. The left most open circle is 70% sucrose (W/W). The mid open and closed circle is 40% sucrose and the two circles on the right are 10% sucrose. Open circles represent measurements carried out in the medium alone. Closed circles represent the same preparations in the P300 beads.

One way to modify the spacing between polymeric or macromolecular diffusion barriers inside cells is to vary the osmotic properties of the cellular bathing medium. Figure 10 shows a plot where derived D_{SL} values are plotted against the reciprocal of osmotic pressure $(1/\pi)$. This data was taken from human embryonic lung cells using TEMPONE as a spin label. The D_{SL} value decreased by a factor of almost three while τ_c changed only slightly. These data indicate that biopolymers capable of restricting diffusional process are crowded closer together as a population of cells are exposed to hyperosmotic stress. Reducing the volume of cytoplasm in this manner effectively isolates and increases the proportion of isolated spin label molecules and therefore reduces ω_{ex} and consequently D_{SL}.

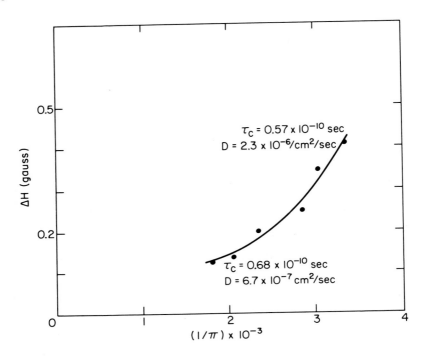

FIGURE 10.   Internal viscosity of cells as a function of osmotic strength.   Human embryonic lung fibroblasts were treated with spin label, NiCl$_2$ at 75 mM, and modulating concentrations of KCl.   The extreme values for mildly hypotonic to hypertonic of both τ_c and   D values are shown.   Under these conditions τ_c values change only slightly while D values change drastically.

Figure 11 shows the modifying effect of Cytocholasin B on BHK cells. Measurements on cells treated with Cytocholasin B (closed circles) show a higher collision frequency for spin labels and consequently the inference is made that barriers to diffusion in the cytoplasm have been reduced as a result of Cytocholasin treatment.

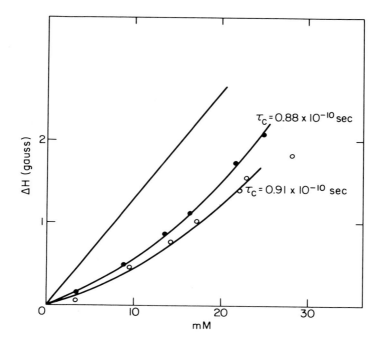

FIGURE 11. Effect of Cytochalasin B on intracellular viscosity. Open circles are BHK cells in growth medium at isotonic concentrations of ions. Closed circles are a split preparation that was treated with Cytochalasin B.

III.   REFERENCES

1.   Abragam, A. "Principles of Nuclear Magnetism" Oxford: Clarendon Press (1961).

2.   Anderson, P. W. *J. Phys. Soc. Japan 9,* 316–339 *(1954).*

3.   Griffith, O. H. and Waggoner, A. S. *Acc. Chem. Res. 2,* 17–24 (1969).

4.   Keith, A., Sharnoff, M. and Cohn, G. *Biochim. Biophys. Acta (300,* 379–419 (Biomem. Rev.) 1973).

5.   Keith, A. D., Snipes, W., Mehlhorn, R. J. and Gunter, T. *Biophys. J., 19,* 205–218 (1977).

6.   McConnell, H. M. *In* "Spin Labeling: Theory and Applications" (L. J. Berliner, ed.). pp. 525–560, Academic Press, New York (1976).

7.   Smith, I. C. P. *In* "Biological Applications of ESR Spectroscopy" (J. R. Bolton and H. Swartz, eds.), pp 483–539, Wiley (Interscience), New York (1972).

TEMPERATURE EFFECTS ON PHOSPHOENOL PYRUVATE CARBOXYLASE FROM CHILLING SENSITIVE AND CHILLING RESISTANT PLANTS

Douglas Graham, Denis G. Hockley and Brian D. Patterson

Plant Physiology Unit
CSIRO Division of Food Research,
and School of Biological Sciences
Macquarie University
North Ryde, N.S.W. 2113, Australia

I.   INTRODUCTION

We have attempted to find enzymes which are affected adversely by chilling temperatures, but which are not membrane-bound or associated with lipids.   PEP-carboxylase may satisfy these criteria.   The temperature coefficient, Q_{10}, or the apparent Arrhenius activation energy, Ea, of this enzyme has been shown to increase below about $10^{\circ}C$ for tropical C4 plants but not for temperate-zone plants (7, 9). Initial work suggested that the detergent Triton X-100 and the enzyme phospholipase C (7) eliminated "break" in activation energy at about $10^{\circ}C$. This would imply an association of the enzyme with lipids; more recent work (13) has shown that, with a highly purified PEP-carboxylase from maize, the break in activation energy was not removed by Triton X-100.

The work quoted refers principally to the enzyme from C4 plants, where it is concerned with photosynthetic CO_2 fixation. In C3 plants, it has an anaplerotic role, replenishing TCA cycle intermediates used for the synthesis of carbon skeletons. This implies an important regulatory role for the enzyme in C3 plants.

We have examined the effect of temperature on PEP-carboxylase from three groups of C3 plants: tropical (chilling-sensitive), temperate and alpine (both chilling resistant).

II.   EXPERIMENTAL

Tomato (*Lycopersicon esculentum*, Mill cv. Rutgers), wheat (*Triticum aestivum* L.), broad bean (*Vicia faba* L.) and purple passionfruit (*Passiflora edulis*, Sims) were grown in a controlled temperature glasshouse (25°C day/18°C night). The Australian alpines, *Cardamine* sp. and *Caltha intraloba* L. collected from 2000m in the Snowy Mountains were maintained in a high light intensity (160 watt m^{2}), controlled environment, growth cabinet (12 h day/12 h night; 15°C/5°C).

Leaf material was extracted at 2°C by grinding in a pestle and mortar with sand and twice the tissue volume of 0.2 M Tris-HCl, pH 7.4, containing 10 mM sodium diethylthiocarbamate (DIECA), 10 mM MgCl$_2$, 2 mM sodium EDTA, 10 mM 2-mercaptoethanol and 1% Polyclar. The extract was filtered through two layers of Miracloth and centrifuged at 12000 g for 30 min. The supernatant was desalted (Sephadex G-25) into 50 mM Bicine pH 7.6, containing 5 mM MgSO$_4$, 1 mM sodium EDTA, and 2 mM DTT. The appropriate eluate was used without further purification for assay of PEP-carboxylase (E.C. 4.1.1.31 activity by following the oxidation of NADH at 340 nm, in a coupled assay with malate dehydrogenase, according to the method of Ting and Osmond (12). Malate dehydrogenase gave linear Arrhenius plots and therefore did not affect our results for PEP carboxylase. Temperature was measured with a fine thermocouple, linked to a recorder. Apparent K$_m$ PEP values were calculated using a GENSTAT computer program based on a method of Cleland (1), involving Lineweaver Burke plots which were linear for the plants tested.

The leaf extracts contained a single PEP-carboxylase as shown by the presence of a single peak of enzymic activity on elution from DEAE Sephacel and a single band of activity after electrophoresis on polyacrylamide gradient gels.

III.   RESULTS

Figure 1 shows an A rhenius plot of the activities at V_{max} for crude preparations of PEP-carboxylases extracted from leaves of the temperate plant, wheat, and the tropical plant, tomato, which both have C3-type photosynthesis. The results for wheat represent a single extract and give an approximately straight line relationship for log enzymic rate versus reciprocal of absolute temperature over the temperature range 1.5° to 38.5°C. The results for tomato are from several experiments in order to verify the non-linear nature of the plot which shows an increasing decline in activity as the temperature decreases below about $10\text{-}12^\circ$C. For the tomato enzyme there is clearly a deviation from the Arrhenius relationship at both low and high (above 30°C) temperatures and the data are fitted well by a curve. The temperature coefficient, Q_{10} for the wheat enzyme is approximately constant at about 2.1

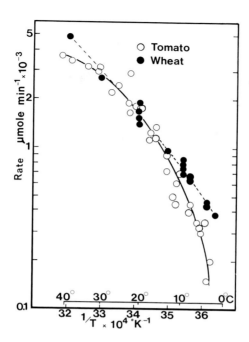

FIGURE 1. Arrhenius plots of wheat and tomato leaf PEP-carboxylase activities from crude preparations after Sephadex G25 treatment. Results have been normalized to 20°C to correct for different extractions of the enzyme. Both were measured at 2 mM PEP. Enzymic rates are μ moles PEP (or NADH oxidized) min⁻¹.

throughout the temperature range whereas the tomato enzyme has a Q_{10} of 1.9 to 2.6 in the temperature range 30° to 10°C. It increases to greater than 5 below 10°C. There is, therefore, a clear distinction in the behaviour of the enzymes from wheat and tomato at low temperatures.

The generality of this finding is illustrated by Fig. 2. The low temperature behaviour of the enzyme from Australian alpine plants (*Cardamine sp.* and *Caltha intraloba*), temperate-zone plants (wheat and broad bean) and tropical plants (tomato and *Passiflora edulis*) is shown. Relative enzymic activity at 1.5°C is plotted as a percentage of that at 20°C. The alpine and temperate species are very similar in their behaviour and give approximately linear Arrhenius plots similar to that of wheat shown in Fig. 1. The tropical species, however, show a relatively much greater decline in enzymic activity at low temperatures.

Relative activity at 1.5$^\cup$ compared to 20°

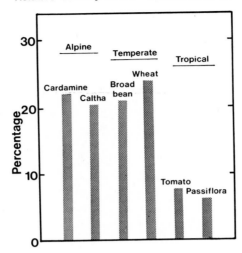

FIGURE 2.   PEP-carboxylase activities from Caltha intraloba, Cardamine sp., Vicia faba (broad bean), wheat, tomato and Passiflora edulis at 1.5°C compared, as a percentage, to the activity at 20°C.

Our work, therefore, confirms that alpine and temperature species can maintain relatively higher rates of PEP-carboxylase activity at low temperatures than can tropical species.

We were interested in the effects of temperature at limiting substrate concentrations since this may better approach the natural metabolic environment of the enzyme. Values for the concentration of PEP in C3 leaves are less than 0.1 mM (5). Accordingly, the enzymes from wheat and tomato were assayed over a wide range of substrate concentrations. In Fig. 3 we show how PEP-carboxylase activity changes with temperature at two different substrate concentrations. For the tropical plant, tomato (Fig. 3a), the activity is reduced much more below 10°C when the substrate concentration is limiting ($\sqrt{}K_m$ PEP) than when it is saturating (>10 times K_m PEP). However, the results in Fig. 3b show that the effect of low temperatures under limiting substrate concentration is not nearly so apparent for the enzyme from wheat, a temperate-zone plant. In all cases initial rates were used in determining the enzymic activities. These observations are amplified by the temperature coefficients, Q_{10}, calculated for the temperature ranges

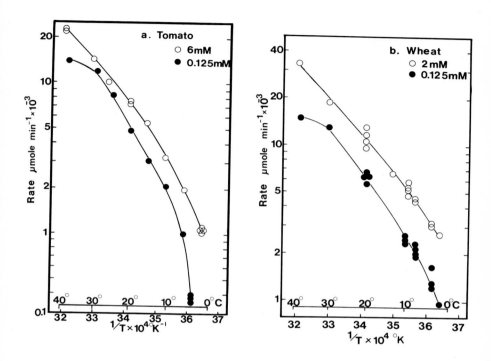

FIGURE 3. Arrhenius plots comparing PEP-carboxylase activities at V_{max} (10 x K_m PEP) and approximately K_m PEP (20°C) for a) tomato, cv. Rutgers, 6 mM and 0.125 mM PEP, and b) wheat, 2 mM and 0.125 mM PEP. Enzymic rates are μmoles PEP (or NADH oxidized) min^{-1}.

30°-20°C and 10°-0°C (Table I) which show a large increase in Q_{10} below 10°C for the tomato enzyme, but only a small change for the wheat enzyme.

We were unable to change the slope of the Arrhenius plots for the tropical species by treatment of the crude preparations with Triton X-100. It is unlikely, therefore, that the marked decline in enzymic activity at chilling temperatures is mediated through lipids associated with the protein. In addition, we could find no convincing evidence for an abrupt increase in activation energy at a definite temperature.

In order to determine whether the enzymic protein undergoes any temperature-induced changes we have examined the kinetic parameter, K_m PEP, which would indicate whether the affinity of the enzyme for one of its substrates is affected by temperature. Changes in K_m are likely to reflect changes in the ordered structure of the protein about the active site. Therefore, apparent K_m PEP was measured at temperatures from 1.3°C to 38.5°C. The apparent K_m PEP values at 1.3° and 20°C for the enzymes from *Caltha* (alpine), wheat (temperate) and tomato (tropical) are compared in Fig. 4. The values for the *Caltha* and wheat enzymes do not change significantly at the lower temperature and remain between 0.1 and 0.2 mM. However, the K_m PEP for the tomato enzyme increases nine-fold from 0.1 mM at 20°C to 0.9 mM at 1.3°C. A minimum value for the tomato enzyme of 0.04 mM occurs at 30°C, representing a 22.5 fold increase at the lower temperature.

TABLE I. *Temperature Coefficients, Q_{10}, of PEP-carboxylase from Wheat and Tomato at Low and High Substrate Concentrations and Temperatures*

		Q_{10} Values	
Plant	*PEP conc.*	30°-20°C	10°-0°C
Wheat	2 mM	1.76	2.10
	0.125 mM	2.09	3.67
Tomato	6 mM	1.89	3.50
	0.125 mM	2.45	>> 6.25

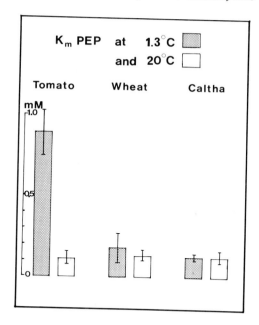

FIGURE 4. Apparent Michaelis constant, K_m PEP, at 1.3° and 20°C for PEP-carboxylases prepared from tomato, cv. Rutgers, wheat and Caltha intraloba. Bars represent 95% confidence limits, derived from statistically analyzed Lineweaver-Burke plots.

We conclude that the active site of the enzyme from this tropical species undergoes changes in its properties with low temperature and that this results in a changed affinity of the enzyme for one of its substrates. This is possibly a direct effect of low temperature on the structure of the protein.

IV. DISCUSSION

The results indicate that the activity of PEP-carboxylase from some tropical (chilling-sensitive) C3 species measured over the temperature range 1.3° - 38.5°C show non-linear Arrhenius plots for V_{max} below about 10°C. Several alpine and temperate (chilling-resistant) species show essentially linear Arrhenius plots for the enzymic activity over the same temperature range. When similar plots are made at low substrate concentrations, around the apparent K_m for PEP, (i.e. at approximately *in vivo* concentrations) the effects of low temperature are accentuated. These results are substantiated by the measurements of apparent K_m PEP for the alpine, temperate and tropical species which show that the tropical species have a substantially increased apparent K_m PEP at low temperatures. It is not suggested that this effect on K_m is wholly responsible for the decline in activity of the enzyme at low temperatures since the enzymic activity was generally measured at V_{max}. However, the finding does indicate that low temperature has a direct effect on the protein, presumably in the vicinity of the active site. Similar effects of temperature on the K_m of several enzymes from animals have been reported (4, 11). In these animals the minimum value for K_m appears to coincide with the normal environmental temperature of the organism and increases outside this range of temperatures.

Possible mechanisms by which low temperature could exert direct effects on the protein are via hydrophobic bonding (4) which is weaker at low temperature, or hydrogen bonding and electrostatic interactions which are stronger at low temperature. Hydrophobic bonding appears to be more important in maintaining enzymic structure (8). Such changes in hydrophobic bonding could influence the kinetic parameters of the enzyme.

Similar effects of low temperature on a number of soluble enzymes such as malate, glutamate, alcohol, glucose-6-phosphate, NADP-isocitrate and glutamate dehydrogenases, from the tropical C3 species, soybean *(Glycine max* (L.) Merr.), have been reported (2). However, it was concluded that the sharp "breaks" in the Arrhenius plots were probably due to lipid association of the enzymes. There is no evidence to support such a conclusion in our studies.

The findings reported in the present paper suggest the possibility that the metabolism of tropical, chilling-sensitive plants may be disrupted by direct effects of low temperature on the anaplerotic enzyme PEP-carboxylase. Preliminary studies in our laboratory indicate that pyruvate metabolism is adversely affected by low temperature in the tropical species studied. While not discounting the possible importance of membrane lipids in accounting for chilling injury in chilling/sensitive species, the present results suggest an alternative, or additional mechanism whereby the plant's metabolism could be disrupted by low temperatures in the chilling range. An imbalance in metabolism could result from decreased activity of PEP-carboxylase, especially since the operation of the TCA cycle as a synthetic pathway is dependent on a continued supply of C4 acids which are principally produced by this enzyme. Such an effect would be alternative, or additional, to the metabolic imbalances which were previously suggested to be a consequence of chilling (6, 10).

V. REFERENCES

1. Cleland. W. W. *Adv. Enzymol. Relat. Areas Mol. Biol. 29,* 1-32 (1967).
2. Duke, S. H., Schrader, L. E., and Miller, M. G. *Plant Physiol. 60,* 716-722 (1977).
3. Hochachka, P. W. *Biochem. J. 143,* 535-539 (1974).
4. Hochachka, P. W., and Somero, G. N. *Comp. Biochem. Physiol. 27,* 649-668 (1968).
5. Latzko, E., Laber, L., and Gibbs, M. *In* "Photosynthesis and Photo-respiration" (M.D. Hatch, C. B Osmond and R. O. Slatyer, eds.), pp. 196-201. John Wiley, New York (1971).
6. Lyons, J. M. *Annu. Rev. Plant Physiol. 24,* 445-466 (1973).
7. McWilliam, J. R., and Ferrar, P. J. *In* "Mechanisms of Regulation of Plant Growth" (R. L. Bieleski, A. R. Ferguson and M. M. Cresswell, eds.) *R. Soc. N.Z. Bull. 12,* 467-476 (1974).
8. Oakenfull, D., and Fenwick, D. E. *Aust. J. Chem. 30,* 741-752 (1977).
9. Phillips, P. J. and McWilliam, J. R. *In* "Photosynthesis and Photo-respiration" (M. D. Hatch, C. B. Osmond and R. O. Slatyer, eds.), pp. 97-104. John Wiley, New York (1971).
10. Raison, J. K. *Symp. Soc. Exp. Biol. 27,* 485-512 (1973).
11. Somero, G. N. *Am. Nat. 103,* 517-530 (1969).
12. Ting, I. P., and Osmond, C. B. *Plant Physiol. 51,* 439-447 (1973).
13. Vedan, K., and Sugiyama, T. *Plant Physiol. 57,* 906-910 (1976).

CELL CULTURE MANIPULATIONS
AS A POTENTIAL BREEDING TOOL

P. J. Dix

Department of Genetics
University of Newcastle upon Tyne
Newcastle upon Tyne
United Kingdom

I. INTRODUCTION

Protagonists of plant tissue and cell culture methods are now commonplace. Applications of these techniques can be found in nearly every field of plant science and they have resulted in many valuable contributions to our knowledge of primary and secondary metabolism, the cell cycle, the regulation of growth and differentiation, and plant/micro-organism interactions. From the agricultural point of view the obvious attractions of tissue culture methods for virus eradication and rapid clonal propagation of commercial varieties have continued to draw a great deal of attention and the number of species for which the basic culture criteria have either been met, or are under intensive study, is now quite extensive (43). Interest in genetic manipulations of plant cell cultures has continued to increase, despite the many obstacles in the way of wide application of these methods. It is therefore expected that those interested in overcoming low temperature stress would also direct some attention to these genetic manipulations.

This review will consider the different methods for genome modification in cell cultures and the problems encountered in the application of these methods. The extent to which these problems have been and can be overcome will be evaluated with particular reference to experiments performed with crop species, and having a potential for crop improvement. In addition the limited progress which has been made in breeding for chilling resistance using tissue cultures will be considered.

II.   THE   CURRENT   STATE   OF   PLANT   CELL   CULTURE TECHNOLOGY

The basic cycle of initiation of undifferentiated cultures, and regeneration of intact plants from them has now been demonstrated for a large number of species and in many of them finely dispersed cell suspension cultures, and protoplast isolation, culture and fusion, have extended the range of methods which can be used to effect genetic modification. In addition anther culture, first used by Guha and Maheshwari (27) has resulted in the availability of haploid material in a far greater range of species than before (50).

It remains true, however, that for a large number of valuable crop species all the essential culture conditions have not been worked out, and for many others they have yet to become sufficiently refined and repeatable as to provide a really useful breeding tool. Certain families, most notably *Solanaceae,* have proved generally more amenable to cell culture manipulations than others. Tree species and cereals have tended to prove recalcitrant although the concentration of effort and resources has begun to yield a measure of success in several cereal species (43). Most commonly the regeneration of plants from cell cultures is the critical stage.

III.   THE USE OF CELL CULTURES IN BREEDING

A.   *Exploitation of Existing Variation*

Anther-derived haploid plants and homozygous diploids derived from them by colchicine treatment (28) provide a means of accelerating the exploitation of existing variation by conventional combination breeding (39). This procedure has already been successfully used to develop new cultivars of tobacco (45, 9) and is certain to be extended to other crop species as the anther culture procedure becomes more widely applicable and reliable.

B.   *Mutation and Selection*

1.   Selection Procedure. Selection in cell cultures is only useful for characters which are likely to be expressed equally well in undifferentiated cells and intact plants. Selection procedures usually involve the exposure of callus, cell suspension, or protoplast, cultures, with or without a prior mutagenesis step, to a suitable selective treatment which kills or inhibits the division of normal cells. Surviving cells give rise to healthy proliferating aggregates which can then be repeatedly exposed to a cycle of selection and regrowth, and plants regenerated from them.

2. Problems

a. Mutagenesis. The number of cells present in a culture make it realistic to select mutants without recourse to a mutagenic treatment, and this was the case with many of the variant lines described here. Effective use of mutagens has been described in some cases (51, 42). It may be wise to restrict the use of mutagens as far as possible when attempting to modify crop plants.

b. Aggregation. Except in the case of protoplasts, cell cultures do not exist predominantly of single cells, but of aggregates of various sizes, and cannot therefore be regarded as microorganisms. Within an aggregate cells may be of different sizes and physiological states, and intercellular connections, together with a variety of gradients across aggregates could interfere with the selection of some kinds of variant.

c. Minimal cell density. Most cell cultures require a minimum cell density for growth which may pose problems when trying to select a few survivors from a large population of dead cells.

d. Chromosomal instability. A well known feature of cell cultures is the chromosomal instability induced by the culture system which over a period gradually gives rise to polyploid or aneuploid cells (48, 11, 1).   This does not occur in all species but appears to be particularly pronounced in haploid cultures. (48). Gross chromosomal changes obviously do not favour the use of cultures in breeding.

e. Loss of morphogenic potential. The potential for the initiation of shoots or embryos in culture is often reduced by an extended culture period, a phenomenon possibly associated with *d.*   This is the most commonly encountered problem in variant selection in cell cultures. The majority of variant cell lines which have been described exist only as cell lines. These two features together mean that expedient use of freshly initiated cultures is likely to remain desirable when crop improvement is the aim.

f. Loss of flowering or fertility. May be further consequences of culture induced incidental genetic changes. Loss of flowering has been found in plants regenerated from a streptomycin resistant cell line of *Nicotiana sylvestris* (37).

g. Epigenetic variation. Phenotypic changes, resulting from causes other than mutation (such as changes in gene expression) can often be found in cell cultures. An example is cycloheximide resistance (35). Some epigenetic variants may be very persistent.

Many of these problems can be avoided, or may not exist for a particular system, and, in spite of them a number of variant lines have been selected in culture. In some of them (38, 33 & 34, 3), sexual transmission has been unequivocally demonstrated. There follows a brief survey of the variant lines which have been selected for characters of potential agricultural interest.

3. *Disease Resistance.* Carlson (5) produced tobacco plants resistant to infection by *Pseudomonas tabaci* by selection for resistance to methionine sulfoximine, an analogue of the toxin, in haploid cell cultures. Gengenbach and Green (22) selected for resistance to the toxin of *Helminthosporium maydis* in Texas male-sterile maize cultures to obtain plants resistant to the pathogen. Certainly the most significant case is the development of sugar cane cultivars resistant to four different pathogens (46).

4. *Herbicide Resistance.* Chaleff and Parsons (6) have demonstrated the sexual transmission, as a dominant allele, of picloram resistance selected in tobacco cultures. Several other cell lines resistant to herbicides have recently been selected (26).

5. *Environmental Stress Resistance*

a. *Low temperature.* Dix and Street (13) obtained cell lines of *Nicotiana sylvestris* and *Capsicum annuum* with enhanced resistance to exposure to $-3^{o}C$ and $5^{o}C$ respectively. Two types of resistant line of *N. sylvestris* differed markedly in their level of resistance. Unfortunately plants could only be regenerated from some of the lines with a lower level of resistance and callus derived from the seedling progeny was sensitive (10). Plants could not be regenerated from resistant or sensitive *C. annuum* cultures. It is concluded that the lower level of chilling resistance in *N. sylvestris* resulted from a fairly stable epigenetic change, persistent through an indefinite number of mitotic divisions in cell culture, but lost during the plants sexual cycle. Alternative explanations for the loss of resistance, such as continual segregation in culture giving rise to chimeral plants cannot be ruled out.

The selection system used in the above work resulted in many lines which survived the first selection, but succumbed to subsequent exposure. This may reflect the physiological spectrum of cells present in the culture simultaneously, and the problems of variations in aggregate size already discussed. These factors may have a greater impact on the selection of this kind of variant than, for example, on selection for resistance to most drugs. This may be the more so since we are looking for survival and growth subsequent to the selection pressure, rather than growth in the presence of a selective agent. Considerable refinement of the selection procedure and a thorough examination of a large number of putative mutants should eventually lead to cultivars with enhanced chilling resistance. It is encouraging that the possibilities are being further investigated using cell cultures of tomato (4) and rice (Xuan and Dix, unpublished).

     b.   *High salinity.* The selection of cell lines with enhanced resistance to growth inhibition by sodium chloride has been described for *Nicotiana sylvestris* (56, 12), *Nicotiana tabacum* (44), *Capsicum annuum* (12), and *Citrus sinensis* (30). In the case of *N. sylvestris* a number of plants have been regenerated from a resistant cell line, and calli initiated from them retain their resistance (Dix, unpublished).

     c.   *Aluminium.* Meredith (41) has reported the selection of cell lines of *Lycopersicon esculentum* with a stable resistance to aluminium toxicity.

     6.   *Amino Acid Overproduction.* A common mechanism for resistance to amino acid analogues is a reduced sensitivity of the feedback control mechanism of the biosynthetic pathway for the appropriate amino acid, resulting in its overproduction. If this could be translated into improved levels of key amino acids in storage organs or seed proteins the potential for crop improvement would be enormous. For this reason more attention has been paid to selection for amino acid analogue resistance in cell cultures than to any other class of mutant, and several recent reviews have covered the subject (54, 55, 32). Many of the analogue resistant lines which have been selected are indeed overproducers of the corresponding amino acids but in none of these has overproduction been shown in regenerated plants, or sexual transmission been demonstrated. S-2-aminoethyl-cysteine (AEC) resistance selected in barley embryos (Bright, personal communication) and methionine sulfoximine resistance selected in tobacco cultures (5) were sexually transmitted as recessive traits, but the mechanism of resistance was not known in these cases.

     The potential of cell cultures for the selection of this kind of mutant has been well illustrated by Widholm (53). Using sequential selection for resistance to four different analogues, he has obtained a carrot cell line simultaneously overproducing lysine, phenylalanine, methionine and tryptophan.

C.   *Somatic Hybridization*

     1.   *Protoplast Isolation, Culture and Fusion.* The main attraction of somatic hybridization lies in the possibility of surmounting the compatibility barrier between species, and the creation of novel hybrids, but protoplasts may also prove the most realistic vector for the transfer of nuclear and cytoplasmic genes from one species to another. The range of species for which the isolation and culture methods have been met, has been covered by several recent reviews (21, 52, 19, 20). Protoplasts can now often be obtained from callus and cell suspension cultures, as well as a wide range of plant organs. Fairly general methods for improving the frequency of fusion between protoplasts have been developed, probably the most popular being the use of polyethylene glycol (PEG) in a method devised by Kao and Michayluk (29).

2. *Hybrid Selection.* The products of protoplast fusion are heterokaryons which may then go on to form hybrids by nuclear fusion. It is then necessary to select out hybrid cells from the large majority of unfused cells, and self-fused cells. Most procedures, well reviewed (7), involve complementation resulting in preferential growth of the hybrids under certain selective conditions. For this the parent species or cell lines must be carefully chosen and mutant cell lines can be particularly useful. Complementation between two non-allelic mutant cell lines has been used, as in nitrate reductase deficient mutants of tobacco (25), but a single mutant line can also be combined with a visually distinct one. For example, Maliga *et al.* (36) fused protoplasts of an albino kanamycin resistant line of *Nicotiana sylvestris* (14) with *N. knightiana* mesophyll protoplasts which divide at a low frequency to give green colonies. Hybrids were selected as green kanamycin resistant colonies.

Where mesophyll protoplasts are fused to those of a cultured cell line, which is normally unpigmented, heterokaryons can often be visually identified and physically separated using a micropipette (40). This may prove a more general procedure for the isolation of hybrids from a mixed population. Confirmation of hybridity is generally sought either by comparison with sexual hybrids (when sexually compatible species have been used), by karyotyping, or by looking for intermediate morphological (17) or biochemical characteristics, such as isoenzyme patterns (36).

3. *Somatic Hybrids*

a. *Intrageneric.* Somatic hybridization within a species or between closely related species could be a very useful breeding tool. Hybridization in species which take a long time to flower could be accelerated, since seedling protoplasts could be used, and there may be incompatibility factors between quite closely related species. In the event of valuable mutant cell lines being obtained, which have lost their capacity for shoot regeneration or flowering, intraspecific, or intrageneric protoplast fusion may provide a means of utilizing the desired phenotype. It may be possible to regenerate fertile plants from the hybrid cells and eliminate chromosome anomalies in subsequent sexual cycles.

b. *Intergeneric.* All indications are that the possibility of producing novel hybrid plants between distantly related species remains remote. The biological barrier involved goes far beyond the physical barrier of the cell wall. There is, however, no special problem in producing heterokaryons and hybrid cells, and the range of viable cell hybrids produced in this way has been recently reviewed by Constabel (8) who also considers the fate of the two sets of chromosomes in the hybrids. Generally hybrid formation is followed, during subsequent divisions, by chromosome elimination, a long familiar feature of animal cell hybrids. In some combinations, such as *Vicia* and *Petunia* (2) and *Arabidopsis* and *Brassica* (24), chromosome elimination seems to be non-specific, chromosomes of either or both parents being lost. In other cases, however, there appears to be specific elimination, of *Petunia* chromosomes from *Parthenocissus* and *Petunia* cells (47) and *Aegopodium* chromosomes from *Daucus* and *Aegopodium* cells (18 and personal communication). This last is of particular interest since plants could be regenerated and although only *Daucus* chromosomes could be detected, certain features of the pigment spectrum of *Aegopodium* were demonstrated. The implication, incorporation of *Aegopodium* genes to *Daucus* chromosomes, could clearly be of greater significance for breeding.

D. Other Methods of Genetic Modification

DNA has been introduced into plant cells and protoplasts in a variety of ways, such as using naked DNA (31), bacteriophage (15), bacteria (23), organelles (23), and animal cells (16). Stable incorporation and expression of exogenous DNA introduced by any of these methods has not been unequivocally demonstrated, and it is difficult to envisage any application to plant breeding in the foreseeable future. More promising is the possibility of transformation using a well characterized plasmid, such as the Ti plasmid of *Agrobacterium tumefaciens,* and this system has been fully reviewed (49).

IV. CONCLUDING REMARKS

To date the increasing use of plant tissue and cell culture methods to achieve agricultural objectives has not been widely extended to the production of new cultivars after genetic modification of cells in culture. Increased experience in tackling the technical difficulties has enhanced the possibility that valuable contributions can be made in this way. It is the view of this author that there are two approaches most likely to be developed as valuable breeding tools used separately or in combination: -
     the direct selection for desirable phenotypes in freshly initiated cultures, followed by plant regeneration, and
     transfer of parts of genomes responsible for desirable traits using intra- or inter-generic protoplast fusion (chromosome elimination may prove an advantage, rather than a problem, for the application of this method).
     With particular reference to the area of interest to this meeting, I know of no serious barrier to the use of these techniques to obtain cultivars of crop species with enhanced resistance to low temperature stress.

V.   REFERENCES

1.   Bayliss, M. W. *Chromosoma 51,* 401-411 (1975).
2.   Binding, H., and Nehls, R. *Molec. gen. Genet. 164,* 137-143 (1978).
3.   Bourgin, J. P. *Molec. gen. Genet. 161,* 225-230 (1978).
4.   Breidenbach, R. W., and Waring, A. J. *Plant Physiol. 60,* 190-192 (1977).
5.   Carlson, P. S. *Science 180,* 1366-1368 (1973).
6.   Chaleff, R. S., and Parsons, M. F. *Proc. Nat. Acad. Sci. 75,* 5104-5107 (1978).
7.   Cocking, E. C. *In* "Frontiers of Plant Tissue Culture 1978" (T. A. Thorpe, ed.), pp. 151-158. IAPTC 1978, Calgary (1978).
8.   Constabel, F. *In* "Frontiers of Plant Tissue Culture 1978" (T. A. Thorpe, ed.), pp. 141-149. IAPTC 1978, Calgary (1978).
9.   Co-Operative Group of Haploid Breeding of Tobacco of Shangtung Institute of Tobacco and Peking Institute of Botany, Academia Sinica. *Acta Bot. Sin. 16* 300-303 (1974).
10.   Dix, P. J. *Z. Pflanzenphysiol. 84,* 223-226 (1977).
11.   Dix, P. J. and Street, H. E. *Plant Sci. Lett. 3,* 283-288 (1974).
12.   Dix, P. J. and Street, H. E. *Plant Sci. Lett. 5,* 231-237 (1975).
13.   Dix, P. J. and Street, H. E. *Ann. Bot. 40,* 903-910 (1976).
14.   Dix, P. J., Joo, F. and Maliga, P. *Molec. gen. Genet. 157,* 285-290 (1977).
15.   Doy, C. H., Gresshoff, P. M. and Rolfe, B. G. *Proc. Nat. Acad. Sci. 70,* 723-726 (1973).
16.   Dudits, D., Rasko, I., Hadlaczky, Gy., and Lima de Faria, A. *Hereditas 82* 121-124 (1976).
17.   Dudits, D., Hadlaczky, G., Levi, E., Fejer, O., Haydu, Zs., and Lazar, G. *Theor. Appl. Genet. 51,* 127-132 (1977).
18.   Dudits, D., Hadlaczky, Gy., Levi, E., Koncz, Cs., Haydu, Zs. and Paal, H. *In* "4th IAPTC Congress, Calgary - Abstracts" pp. 67 (1978).
19.   Eriksson, T. *In* "Plant Tissue Culture and its Bio-technological Applications" (W. Barz, E. Reinhard, M. H. Zenk, eds.), pp. 313-322, Springer-Verlag, Berlin, Heidelberg (1977).
20.   Eriksson, T., Glimelius, K. and Wallin, A. *In* "Frontiers of Plant Tissue Culture 1978" (T. A. Thorpe, ed.), pp. 131-139, IAPTC 1978, Calgary (1978).
21.   Gamborg, O. L. *In* "Cell Genetics in Higher Plants" (D. Dudits, G. L. Farkas, P. Maliga, eds.), pp. 107-127, Akademiai Kiado, Budapest (1976).
22.   Gengenbach, B. G., and Green, C. E. *Crop Sci. 15,* 645-649 (1975).
23.   Giles, K. L. *In* "Frontiers of Plant Tissue Culture 1978" (T. A. Thorpe, ed.) pp. 67-74. IAPTC 1978, Calgary (1978).
24.   Gleba, Y. Y., and Hoffmann, F. *Molec. gen. Genet. 165,* 257-264, (1978).
25.   Glimelius, K., Eriksson, T., Grafe, R., and Muller, A. J. *Physiolo-*

gia Plant. 44, 273-277 (1978).
26. Gressel, J., Zilkah, S., and Ezra, G. In "Frontiers of Plant Tissue Cultures 1978" (T. A. Thorpe, ed.), pp. 427-436. IAPTC 1978, Calgary (1978).
27. Guha, S., and Maheshwari, S. C. Nature 204, 497 (1964).
28. Jensen, C. J. In "Haploids in Higher Plants" (K. J. Kasha, ed.), pp. 153-190. The University of Guelph, Guelph (1974).
29. Kao, K. N., and Michayluk, M. R. Planta 115, 355-367 (1974).
30. Kochba, J., Spiegel-Roy, P., and Saad, S., IAEA publ., (1978).
31. Ledoux, L. "Genetic Modifications with Plant Materials." Academic Press, New York (1975).
32. Maliga, P. In "Frontiers of Plant Tissue Culture 1978" (T. A. Thorpe, ed.), pp. 381-392. IAPTC 1978, Calgary (1978).
33. Maliga, P., Sz-Breznovits, A., and Marton, L. Nature New Biol. 244, 29-30 (1973).
34. Maliga, P., Sz-Breznovits, A., Marton, L., and Joo, F. Nature 255, 401-402 (1975).
35. Maliga, P., Lazar, G., Svab, Z., and Nagy, F. Molec. gen. Genet. 149, 267-271 (1976).
36. Maliga, P., Lazar, G., Joo, F., H-Nagy, A., and Menczel, L. Molec. gen. Genet. 157, 291-296 (1977).
37. Maliga, P., R.-Kiss, Zs., Dix, P. J., and Lazar, G. Molec. gen. Genet. 172,13-15, 1979).
39. Melchers, G. Plant Res. and Dev. 5, 86-110 (1977).
40. Menczel, L., Lazar, G. and Maliga, P. Planta 143, 29-32 (1978).
41. Meredith, C. P. Plant Sci. Lett. 12, 25-34 (1978).
42. Muller, A. J., and Grafe, R. Molec. gen. Genet. 161, 67-76 (1978).
43. Murashige, T. In "Frontiers of Plant Tissue Culture 1978" (T. A. Thorpe, ed.), pp. 15-26 IAPTC 1978, Calgary (1978).
44. Nabors, M. V., Daniels, A., Nadolny, L., and Brown, C. Plant Sci. Lett. 4, 155-159 (1975).
45. Nakamura, A., Yamada, T., Kadotani, N. and Itagaki, R. In "Haploids in Higher Plants" (K. J. Kasha, ed.), pp. 277-278. The University of Guelph, Guelph (1974).
46. Nickell, L. G. Crop Sci. 17, 717-719 (1977).
47. Power, J. B., Frearson, E. M., Hayward, C. and Cocking, E. C. Plant Sci. Lett. 5, 197-207 (1975).
48. Sacristan, M. D. Chromosoma 33, 273-283 (1971).
49. Schilperoort, R. A., Klapwijk, P. M., Hooykaas, P. J. J., Kockman, B. P., Ooms, G., Otten, L. A. B. M., Wurzer-Figurelli, E. M., Wullems, G. J., and Rorsch, A. In "Frontiers of Plant Tissue Culture 1978" (T. A. Thorpe, ed.), pp. 85-94. IAPTC 1978, Calgary (1978).
50. Sunderland, N. In "Haploids in Higher Plants" (K. J. Kasha, ed.), pp. 91-122. The University of Guelph, Guelph (1974).
51. Sung, Z. R. Genetics 84, 51-57 (1976).
52. Vasil, I. K. In "Advances in Agronomy" 28, pp. 119-160. Academic Press, New York, San Francisco and London (1976).
53. Widholm, J. M. Can. J. Bot. 54, 1523-1529 (1976).

54. Widholm, J. M. *In* "Plant Tissue Culture and its Biotechnological Applications" (W. Barz, E. Reinhard, and M. H. Zenk, eds.), pp. 112-122. Springer, Berlin, Heidelberg (1977a).
55. Widholm, J. M. *Crop Sci. 17,* 597-600 (1977b).
56. Zenk, M. H. *In* "Haploids in Higher Plants" (K. J. Kasha, ed.), pp. 339-353. The University of Guelph, Guelph (1974).

GENETIC DIVERSITY OF PLANTS FOR RESPONSE
TO LOW TEMPERATURES AND ITS POTENTIAL
USE IN CROP PLANTS

C. Eduardo Vallejos

Department of Vegetable Crops
University of California, Davis

I.   INTRODUCTION

Genetic diversity for response to temperatures can be observed at almost all levels of organization in the plant kingdom. This diversity has probably evolved as a result of the climatic variation existent on this planet. Plants, evolved to adapt to a certain range of environmental conditions, are stressed if any environmental component extends beyond that range. Stress then is a relative term. For example, *Tidestromia oblongifolia,* a bush from Death Valley, California has an optimum temperature for photosynthesis of $45^{\circ}C$, but at $20^{\circ}C$ or less photosynthesis stops and dormancy is induced (5). On the other hand there are the alpine and arctic plants such as *Mimulus lewisii* (25) and *Oxyria digyna* (2) which are capable of withstanding very low temperatures. So, low temperature stress has a different meaning depending on the group of plants, and it is necessary to differentiate between freezing (below $0^{\circ}C$) and chilling (low temperatures above $0^{\circ}C$).

While considerable information has been reported in regard to the phenomenon of chilling injury, little is known about the exact sequence and timing of events at the cellular and molecular level that lead to the injury and the subsequent expression of the symptoms. One line of evidence accumulated in chilling sensitive species suggests that the primary effectors of the damage are the cellular and subcellular membranes (30). The physiological disorder caused by low temperatures is a complex phenomenon. Several important factors affect the degree of injury including: a) the temperature, b) time of exposure, c) the organ of the plant or the tissue exposed, d) its physiological stage, and e) the temperature at which the organism has been growing.

Chilling sensitive species are in general from tropical and sub-tropical origin and they can be found in taxa as diverse as *Solanaceae* (dicot) and *Musaseae* (monocot). Variability in response to low

473

temperatures, both chilling and freezing, is found at every level of taxonomic organization in the plant kingdom, from within a phylum to within a genus such as *Lycopersicon* and a species such as *Zea mays,* corn. Although tomato and corn are evolutionarily distant, they share a trait or number of traits that render them chilling sensitive.

It is also possible to see a similar kind of variability from a different perspective, even within a single chilling sensitive plant. Thus variation in sensitivity of different tissues and/or organs can be observed in a plant of a given age when this plant is exposed to chilling temperatures (28); also variation in sensitivity of a particular tissue/organ at different stages of development (30, 28). The last and by no means the least important is the variation in sensitivity of different organelles and other subcellular structures within a cell at a given time (see R. Ilker in this volume).

Each structure in a plant has at least one main physiological function, a certain physiological process which in turn involves some biochemical reactions or pathway. The extent to which this pathway is sensitive to low temperatures relative to other pathways will impart sensitivity to the physiological process and to the tissue or organ in which this process is the most important.

These distinct levels of variability mentioned above result from the fact that a plant, chilling sensitive or not, is a dynamic entity which is continuously growing and developing. Changes in sensitive sites represent the dynamic nature of the plant and the process of differentiation, which is an ordered sequence of gene expression and repression, and is regulated to some extent by the environment. Thus, in a loose sense, variability in sensitivity within a plant (with time and stage of development) can be considered a form of genetic variability. This is a very important concept for studies in comparative physiology.

II.   THE GENETIC APPROACH TO STUDY CHILLING INJURY

A systematic study of a simple system or mechanism will most likely involve direct measurement of its components and the factors that directly affect the system. However, for more complex systems, especially those with many unknown components, a different approach should be taken. One likely approach is to vary the external factors that affect the system and measure the effects of the changes that occur. Identification of the unknown components responsible for a particular response can be achieved by analytical comparison of two systems which differ only for that response; for example, two different ecotypes from the same species.

To utilize genetically controlled differences for studies in comparative physiology in plant systems, especially of an unknown mechanism, one has to rely on induced or natural variability. Although many studies have utilized variants induced by chemical mutagens, it is often not possible to induce the kind of variation desired and in these cases the only alternative is to rely on natural genetic variation. The amount of variability found in cultivated varieties of crop plants is usually minuscle compared to that found in wild populations or closely related species. Many crop plant species have been domesticated by man to the point that no wild forms of the species exist, whereas others do have wild forms or close intercrossable relatives growing in the wild; it is in these wild forms that great variability can be found. Although attempts with plant cell suspension culture have had some degree of success in recovering useful variants, it will be a long time before the technique becomes a standard procedure to aid in genetically defining a physiological process (see Dix in this volume).

An exhaustive survey of variability for temperature response in the plant kingdom will yield a large list of species. These species can be classified in three groups. First, those crop plants for which no relatives from the same genus can be found in the wild such as *Zea mays,* corn or maize. Second, crop plants whose close relatives can be found in the wild, as in the case of the genus *Lycopersicon,* tomato. And third, wild species like those belonging to the genus *Atriplex.* While *Atriplex* has a worldwide distribution, corn and tomato are species from the new world. A geographical region in this continent with a highly complex physiography, such as Peru, provides excellent conditions for the evolution of divergent genetic variants with specific physiological mechanisms adapted to different environmental conditions.

It is in the geographical context of Peru where I want to describe examples of genotypic variability within (corn) or between closely related species (tomato). A brief description of the geographical location and orographic system of Peru here will be helpful in order to understand the nature of its ecological diversity (Figure 1). Its geographical location is essentially equatorial, the northernmost point is 0° 2' S and the southernmost point is 18° 12' S. The Andean Ranges of mountains (Cordilleras) run parallel and very close to the desertic Pacific coast (52). The orographic description of the Peruvian Andean system proceeding from south to north is as follows: Two ranges of mountains enter from Chile and Bolivia; these are the "Cordillera Volcanica" (from Chile) and the "Cordillera Real" (from Bolivia), which converge south of Cuzco to form the Vilcanota Knot. Three "Cordilleras", "Oriental", "Central", and "Occidental" surge northward from the Vilcanota Knot, meeting at the Pasco Knot in Central Peru. Three new "Cordilleras", "Oriental", "Central" and "Occidental" proceed from this point northward until they join in the Loja Knot in Ecuador. The Occidental Cordillera, which has the highest peaks in Peru, is the Andean Continental Divide, it separates the Pacific and Atlantic watersheds. The country is divided by the Andes into three main natural regions: the coast, a long and narrow strip of desertic land between the Andes and the Pacific Ocean, a number of rivers transect the coast carving deep and parallel valleys into the western slopes; the highlands, comprising the mountains with many interAndean valleys and high plateaus; and the jungle on the eastern slopes of the Andes, comprising the lowlands with tropical and

subtropical rainy forest (52). Thirty-five different tropical climates
have been identified in Peru, associated with differences in mean
temperature, annual precipitation, humidity, and latitude (52).

A. Zea mays

Corn is a crop plant from the new world. It developed under
extremely diverse environmental conditions found in North, Central, and
South America. This is reflected in the tremendous variability found in
this species. There are 305 races of corn listed in 11 different volumes
published by the National Academy of Sciences - National Research
Council. They represent almost a complete inventory of the new world
maize (31).

Peru has been postulated to be an independent center of maize
domestication (23). At least 42 races of maize can be found in Peru
coming from areas of diverse ecological conditions having been produced
by hybridization and selection along with naturally occurring mutation.
Grobman *et al.* (23) have proposed that domestication of corn in Peru
occurred in the low to middle altitudes of the Andes resulting in the
formation of primitive races and extension of the original range of
adaptation to higher and lower altitudes as well as more northern and
southern latitudes. The presence of genetic variability in corn allowed an
effective artificial selection for adaptation during the different stages
of domestication of corn and hence has extended the range of
geographical distribution of maize to habitats with extreme diverse
conditions for plant growth and development, from sea level in the
desertic coast to an elevation of 4,000 meters in the Andes. Grobman
et al. (23) stated that "corn populations were then subjected to multiple
pressures of natural selection interacting with fixed artificial selection-
pressure, producing in the end a number of morphological genetically and
adaptively distinct populations." In these distinct populations,
morphological and physiological characters have a certain degree of
coherence; however, they do not appear to be genetically linked, for
reductions in recombination in hybrids of interpopulation crosses have
not been observed. The concomitant presence of these two sets of
characters in a given population will be maintained through continuous
selection. The development of these defined ecological races was
followed by expansion of distribution. In this process certain racial
characteristics were maintained at a low level of variability, while
others had to vary in order to adapt to the new environment. It should
also be mentioned that some of the existing races arose from
hybridization.

Grobman *et al.* (23) classified the existing races of maize from Peru
into six main groups according to their chronological origin:

1) Primitive Races (of greatest antiquity)
2) Anciently Derived or Primary Races (derived from Primitive
Races)
3) Lately Derived or Secondary Races (derived from Primary
Races)

4) Introduced Races
5) Incipient Races (restricted geographic distribution)
6) Imperfectly Defined Races

In the following paragraphs I will list some of the corn races from groups 1 and 2, briefly describe their characteristics, origin and relationships. A discussion on the possibilities of studying the inheritance of adaptation mechanisms will follow in subsequent pages. The classification of the Peruvian Races was based on archaeological evidence for the prehistoric races and morphological and cytological characters of the existing races.

1. Primitive Races: All primitive races are confites or popcorns.

a) Confite Morocho. A short plant, early in maturity. It is found at intermediate altitudes 2,500–3,000 meters above sea level. This corn easily adapts to the coast showing resistance to extreme environmental fluctuations due to a lack of environmental specificity. It represents an evolved form of Proto Confite Morocho, a prehistoric race.

b) Confite Puntiagudo. Also an early maturing short plant (117 cm.). Collected at altitudes of 2,500–3,500 m, it is widely distributed along the Andes. Derived directly from Confite Morocho with some introgression of *Tripsacum* probably through a race called "Enano" from the lowlands east of the Andes.

c) Confite Puneno. The earliest maturing race in the highlands and also the shortest of all (56 cm). This race grows at the highest altitudes in the world. It is found around Lake Titicaca at altitudes of 3,600 to 3,900 m. Ramirez *et al.* (45) reported the highest elevation for this corn in Bolivia to be 3,993 m. It is proposed that this corn evolved when the prehistorical race Confite Chavinense was forced by early farmers into the higher lands around Lake Titicaca.

d) Kculli. A very short plant, (92 cm) found at between 2,300 and 3,300 m. although the majority of typical collections come from above 3,000 m. It was extensively cultivated in both the coast and the highland in pre-Columbian times. Thought to be originated from Proto Kculli, an ancient popcorn race of the Andes.

e) Enano. A short plant found in the lowlands and jungle east of the Andes in the southern part of Peru. This corn was collected at 270 m. and can be found eastward at lower elevations. It is suspected to have derived from "Confite Chavinense" and have *Tripsacum* introgression.

It has been established that of the primitive races, "Confite Morocho" and "Confite Puntiagudo" have a wide range of adaptation. It would be interesting to know if adaptation of "Confite Puneno" to the highest elevation and "Enano" to the lowlands in the tropical jungle resulted in a loss of their ability to stand the opposite extreme conditions.

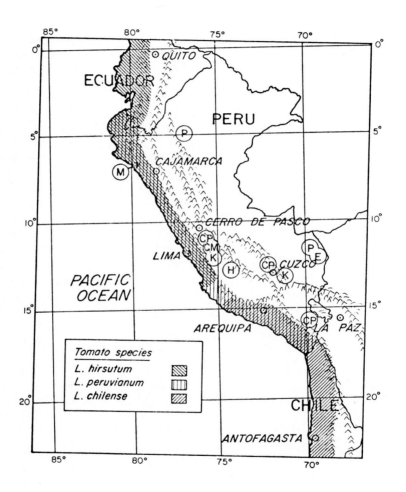

FIGURE 1. *Orographic map of Peru. It shows the distribution of 3 species of the genus Lycopersicon, and the Peruvian races of corn. Confite Morocho (CM); Confite Puntiagudo (CPo); Confite Puneno (CP); Kculli (K); Enano (E); Mochero (M); Huancavelicano (H); Piricinco (P).*

2.   *Anciently Derived Races.* These are racial selections from simple as well as complex hybrid populations that resulted from the intercrossing that took place among the primitive popcorns.

a)   *Mochero.* A short plant endemic to the north coast at altitudes below 50 m.   There is ample evidence pointing at "Proto-Iqueno", a coastal prehistoric popcorn, as the ancestor of "Mochero."

b)   *Huancavelicano.* Plants with an average height of 1.32 m. and altitudinal distribution limited to lands between 2,200 to 3,500 m, although 80% of its collections come from elevations above 2,800 m. This race is found in the southern sierras of Peru. The primitive races "Confite Morocho" and "Kculli" are proposed to be the ancestors of this race.

c)   *Piricinco.* A plant with a very long and slender ear, grows on the subtropical and tropical lowlands east of the Andes at altitudes between 150 and 940 m.   This race is proposed to be a hybrid between "Enano" (primitive race from the tropics) and a middle altitude race from the highlands which in turn is a descendant of Confite Morocho. This race shows strong tripsacoid characters arising from introgression with *Tripsacum australe,* a tropical tripsacum from South America.

These races arose from both hybridization and selection of older races so one expects them to be better adapted to certain specific environments.   Here, it will also be interesting to know if this latter selection for more specific habitats narrowed the range of adaptability displayed by the ancestors of this group.

B.   Lycopersicon

The 8-10 species belonging to this genus have a natural area of distribution limited to western South America.   This area comprises northern Chile, Peru, Ecuador, including the Galapagos Islands and southern Colombia (47). There is a varying degree of intercompatibility among the species.   Some combinations are fully compatible whereas hybrids of other combinations can be attained only with some difficulty (48).   All the species have 12 pairs of chromosomes (2n=2x=24) and cytological studies of F_1 hybrids have shown there is very little if any structural differences between sets of chromosomes (48).

Holle *et al.* (26, 27) have recently published a catalog of collections of green fruited species.   There are three green fruited species of interest (due to their wide range of altitudinal distribution) found on the western slopes of the Andes: *L. chilense,* which inhabits one of the most arid of the world's temperate deserts in southern Peru and northern Chile (47), with some populations also found on the hillsides at higher elevations; *L. peruvianum,* a species extremely polymorphic both within

and between populations (46), which has a range of distribution from
northern Peru to the northernmost coast of Chile, and its altitude range
is from sea level to 3,000 m; and *L. hirsutum,* which is native to an
area that extends from central Peru to northern Ecuador; the area of
distribution is divided in 4 main regions, three on the western side of the
Continental Divide and one on the eastern side. The altitudinal range in
Peru extends from 500 m to 3,300 m*, the highest limit known for any
tomato species. However, in Ecuador, a subspecific taxon *f. glabratum*
inhabits coastal lowlands. The Ecuadorean form of *hirsutum* has a
much smaller corolla, slender calyx and considerably less hairiness in
leaves and stem than its Peruvian counterpart *f. typicum* (49).

Populations of *L. hirsutum* from the valleys in northern Peru the
center of distribution, are self-incompatible and highly variable for
allozymic and morphological characters, however, populations from
northern (Ecuador) and southern (Central Peru) limits of distribution are
self-compatible and tend to be less polymorphic but extensively different
from each other (49).

C.   Atriplex

The numerous species in this genus belong to two subgenera – Obione
and Euatriplex which have diverged profoundly from each other during
evolution (41).

The species of this genus are widely distributed in Europe, North
America, and Australia. Some species are typical of cool coastal areas
whereas others are from warm interior valleys or warm deserts.
Interestingly there are some species, such as *A. lentiformis,* that have
ecotypes well adapted to both environments (5, 6, 43). Species with C_3
metabolism are found in both thermal environments, the same is true for
species with C_4 metabolism (41). *A. lentiformis* seems to be the most
suitable species for comparative physiology and genetic studies of
temperature adaptation. The species of this genus, even within each
subgenus, have become highly differentiated genetically and that poses a
problem for interspecific genetic analysis (41). Although a complete
understanding of the genetic relationships among the species has not
been attained, some information is available in the literature. Within
Euatriplex the species *A. rosea* (from warm areas) and *A. sabulosa* are
intercrossable with some degree of success, both have C_4 metabolism
(41). It has been shown that *A. vesicaria,* a warm area species, can be
cold tolerant (9).   This species is capable of growing and

---

*Seeds of L. hirsutum in this altitudinal range were collected by the
author during an expedition in 1976.*

photosynthesizing when grown in a temperature regime $8°C$ day/$6°C$ night. *A. vesicaria* and *A. confertifolia* showed evidence of some degree of photosynthetic acclimation to growth temperature, mainly through some kinetic adjustments. The authors suggested that cold tolerance stems more from maintaining membrane integrity at low temperatures rather than the kinetic adjustments during acclimation, this remains to be determined.

III. UNDERSTANDING SENSITIVITY, TOLERANCE, AND ADAPTA- TION TO LOW TEMPERATURE

There is a basic need to understand the mechanisms of sensitivity and resistance to low temperature stress if one intends to introduce the latter trait into sensitive crop plants. Pursuant to this model, one needs systems such as those with species or genera that have interbreeding populations adapted to extreme thermal environments [4]. A few examples are mentioned elsewhere in this chapter. The selection of wild species (tomatoes) or primitive forms of cultivated plants (such as corn) has the advantage of providing genetic variability. As Björkman [4] pointed out, the use of wild species spares us the problem of dealing with cultivated plants which have been selected for uniformity and bred for desirable characteristics with the exclusion or alteration of adaptation mechanisms essential in stress environments.

Hiesey *et al.* [24] outlined the basic principles for studies of comparative physiology of ecological races. An extensive genetic, ecological, and physiological characterization of the races or ecotypes from contrasting environments is required in a "...program aimed at discovering basic physiological mechanisms that determine the performance of plants in definite environments..."

Genetic and ecophysiological studies have been carried out on altitudinal races of *Potentilla glandulosa* by Clausen and Heisey [14] and of altitudinal as well as latitudinal races of *Mimulus* spp. by Hiesey *et al.* [25]. With both ecotypes, certain morphological and physiological characters were genetically characterized for races from contrasting environments (Stanford, Timberline), and for their F_1, F_2 and F_3 progenies. In all cases, the individuals were diploid, freely interfertile and with normal pairing of chromosomes. Clear morphological marker characters distinguishing well-defined contrasting ecotypes were chosen. Each morphological character and the ability to survive in contrasting environments were graded from one to nine according to the resemblance of the character to either of the parents. Regressions and correlations were obtained for all possible combinations of two characters. Analysis of these results showed that there was a partial correlation between independent non-linked characters that distinguish a species, subspecies, race or ecotype.

Anderson (1) and Mather (34) had previously recognized the fact that most characters that distinguish species and races are governed by systems of genes consisting of several components referred to as *polygenes* which show complex segregation. Hiesey *et al.* (25) then coined the term 'Genetic Coherence'* to define "the tendency of F_2 and later generation progeny to inherit certain combinations of characters of the parents more frequently than would be expected on the basis of free random recombination." Genetic coherence is evaluated as the range of correlation coefficients with the highest frequency. When clones of parents and F_1, F_2 and F_3 progenies were grown at three different elevations (Stanford 30 m, Mather 1,400 m, Timberline 3,100 m) it was discovered in the segregating generations that there were significant correlations between many of the morphological markers and the ability of individuals to survive at each elevation. In other words, there was a marked tendency for the types most resembling the parental ecotypes to be better fit for survival, growth and development in the environments normally occupied by the putative parent ecotype.

The expression of genetic coherence is enhanced when partial incompatilibity exists between the parental forms. That is, the group of correlation coefficients with highest frequency has a higher value. The enhanced genetic coherence is interpreted as a result of the elimination of gametes and zygotes differing markedly in genic composition from the original parental combination. Some of the inter- and intraspecific crosses of *Mimulus* (25) showed heterosis. Parental ecotypes unable to survive at certain elevations yielded a vigorous progeny capable of surviving at those elevations. A case was also observed in which a hybrid between high and low altitude yielded a first generation hybrid with better frost resistance than the alpine parent.

Based on responses obtained for altitudinal races of *Potentilla glandulosa* at contrasting altitudes, Clausen and Hiesey (14) concluded that evolution of natural races for contrasting environments operates on two basic principles. First, the presence of a mechanism that enables each race to attain a certain degree of genetic coherence, which allows it to exist as a defined biological entity capable of surviving in its respective ecological niche. Second, this genetic coherence is not rigid enough as to prevent extensive recombination of the germplasm. The potential for increasing genetic diversity, through hybridization of different ecological entities, as a response to changes in selective pressures, lies in those two principles.

All the basic concepts and principles on the genetics and ecophysiology of altitudinal races pioneered earlier by Anderson and Mather and later by Clausen and Hiesey (14) and Hiesey *et al.* (25) can be used to study the genetic characters involved in the phenomenon of

---

*For a more extensive explanation of this term see Heisey et al. (25) pp. 78-82.

chilling injury. Here I will discuss the advantageous aspects of using two systems as working models: A species *Zea mays* (corn, *Graminneae*, a monocot); and a genus, *Lycopersicon* spp., (tomato, *Solanaceae*, a dicot).

A working model to explain the basic mechanism in chilling injury has been proposed by Lyons (30), in which cellular and subcellular membranes of chilling sensitive plants are assigned a major role. Here I suggest the use of the Peruvian races of corn and the genus *Lycopersicon* as model systems to undertake studies toward elucidating the mechanism for chilling injury. These two crop plants share a number of characteristics that make them useful for genetic and comparative physiological studies of adaptation to different environments. Both corn and tomato show great genetic variability: in corn it has been brought about by extensive cross pollination, wide distribution and selection for different environments by man (39); whereas variability in the tomato is found in the natural populations of wild species. Both species have relatively short life cycles and produce large numbers of seeds. The maize genetic strains are available from the Maize Genetics Cooperative (39) and the tomato accessions from the Tomato Genetics Cooperative (26, 27). These two species are among the best genetically characterized in the plant kingdom. For tomato, more than 1,100 genes have been identified and an extensive linkage map has been developed (48).

In order to better understand the processes of natural selection and adaptation it is necessary to have information on the phenotypic expression of genes or gene combinations over a range of environments (14) and establish a direct relationship between the functional characters of an ecotype and its natural environment (3).

A number of comparative studies (3, 4, 5, 6, 9, 22, 25, 35, 36, and 43) have shown the optimal temperature for photosynthesis of plants native to cold environments is lower than that of their counterparts from warmer environments. Björkman (3) recognizes three distinctive processes in photosynthesis: photochemical, CO_2-diffusion, and biochemical; and each one can be affected differently by the diverse factors present in the surrounding environment. Since any of these three processes can be limiting under different environments, plants have most likely evolved mechanisms of adaptation to overcome a particular limitation. So far the major concern in comparative studies of altitudinal ecotypes has been temperature as the principal determinant. However, caution should be taken to consider other determinants related to high altitudes, e.g. the lower partial pressure of CO_2 at high elevations. In this regard Hiesey *et al.* (25) and Billings *et al.* (2) found that ecotypes from higher elevations were very efficient in absorbing CO_2 at low concentrations and that their saturation point for CO_2 was lower than that of low altitude ecotypes.

Temperature can affect the three processes of photosynthesis in different ways. Although photochemical reactions are independent of temperature, at least within the range of temperature optimum for life, this process may be affected by alteration of the fluidity or molecular ordering of thylakoid membranes where the two photosystems, the electron transport chain and the energy transducing system are imbedded (18, 37, 38). The diffusion of CO_2 from the atmospheric air to the site of carboxylation is affected by temperature, for diffusion is a temperature dependent parameter. Additionally there is the effect of temperature on the resistances that CO_2 molecules encounter on their way to the site of carboxylation, especially those resistances that involve cellular membranes such as the plasmalemma of guard cells (55, 56) of mesophyll cells and chloroplast membranes. The rate of the biochemical reactions from the Calvin cycle also depends directly on temperature and on CO_2 concentration at the reaction site. Thus, it appears that the effects of temperature on the three processes involved in photosynthesis can be transduced in one way or another by cellular and subcellular membranes.

Studies on adaptation of photosynthetic apparatus of species from contrasting environments have led Björkman (4) to the conclusion that a single genotype has a potential for phenotypic adjustment of its photosynthetic characteristics to changes in the environment, but he pointed out that the range of such adjustment is limited and reflects a genotypic adaptation to the particular conditions prevailing in its native habitat.

Comparative studies of tomato ecotypes and the races of corn from contrasting environments should be most helpful to understand, as stated earlier by Björkman (4), "The environmental and evolutionary limits of adaptation. . .and the physical, structural and molecular mechanisms involved."

What is the importance of the physical properties of membranes (as they are affected by temperature) in the evolution and adaptation of the species considered in this chapter? In preliminary experiments (53), both high altitude (3,300 m) and low altititude (100 m)* ecotypes of L. hirsutum were grown in hydroponic culture in a greenhouse and later the adult plants were transferred to a growth chamber with a $13^{\circ}/5^{\circ}C$ day/night temperature regime for a two-week period. At the end of the experiment the low altitude ecotype had died whereas the other was not only alive but without any symptoms of injury. It should also be mentioned that both ecotypes grow normally in the greenhouse at high temperatures up to $35^{\circ}C$. These results suggest that ecotypes from the region of high variability in northern Peru (49) have greater phenotypic plasticity in their ability to stand a wider range of temperatures than their Ecuadorian counterparts which show less variability. This could also be interpreted as a greater thermal stability of membrane structures in the ecotype from the region of higher variability, but this needs to be proven experimentally. Thus, the origin of the Ecuadorian ecotypes might be explained in terms of the "Baldwin effect" (22) of organic selection, a process in which phenotypic reactions permit a population to exist in an environment to which it is not well adapted and give it time to

*Seeds of this ecotype were kindly provided by Dr. C. M. Rick.

acquire, by a more or less random process of gene mutations, genotypes which are adapted to the new environment. Loss of its ability to survive at low temperatures is probably the price this ecotype had to pay in order to adapt to the Ecuadorian tropical forest. This can also be supported by the evidence accumulated by Björkman (4), in studies of genotypic adaptation of plants from contrasting environments. He found that genotypic adaptations which enable the plant to photosynthesize with an unusually high efficiency under one environmental extreme of light or temperature require a specialization of its photosynthetic machinery which precludes a high photosynthetic efficiency at the other extreme. On the other hand, there are the populations from northwestern Peru which Rick (49) has found to be the region of highest variability and greatest outcrossing which are characteristic of an ancestral species in that region of great ecological diversity. For corn a similarity could be drawn between the primitive race of Peruvian corn and the populations of *L. hirsutum* from northwestern Peru, both with a wide range of temperature adaptation on one hand the the marginal populations of *L. hirsutum* from Ecuador and the modern and climate specialized races and selections of corn, which have a diminished range of adaptation.

Morphological and allozymic characters have been recorded extensively for ecotypes of *L. hirsutum* from contrasting environments (49). Many of these characters can be used as genetic markers to follow up in future generations. Clausen and Hiesey (14) pointed out such markers themselves may not facilitate ecological adjustment of their ecotypes to the environment but may be linked with genes that control the subcellular structures and physiological characters of importance in determining adjustments to the environment. Establishing such correlations would be extremely helpful for screening purposes in the future.

It has been suggested in the past that cold tolerance is maternally inherited to some extent, at least for seed germination, but testing that characteristic turns out to be of some difficulty using the Peruvian and Ecuadorian ecotypes due to the existence of unilateral incompatibility (33). The only successful way to make the cross is to use the Ecuadorian type as the pistillate parent. The use of some extreme variants from the lowest elevations in Peru (500 m) which are fully compatible with the population from the highest altitude (3,300 m) will help to elucidate this problem. Full compatibility can be found in the races of corn.

IV.   THE SEARCH FOR "COLD TOLERANT GENES" IN CROP PLANTS

Some effort has been made towards introducing and increasing cold tolerance of chilling sensitive cultivated plants. In general the accomplishments in these efforts have not been spectacular. Plant breeders have focused on seed germination at low temperatures to enable the early planting of crops. Selection for seed germination at low temperatures had been made in cotton (12, 13, 15, 32), soybean (42), beans (16), corn (7, 8, 44) and tomato (10, 17, 19, 40, 50). In almost all cases, selection was applied to cultivars from each crop. Once a cold tolerant line was selected, it was crossed with a sensitive line and from study of their progenies, conclusions were drawn about the nature and complexity of the inheritance for cold tolerance. In general the results are quite contradictory. It has been claimed that cold tolerance for seed germination is monogenic (10, 19). It has also been claimed polygenic in most cases and maternally inherited as well (12, 13, 32, 40). Some of the problems plant breeders have to face are selection criteria and a meaningful way to quantitatively measure the response of breeding lines in field trials.

Comparisons and selections of cultivars for their tolerance to low temperatures during different stages of development (other than germination) have been obtained with a varying degree of success in: tomatoes, during a whole life cycle (28, 29) and fruit set (11); corn, growth in general (7); and cotton, growth, development and boll maturation (20, 21). At the present time efforts are being made to introduce cold tolerance into sorghum using races native to the highlands of Uganda and Ethiopia (54). The selection in cotton, G. hirsutum, line M-8 is of interest since it seems to be a line with a variant for desaturase Δ15, directly involved in the synthesis of linolenic acid, which might be activated at low temperatures (51).

In all the examples listed above, selection has been used only in commercial cultivars. There have not been any reports of the use of wild species or primitive forms of a species as a source of cold tolerance. The main reason for the absence of this attempt is the time involved in recovering the trait in a commercially acceptable type. Since most present day cultivars are the result of a long period of selection for uniformity, high yields and adaptation to mild environments, the probability of finding striking variants within these populations is extremely low and at best any accomplishments will be limited.

Many useful characters from wild species have been successfully transferred to the cultivated tomato (20, 21). At the present time a joint effort is being made with Dr. C. M. Rick, Dr. A. Stevens, Dr. R. Jones, S. D. Tanksley and the author to introduce cold tolerance into the cultivated tomato.

According to Mangelsdorf (31), the use of exotic germplasm of corn in the USA has been given little attention. The reasons are, (a) the possibility for improvement of existing cultivated varieties is still far from exhausted, and (b) the presence in exotic races of some undesirable characters such as some responsiveness to photoperiod, luxurious growth, late flowering, etc. Mangelsdorf suggests that the most promising method of utilizing exotic races in the U.S. is by modifying one or more inbred lines. Hybrids of native inbred lines with exotic races (usually not well adapted to conditions of the U.S.) followed by two backcrosses with their inbred parent yield vigorous progeny with most of the undesirable characteristics of the exotic race masked by the inbred. One of the drawbacks of the cold tolerance Peruvian race "Confite Puneno" is its high susceptibility to smut. The main concern of American corn breeders in the past years has been yield; but today, with increasing demands for food by a hungry world, there is a need to conquer new land in marginal areas for crop production. Areas which are presently limited in use because of cool climates may some day support new varieties of existing crops bred to grow and develop normally in the cold. This possibility represents a challenge which has to be met with new genetic resources.

ACKNOWLEDGMENTS

I express my gratitude to Dr. Charles M. Rick, Dr. M. Allen Stevens, Dr. Richard Jones, Steve Tanksley, and Karen Koevary for reviewing this manuscript as well as for their valuable suggestions and comments. I also thank Dawn Nichols for typing the initial drafts and Moira Tanaka for the art work.

IV. REFERENCES

1. Anderson, E. Genetics 24, 668-698 (1939).
2. Billings, W. D., Clebach, E. E. C., and Mooney, H. A. Science 133, 1834 (1961).
3. Björkman, O. Brittonia 18, 214-224 (1966).
4. Björkman, O. In (Marcelle, R. ed.) "Environmental and Biological Control of Photosynthesis" pp. 1-16. Dr. W. Junk B.V., Publishers, The Hague, 1975 (1975).
5. Björkman, O., Mooney, H. A., and Ehleringer, J. Carnegie Institution of Washington, Yearbook 74, 743-759 (1975).
6. Björkman, O., Nobs, M., Mooney, H., Troughton, J., Berry, J., Nicholson, F., and Ward, W. Carnegie Institution of Washington Yearbook 73, 748-767 (1974).
7. Cal, J. P. and Obendorf, R. L. Crop Sci. 12, 572-(1972a).
8. Cal, J. P. and Obendorf, R. L. Crop Science 12, 369-373 (1972b).
9. Caldwell, M. M., Osmond, C. B., and Nott, D. L. Plant Physiology 60, 157-164 (1977).
10. Cannon, O. Utah Science 32, 8 (1971).
11. Charles, W. B. and Harris, R. E. Canadian J. of Plant Sci. 52, 497-506 (1972).
12. Christiansen, M. N. Proceedings Beltwide Cotton Production Research Conference 1972, p. 32 (1972).
13. Christiansen, M. N. and Lewis, C. F. Crop Sci. 13, 210-212 (1973).
14. Clausen, J. and Hiesey, W. M. Carnegie Institution of Washington, Publication 615, Washington, DC (1958).
15. Cole, D. F. and Christiansen, M. N. Crop Sci. 15, 410-412 (1975).
16. Dickson, M. H. Crop Science 11, 848-850 (1971).

17. El Sayed, M. N. and John, C. A. *J. Amer. Soc. Hort. Sci. 98,* 440-443 (1973).
18. Fork, D. C., Murata, N., and Auron, M. *Carnegie   Institution   of Washington, Yearbook 76,* 220-235 (1977).
19. Gatherum, D. M., Miles, W. G., and Cannon, O. S. *Utah   Acad.   of Sci. Arts and Lett. Proc. 47,* 278 (1970).
20. Gipson, G. R. and Ray, L. L. *Proceedings Beltwide Cotton Production Research Conference 1975,* p. 72 (1975).
21. Gipson, J. R. and Ray, L. L. *Proceedings, Beltwide Cotton Production Research Conference, 1977,* p. 62 (1977).
22. Grant, V. "The Origin of Adaptations." Columbia University Press, 606 pp., New York and London (1963).
23. Grobman, A., Salhauna, W., and Sevilla, R. in collaboration with Mangelsdorf, P. C. *Natl. Acad. Sci.-Natl. Res. Council Publ. No. 915* (1961).
24. Hiesey, W. M., Milner, H. W., and Nobs, M. *Carnegie Institution of Washington Yearbook 58,* 344-346 (1958-59).
25. Hiesey, W. M., Nobs, M. A., and Björkman, O. *Carnegie Institution of Washington Publication 628,* Washington DC (1971).
26. Holle, M., Rick, C. M., and Hunt, D. G. *Tomato Genetics Cooperative 28,* 49-78 (1978).
27. Holle, M., Rick, C. M., and Hunt, D. G. *Tomato Genetics Cooperative 29,* 69-91 (1979).
28. Kemp, G. *Canadian J. Plant Sci. 48,* 281-286 (1968).
29. Li, S-C. "Genetic studies of earliness and growth stages of *Lycopersicon esculentum.*" Thesis. The University of British Columbia (Canada). Dissertation Abstracts 36B, 3143 (1975).
30. Lyons, J. M. *Annual Review in Plant Physiology 24,* 445-466 (1973).
31. Mangelsdorf, P. C.. "Corn, Its Origin, Evolution, and Improvement." The Belknap Press of Harvard University Press, Cambridge, Massachusetts (1947).
32. Marani, A. and Dag, J. *Crop Sci. 3,* 243-245 (1963).
33. Martin, F. W. *Evolution 17,* 519-528 (1963).
34. Mather, K. *Biological Reviews 18,* 32-64 (1943).
35. Mooney, H. A. and Billings, W. D. *Ecological Monographs 31,1-28* (1961).
36. Mooney, H. A. and Billings, W. D. *Ecology 44, 812-816* (1963).
37. Murata, N. and Fork, D. C. *Plant Physiology 56,* 791-796 (1973).
38. Murata, N., Throughton, J. H., and Fork, D. C. *Plant Physiol 56,* 508-517 (1975).
39. Neuffer, M. H. and Coe, E. H., Jr. In: King, R. C. *Handbook of Genetics 2,* 3-30 (1974).
40. Ng, T. G. and Tigchelar, E. C. *J. Amer. Soc. Hort. Sci. 98,* 314-316 (1973).
41. Nobs, M. A. *Carnegie Institution of Washington Yearbook 75,* 2121-2123 (1976).
42. Obendorf, R. L. and Hobs, P. R. *Crop Sci. 10,* 563-566 (1970).
43. Pearcy, R. W. and Harrison, A. T. *Ecology 55,* 1104-1111 (1974).

44. Pinnell, E. *Agronomy Journal 41,* 563–568 (1949).
45. Ramirez, R., Timothy, D. H., Diaz, E., and Grant, U. J. in collaboration with Nicholson, G. E., Anderson, E., and Brown, W. L. *Natl. Acad. Sci., Natl. Res. Council Publ. No. 747 (1960).*
46. Rick, C. M. *Evolution 17,* 216–232 (1963).
47. Rick, C. M. In: Srb, A. M. *Genes, Enzymes, and Populations,* pp. *255–269 (1973).*
48. Rick, C. M. In: King, R. C. *Handbook of Genetics 2,* 247–280 (1974).
49. Rick, C. M., Fobes, J. F., and Tanksley, S. D. *Plant Systematics and Evolution* (In Press) (1979).
50. Smith, P. G. and Millet, A. H. *Proceedings of the American Soc. Hort. Sci. 84,* 480–484 (1959).
51. St. John, J. and Christiansen, M. N. *Plant Physiology 57,* 257–259 (1976).
52. Tosi, J. *Boletin Tecnico No. 5, Zona Andina, IICA, OEA* (1960).
53. Vallejos, C. E., Lyons, J. M., and Breidenbach, R. W. (unpublished results) (1977).
54. Van Arkel, H. *Netherlands J. of Agric. Sci. 25,* 135–150 (1977).
55. Wilson, J. M. *New Phytologyst. 76,* 257–270 (1976).
56. Wright, M. *Planta 120,* 63–69 (1974).

ADAPTATION TO CHILLING STRESS IN SORGHUM

J. R. McWilliam, W. Manokaran, and T. Kipnis[1]

University of New England
Armidale, N.S.W., 2351, Australia

I.   INTRODUCTION

Many annual plants of tropical and subtropical origin are now widely cultivated as summer crops in temperate environments. This has been achieved by adjusting the phenology of the crops so that their development takes place during the period of favorable summer temperatures and the exposure to chilling temperatures in either air or soil is minimized.

Despite this long history of temperate cultivation there is little evidence that any of these crops have acquired a substantial degree of chilling-resistance and as a result they regularly suffer from chilling damage when spring-sown in temperature environments.   Common symptoms include failure to germinate or retarded germination, radicle inury or death, slow emergence of the plumule or cotyledons and chlorosis and accompanying necrosis of newly formed photosynthetic tissue (2, 4, 5, 7, 15, 18, 22). Variation in chilling-sensitivity has been reported in a number of crops of tropical origin. Genetic differences in germination and early growth of inbred and hybrid corn grown under chilling temperatures have been reported (4, 5, 20).   Also varietal differences in the minimum temperature required for the germination of tropical rice have been shown by Oka (17). Patterson *et al.* (19) have shown a good correlation between the chilling tolerance of *Passiflora* species and the temperature of the environment in which they occur and have also recorded variation in germination and greening of ecotypes of wild tomato which is closely related to the altitude of their origin in the Andean region of South America.   There is also evidence that the movement of a number of tropical grass species north and south into

[1]*Permanent address: Agricultural Research Organization, The Volcani Centre, Bet Dagan, Israel*

more temperature environments has been accompanied by the acquisition of a greater degree of chilling tolerance (12).

This paper provides a comparative analysis of the germination, early seedling development and chlorophyll synthesis of a number of sorghum species adapted to contrasting thermal environments, when exposed to a wide range of chilling temperatures. A better understanding of the effects of chilling stress on these important sequential steps in the process of establishment and evidence for variation in these responses may suggest more efficient ways of selecting for increased chilling tolerance in this and other important crops of tropical origin.

II.   MATERIALS AND METHODS

A.   Plant Material

For the majority of experiments three species of *Sorghum* were used (1) *S. bicolor,* a North American commercial sorghum cultivar with a Hegari type background; (2) *S. verticilliflorum,* a tropical wild sorghum (shatter cane) now a widespread weed of the tropics, collected (latitude 17°S) in coastal north Queensland, and (3) *S. leiocladum,* a summer growing wild perennial species, native of eastern and northern Australia, collected (latitude 31°S) in northern New South Wales at an elevation of 1000 m. Although *Sorghum* is primarily a genus of tropical origin, the species does occur naturally and is grown in temperate environments. These three species were chosen to represent material with potentially contrasting thermal adaptations. In all experiments the behaviour of these three chilling-sensitive species has been compared with barley *(Hordeum vulgare* cv Abyssinian), a chilling-resistant temperate cereal. All seed used was approximately the same age, collected within six months of undertaking the experiments. Seed viability was high (>95%) in all species with the exception of *S. leiocladum* which was lower because of the difficulty of identifying empty caryopses. All seed was treated during storage with a fungicide to minimize fungal contamination.

B.   Temperature Control

Most responses were measured over a temperature range from 24°-4°C, although there was little measurable activity observed in the sorghums below 8°C. Incubators were used to obtain the various temperature regimes. These were maintained in a cold room at 4°C to minimize temperature fluctuations which varied within ±0.25°C of the stipulated temperature. In the case of the respiration experiment, the water bath temperatures for the manometers was maintained at ±0.1°C. In those experiments where young seedlings were exposed to light during greening at chilling temperatures, refrigeration was provided in incubators to remove heat generated by cool beam lamps. Temperature was stepped by $2°C$ intervals between $24°$ and $12°C$ and by $1°C$ intervals below $12°C$.

C. Experimental Procedures

1. Germination. Seeds were germinated at the various temperatures in the dark in petri dishes on moist filter pads. Seeds were considered to have germinated when the length of the radicle exceeded the small diameter of the seed. The experiment was terminated after 50 days.   Rates of germination were based on time in days to 50% germination (G_{50}) calculated by means of probit analysis (8).

2. Respiration. Respiration rates of seed germinated at $24°C$ were measured over the temperature range, using Warburg equipment (25). Oxygen uptake ($\mu l\ O_2g^{-1}h^{-1}$) was measured on replicate samples of seedlings of known weight at a similar stage of germination (emergence and active extension of radicle and plumule). All seed was germinated at $24°C$ and then equilibrated for 24 h at the temperature at which respiration was to be measured. Five seedlings were used for each measurement and after equilibrating for 30 min. in the flasks, respiration measurements were taken over a period of two hours under steady state conditions.

3. Extension. Rates of extension of the subcoleoptile internode (mesocotyl) were measured on seedlings that had been germinated at $24°C$ and then grown without exposure to light, on agar medium in single vials. The mesocotyls were permitted to elongate up through glass tubing with a diameter slightly larger than the widest diameter of the coleoptile. Extension measurements (mm h^{-1}) were made over the range of temperatures without disturbing the seedlings in green (safe) light at 6 or 12 hr intervals depending on growth rate, using a binocular microscope. Five successive readings were taken on each of 15 seedlings of each species after a 12 h equilibration period at the designated temperature.

4. Arrhenius Plots. Rate functions for germination, respiration and extension growth are presented in the form of Arrhenius plots by plotting log rate against the reciprocal of the absolute temperature. The slopes of the curves were obtained by fitting regression to points derived from the averages of three replications and selecting those giving the best fit. Where discontinuities were suspected, regressions were fitted to all combinations above and below the apparent break and the partition with the minimum sum of square was selected. Lines of best fit were then drawn for each partition and from the intersection of the regressions an estimate of the break point (transition temperature) was made.

5.   *Carry-Over Effects*. The effects of exposure to chilling temperature on subsequent development at a favourable temperature was examined in relation to germination and extension growth of *S. bicolor*. Seed was either imbibed at 8° for periods up to 20 days or germinated at 24°C and then stored as an elongating seedling at 8° for up to 10 days. Rates of germination (G_{50}) and rates of mesocotyl extension were then measured as described at 24°C to detect any carry-over of the previous temperature storage.

6.   *Chlorophyll Synthesis*. The capacity of young etiolated seedlings to synthesize chlorophyll in the light at the various temperatures was measured by exposing seedlings, which had been grown to the first leaf stage in the dark at 24°C, to a light flux of either 250 or 25 μE m^{-2}s^{-1} of PAR for 24 h.[a] Prior to exposure, all seedlings were pre-conditioned for 24 h at the designated temperature. Low vapour pressure deficits were maintained throughout. The extent of greening was determined by measuring the total chlorophyll (a+b) concentration of the first true leaf (μg chlorophyll g^{-1} FW) spectrophotometrically (1) after extraction in 80% (w/v) acetone.

7.   *Electron Microscopy*. Leaf segments (1 cm) were fixed in cold 3% (w/v) glutaraldehyde in 0.01 M Sørensen's phosphate buffer, post fixed with osmium tetroxide and after dehydration, embedded in Spurr's low viscosity resin (24). Thin sections were cut on an LBK ultramicrotome and examined under the electron microscope after contrasting with uranyl acetate and lead citrate.

III. RESULTS AND DISCUSSION

A.   *Germination and Early Seedling Development*

The rates of the three germination responses studied, initial germination, seedling-respiration and mesocotyl extension all declined as the temperature was reduced from 24° down to 8°C. The rate of decline, however, varied considerably between species, especially in the lower part of the temperature range below about 12°C (Fig. 1a-c).

A single discontinuity was observed in the Arrhenius plots for all three responses, both in the chilling-sensitive sorghum species and in the chilling-resistant barley control. The plots in all cases were linear above and below the break, which with one exception (germination of *S. verticilliflorum*) occurred at a temperature (transition temperature) around 12°C which lies in the temperature range (6-14°C) reported for most chilling-sensitive plants. Similar breaks in the Arrhenius plots of

[a]*Lambda Quantum Meter L185. Measures quantum flux density of PAR (400-700 nm).*

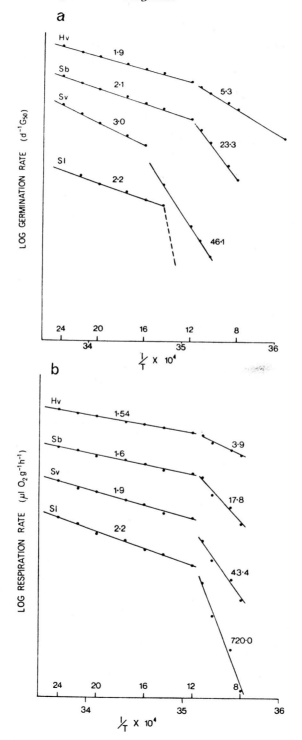

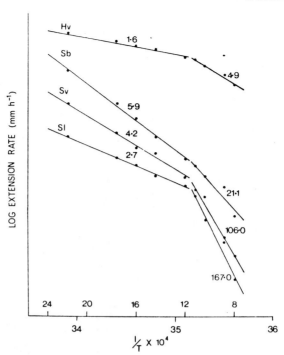

FIGURE 1.   Arrhenius plots for three processes involving the germination of chilling-resistant barley and three chilling-sensitive sorghum species adapted to different temperature environments. Q_{10} values derived from regressions are indicated for each slope.   (a) Germination rate (reciprocal of days to 50% germination; (b) Respiration rate $\mu \ell\ O_2\ g^{-1}h^{-1}$); (c) Mesocotyl extension rate (mm h^{-1}; H.v. Hordeum vulgare; S.b. Sorghum bicolor; S.v. Sorghum verticilliflorum; S.l. Sorghum leiocladum.

various aspects of germination activity have also been reported in other chilling-sensitive species including cucumber and mung bean (22), castor bean (27) and rice (15). In the case of tropical indica rice varieties, the breaks occurred at a temperature of 16°-17°C which is in the range found for the tropical ecotype of S. verticilliflorum used in this study.

Discontinuities in the Arrhenius plots of all three responses in chilling-resistant barley, although not as pronounced as in the sorghums, do represent a significant change in the temperature coefficients of these reactions. Similar breaks in Arrhenius plots for chloroplast development, Hill reaction activity and leaf elongation of barley at around the same temperature have been reported (16, 23) and also for respiration in chilling-resistant wheat and rye seedlings (21).

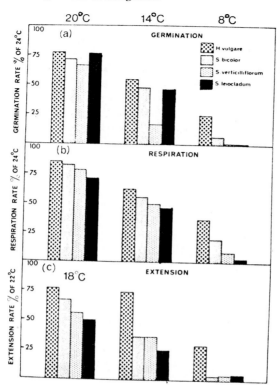

FIGURE 2. *Relative rates of (a) germination, (b) respiration and (c) mesocotyl extension, of chilling-resistant barley and three chilling-sensitive sorghum species adapted to different temperature environments. Values are given for three temperatures, one above and two in the chilling range.*

In the three chilling-sensitive sorghum species, although in most cases the breaks in the Arrhenius plots occurred at or about the same temperature (around 12°C) as in barley, the slopes of the regressions as measured by the temperature coefficients (Q_{10}'s) were greater especially below the transition temperature. The enormous increases in the Q_{10} values below 12°C in the most chilling-sensitive sorghums indicate extremely high activation energies and help explain the poor germination and low rates of extension growth at these temperatures. The relative differences in the rates of germination, respiration and extension growth above and below the chilling temperature range are more clearly illustrated in Figure 2. The sorghums as a group are relatively more depressed than barley immediately above the chilling temperature range (20-14°C) and dramatically so below 10°C.

These results suggest that a discontinuity in the Arrhenius plots of these various germination processes may not be a unique or distinguishing feature of chilling-sensitive species, but that the differences in the apparent activation energies or the Q_{10}'s above, and especially below, the transition temperature may be a much more useful diagnostic feature, (Fig. 1a-c).

The rates of Q_{10}'s of the various responses below $12^{\circ}C$ may also provide a measure of the relative chilling-tolerance within a group of chilling-sensitive species. The Q_{10} of the three sorghums species used in this study below 12° differed quite substantially. In all cases the Q_{10} for the tropical *S. verticilliflorum* was higher than *S. bicolor,* which is to be expected, because of greater chilling-tolerance associated with the commercial Hegari type sorghums. The highest Q_{10}, suggesting the greatest chilling-sensitivity, were found for *S. leiocladum* which is an ecotype adapted to a relatively temperate environment. This species is a wild perennial and its greater chilling-sensitivity may represent an adaptation to escape the effects of low temperature in the spring and autumn, as the species remains dormant during this period and grows only during the warmer summer months.

In addition to differences in chilling-sensitivity in the germination response of different sorghum species, there is also some evidence that differences exist within cultivars of *S. bicolor.* G_{50} values for three groups of commercial sorghum, one from U.S.A., one from tropical Africa and a third consisting of hybrids between the two, were compared over the temperature range. The U.S. cultivars and their hybrids, possibly because of their history of selection for adaptation to cooler environments, had significantly faster rates of germination below $12^{\circ}C$ but not at temperatures above the chilling range.

B.   Carry-Over Effects of Chilling Exposure

The effect of storing either imbibed seed or germinated seedlings of *S. bicolor* for varying periods at a chilling temperature on their subsequent performance at a favourable temperature $(24^{\circ}C)$ is illustrated in Figure 3. Storing imbibed seed for different periods up to 20 days at $8^{\circ}C$ caused a small delay in subsequent germination rate (G_{50}) but the differences observed after different periods of chilling were not significant (Fig. 3a). The rates of germination were generally enhanced by low temperature storage by comparison with the unchilled control, but again the differences were not significant and can be accounted for by the extra time required to complete inhibition in non-chilled seeds.

The absence of a marked response to chilling during imbibition and the early stages of germination in sorghum supports the finding of Nishiyama (15) that the water uptake stage in rice has a low Q_{10} and is much less temperature dependent than the subsequent germination phase. It is at variance, however, with the results of Wiles and Downs (28) who found that even short periods of chilling were harmful in cotton, if given after the seeds had begun hydration.

Once germination is well advanced in sorghum the rates of mesocotyl extension at a favourable temperature were significantly

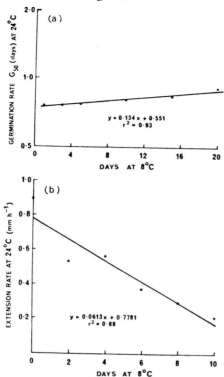

FIGURE 3. *Carryover effects of storing either imbibed seed or germinating seedlings for periods up to 20 days at a chilling temperature of* $8^\circ C$*, on the subsequent (a) germinatioan rate (days to 50% germination) and (b) rate of mesocotyl extension (mm* h^{-1}*) of young seedlings. Both rates were measured at* $24^\circ C$.

depressed by prior storage of the germinated seedlings at $8^\circ C$ (Fig. 3b). The response was linear and after 10 days, the maximum period of exposure to chilling stress, the extension rate was only 25% of the control rate. Similar results have been reported in cotton (3, 6).

These results suggest that the indirect effects of chilling stress are more serious in actively growing tissue which is undergoing cell extension and/or cell division. The damage may be initiated in the cell membrane causing an increase in permeability and leakage of cell solutes and ultimately leading to metabolite disturbance and the accumulation of toxic products. All these lesions have been reported previously (10) but none were specifically identified in this experiment.

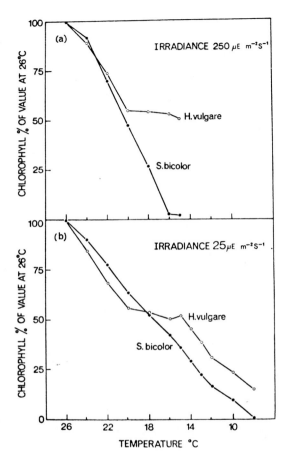

FIGURE 4. *Greening of etiolated barley (H. vulgare) and sorghum (S. bicolor) seedlings produced at 24°C and then transferred to a range of temperatures from 26°-8°C and two levels of irradiance (400-700 nm) (a) 250 $\mu E\ m^{-2}s^{-1}$, (b) 25 $\mu E\ m^{-2}s^{-1}$.*

D. Chlorophyll Synthesis in Etiolated Seedlings

The extent of greening of high temperature grown etiolated leaf tissue of barley and sorghum *(S. bicolor)* when exposed for 24 h to a range of temperatures from 26 to 8°C at two levels of irradiance is illustrated in Figure 4. All values of chlorophyll have been expressed as a percentage of the concentration found at 26°C.

The amount of chlorophyll synthesized in both species declined as the temperature was reduced, but below 17° in the chilling temperature range the ability of sorghum to accumulate chlorophyll was dependent on the level of irradiance. At the low flux density (25 μE m^{-2}s^{-1}) sorghum was able to synthesize and accumulate chlorophyll down to 8°C but at the higher flux (250 μE m^{-2}s^{-1}, approximately 10% of full sunlight) virtually no chlorophyll accumulated in sorghum at 16°C and below (Fig. 4a). By contrast, greening in barley was unaffected by the level of irradiance down to 15°C and where data are available at low irradiance, the relative concentration of chlorophyll was higher than in sorghum.

These results suggest that chlorophyll synthesis in etiolated tissue of a chilling-sensitive species such as sorghum, which had developed at high temperature, is possible at temperatures down to 8°C under conditions of low light flux, but below 16°C at higher levels of irradiance the rate of photo-oxidation or chlorophyll exceeds its rate of synthesis. Sensitivity to photo-oxidation under these conditions appears to be increased if chilling-sensitive plants are first etiolated at lower temperatures (13), or if chloroplast development is defective as in chlorophyll mutants of barley (24) and maize (14). Evidence from electron micrographs of sorghum leaf tissue after 24 h exposure at 15°C to either 25 or 250 μ μEm^{-2}s^{-1}, indicate that the failure to develop chlorophyll at the higher irradiance is due to the arrested development of the membrane system of the developing plastids which consisted largely of primary lamella layers with no evidence of grana development (Fig. 5a,b). A similar situation has been described in zantha mutants of barley at low temperature (24). Under the same condition, but at a lower irradiance, the sorghum chloroplasts were more comparable to those developed at the higher temperature (Fig. 5c,d) although the stroma thylakoids and grana were less developed in the mesophyll chloroplasts and there were no starch grains observed in the bundle sheath chloroplasts.

The presence of chlorophyll and the abundance of starch grains in the mesophyll chloroplasts, however, suggest that the chloroplasts were fully functional. This indicates that the photo-oxidation of chlorophyll and other leaf pigments such as carotene (11, 26) may prevent the development of a fully functional grana and stroma thylakoid system in the chloroplast and if prolonged, may damage membranes and result in permanent chlorosis of leaf tissue as is commonly found in many crops and grasses of tropical origin when grown at chilling temperatures (9).

The sensitivity of etiolated tissue to photo-oxidation in the light is a function of the light flux and appears to be a general response in chilling-sensitive plants and may reflect a temperature induced decline in the rate of a key enzyme(s) responsible for the synthesis of chlorophyll and its associated pigments. As the extent of greening can be readily observed or measured, it may be possible to use this response to identify variation in the levels of chilling-tolerance in chilling-sensitive crop plants.

As a preliminary test of this hypothesis a range of known chilling-tolerant and sensitive species were etiolated as described previously at 24°C, then after 24 h equilibration at 17°C were greened for 24 h at 17°C at an irradiance of 250 μE cm^{-1}s^{-1}. The results of this trial are given in Table 1. The chilling-resistant species are immediately identifiable,

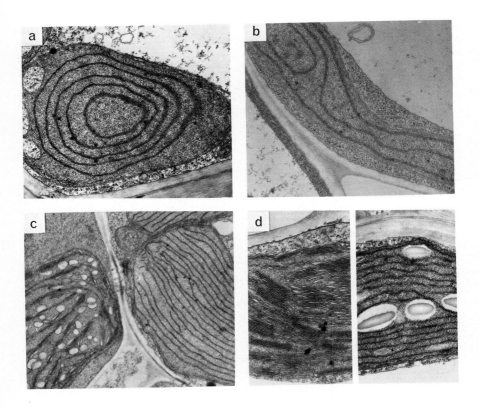

FIGURE 5. Chloroplasts of sorghum (S. bicolor) following exposure
to different levels of irradiance at two temperatures. (a,b) Developing
plastids from sorghum exposed at 15^0C to an irradiance of 250
$\mu E\ m^{-2}s^{-1}$, containing only primary lamella layers, often in concentric
rings. No grana development or even thylakoid extensions suggesting
incipient grana development were present.   15000 and 21000 X
respectively. (c) Mesophyll and bundle sheath chloroplasts greened at
15^0C at an irradiance of 25 $\mu E\ m^{-2}s^{-1}$. Lamella system of mesophyll
chloroplasts is less developed than at 24^0C. Numerous small starch
grains are present in mesophyll, but absent from bundle sheath
chloroplasts.   15000 X. (d) Mesophyll and bundle sheath chloroplasts
(showing prominent starch grains) greened at 24^0C at 250 $\mu Em^{-2}s^{-1}$.
21000 X.

as they synthesize appreciable amounts of chlorophyll under these
conditions. Amongst the chilling-sensitive group the relative amounts of
chlorophyll synthesized do correlate reasonably well with the known or
suspected chilling-sensitivity of these species.

TABLE I.   Chlorophyll content of the first leaf or cotyledons of etiolated seedlings of two groups of chilling-resistant and chilling-sensitive plants greened for 24 h at $17^{\circ}C$ at an irradiance of 250 $\mu E\ m^{-2}\ s^{-1}$. All values expressed as % of chlorophyll present in the tissue greened under similar conditions at $24^{\circ}C$.

Chilling-resistant	% Chlorophyll	Chilling-sensitive	% Chlorophyll
Cauliflower	83	Pearl Millet	28
Cabbage	62	Rice (Indica)	23
Radish	60	Cantaloupe	22
Barley*	57	Cotton	21
Wheat*	45	Sorghum	19
Oats*	43	Maize	19

*Spring varieties.

IV. CONCLUSIONS

Most plants adapted to tropical and subtropical environments respond in a similar fashion when subjected to chilling temperatures and develop comparable physiological damage when this exposure is prolonged. This is consistent with the hypothesis that the primary event initiating these responses is a temperature induced physical change in their membranes.

One common feature of many responses of chilling-sensitive species is a discontinuity in the Arrhenius plots which occur over a fairly characteristic temperature range, as found for the sorghum species used in this study. The existence of such breaks in a chilling-resistant species such as barley indicates that this phenomenon is not unique to chilling-sensitive species and suggests that the magnitude of the temperature coefficients above and below the break, or transition temperature, is more significant in identifying chilling-sensitive species than the breaks *per se,* or the temperatures at which they occur.

Variation in chilling-sensitivity between members of related species and within ecotypes of a single species was apparent in sorghum and has been reported in a number of other crop plants. This appears to be an adaptive response and under genetic control. One approach which appears to provide a means of identifying such variability and one which could provide an index of chilling-sensitivity, is the greening response of etiolated leaves. A further study has been initiated with a range of chilling-sensitive crop species using small leaf segments or cotyledons to determine if this technique could be developed as a rapid mass screening procedure to assist in selecting for greater chilling-tolerance.

ACKNOWLEDGMENTS

We are grateful to Mr. B. Ellem of the Division of Mathematical Statistics, CSIRO for assistance with probit analyses and for access to CIRONET the CSIRO computer, also to Dr. Hazel Harris and Dr. V. J. Bofinger, University of New England, Armidale, NSW, for statistical advice and helpful criticism.

V.   REFERENCES

1. Arnon, D. *Plant Physiol. 25,* 1-15 (1949.
2. Breidenbach, R. W., Wade, N. L. and Lyons, J. M. *Plant   Physiol. 54,* 324-327 (1974).
3. Buxton, D. R., Sprenger, J. P., and Pegelow, E. J. *Crop Sci. 16,* 471-474 (1976).
4. Cal, J. P. and Obendorf, R. L. *Crop Sci. 12,* 369-373 (1972a).
5. Cal, J. P. and Obendorf, R. L. *Crop Sci. 12,* 572-575 (1972b).
6. Christiansen, M. N. *Plant Physiol. 42,* 431-433 (1967).
7. Creencia, R. P. and Bramlage, W. J. *Plant Physiol. 47,* 389-392 (1971).
8. Finney, D. J. "Statistical Method in Biological Assay", 2nd Ed., Hafner, N.Y. (1964).
9. Kawanabe, S. *Proc. Japan Soc. Plant Taxonomists 2,* 17-20 (1968).
10. Levitt, J. *In* "Responses of Plants to Environmental Stress" (T. T. Kozlowski, ed.) pp. 27-43, Academic Press, New York.
11. McWilliam, J. R. and Naylor, A. W. *Plant Physiol. 42,* 1711-1715 (1967).
12. McWilliam, J. R. and Ferrar, P. J. *In* "Mechanisms of Regulation of Plant Growth" (R. L. Bieleski, A. R. Ferguson, M. M. Cresswell, Eds.), pp. 467-476. Bulletin 12, The Royal Society of New Zealand, Wellington (1974).
13. Millerd, A. and McWilliam, J. R. *Plant Physiol. 43,* 1967-1972 (1968).
14. Millerd, A., Goodchild, D., and Spencer, D. *Plant Physiol. 44,* 567-583 (1969).
15. Nishiyama, I. *Plant and Cell Physiol. 16,* 533-536 (1975).
16. Nolan, W. G. and Smillie, R. M. *Biochimica et Biophysica Acta 440,* 461-475 (1976).
17. Oka, H. *Jap. J. Breeding 4,* 140-144 (1954).

18. Obendorf, R. L. and Hobbs, P. R. *Crop Sci. 10,* 563-566 (1970).
19. Patterson, B. D., Murata, T., and Graham, D. *Aust. J. Plant Physiol. 3,* 435-442 (1976).
20. Pinnell, E. L. *Agron. J. 41,* 562-568 (1949).
21. Pomeroy, M. K. and Andrews, G. J. *Plant Physiol. 56,* 703-706 (1975).
22. Simon, E. W., Minchin, A., McMenamin, M. M. and Smith, J. *New Phytol. 77,* 301-311 (1976).
23. Smillie, R. M. *In* "Genetics and Biogenesis of Chloroplasts and Mitochondria" (Th. Bucher *et al.* Eds.), pp. 103-110. Elsevier, Amsterdam (1976).
24. Smillie, R. M., Nielsen, N. C., Henningsen, K. W., and von Wettstein, D. *Aust. J. Plant Physiol. 4,* 439-449 (1977).
25. Umbreit, W. W. *In* "Manometric Techniques" (W. W. Umbreit, R. H. Burris and J. F. Stauffer, Eds.), pp. 1-17. Burgess Publ. Co., Minnesota (1964).
26. van Hasselt, Ph. R. *Acta Bot. Neerl. 21,* 539-548 (1972).
27. Wade, N. L., Breidenbach, R. W., and Lyons, J. M. *Plant Physiol. 54,* 320-323 (1974).
28. Wiles, E. L., and Downs, R. J. *Seed Sci. and Tech. 5,* 649-657 (1977).

CHILLING INJURY ASSAYS FOR PLANT BREEDING

R. E. Paull

Department of Botany
University of Hawaii at Manoa
Honolulu, Hawaii

B. D. Patterson and D. Graham

Plant Physiology Unit
CSIRO Division of Food Research
and School of Biological Sciences
Macquarie University
North Ryde, N.S.W., Australia 2113

I.   INTRODUCTION

Plants from tropical and subtropical regions show considerable genetic diversity in their ability to tolerate chilling. The wild green-fruited tomato, *Lycopersicon hirsutum* Humb. & Bonpl., is found from sea level in Ecuador under tropical rain forest conditions to 3,300m in the Andes in northern Peru, well into the zone of chilling temperatures (65). For geographical varieties of this species, chilling tolerance increases with altitude of origin (59, 60). Similar variations are shown by cultivated species of the genus *Passiflora* (58). The diversity in wild populations may be ascribed to variations in the intensity of selection pressures which remove the more chilling sensitive members from the population. There is considerable interest in the development of a selection procedure for determining chilling tolerance or resistance for use in crop breeding (18, 37, 66, 52). Much of the published work on chilling injury does not suggest methods for quantifying small differences in chilling tolerance, because it concerns comparisons between unrelated plants. Therefore, in this review, studies using closely related species,

[a]*Journal Series #2369 of the Hawaii Agricultural Experiment Station.*

507

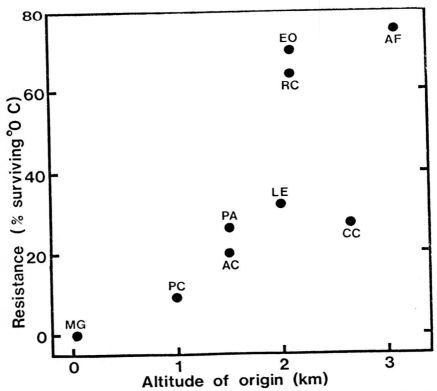

FIGURE 1.   *Survival of altitudinal variants (L. hirsutum) after a chilling stress.   Seedlings were chilled for 7 days at 0°C after the appearance of the first true leaf.   Survival was assessed after a further week at 18/22°C in a growth cabinet (60).*

cultivars or races within the same genera have been emphasized. Chilling injury is usually manifested as visual symptoms which are difficult to quantify.   This has meant that progress has been slow in developing quantitative measures of chilling tolerance. It has often been noted that different criteria may measure different physiological parameters at different developmental stages.   This is true for instance for bananas (57) and rice (9).   At some stages plants can be hardened against chilling (79); this must also be taken into account when measuring sensitivity to chilling.

An assay for use in a breeding program should be i) rapid, ii) reproducible, iii) non-destructive to the whole plant, iv) quantitative or at least able to rank material, and v) have a low labor cost per assay.

II.   SELECTION METHODS

A.   *Assays Under Natural or Semi-Natural Conditions*

These tests integrate the plant's ability to survive, grow and develop under chilling conditions.   A program based on field testing in areas which are marginally chilling is dependent upon year to year variation. However, unpredictable field conditions can be replaced by controlled conditions. Holbert *et al.* (31) designed a portable refrigerated chamber to study cold resistance of adult corn plants in the field. The method gave a relative ranking, but was slow. To overcome some of these problems with whole plants, parts of plants or seedlings have been used. Patterson *et al.* (60) tested seedlings of the wild altitudinal collection of tomatoes *L. hirsutum* by chilling at 0°C. Survival was related to the altitude of origin (Figure 1, 2). In this type of selection procedure the genetic material whose tolerance falls between the arbitrary temperature chosen and temperature below which injury normally occurs (less than 10°C to 12°C) would be lost. However, it could probably be used to select the most chilling resistant seedlings from a genetically diverse population.

TABLE I. Lag Time to 10% Germination at 15°C, Lowest Temperature at which 20% Rate Per Day of Germination was Attained and the Minimum Germination Temperature at which 50% Germination has Attained for an Altitudinal Collection of Lycopersicon hirsutum, L. esculentum and their Hybrids. For Details of the Altitudinal Varieties of L. hirsutum see (60).

Species	Altitude of origin (m)	Lag time (days) (to 10% germination at 15°C)	Temperature ($^{\circ}$C) (20%) rate of germination attained	Minimum germination temperature ($^{\circ}$C) (50% germination)
L. hirsutum				
race - BG	400	8.3	17.0	15.2
PC	1000	5.0	18.2	13.0
AC	1500	4.4	12.2	8.8
LE	2000	5.6	10.4	10.0
EO	2100	2.7	21.8	7.1
CC	2650	5.7	13.9	9.2
AF	3100	3.6	11.4	5.9
Ric	3300	-	-	7.4
L. esculentum				
Rutgers		5.5	16.0	9.1
PI-341985		2.0	12.6	5.5
Rutgers x BG		3.7	17.0	9.1
Rutgers x AF		8.0	21.7	7.0

Tests based on seed germination at marginally chilling temperatures ($10°$ to $12°C$) have been applied to crops such as: beans (18, 19, 38), lima beans (75), cotton (13, 42, 46) and tomatoes (35). In altitudinal varieties of *Lycopersicon hirsutum* there was considerable variation in lag time before radical emergence and the rate of germination, and the ability to germinate at low temperature did not necessarily correlate with the ability to survive chilling at $0°C$ (Table I). The lag time before germination was more dependent on temperature than was germination rate.

Methods based upon growth and development can be applied to parental selection and bulk breeding lines. Selection methods can be based on growth rate (1, 3, 35, 56, 70, 71), dry weight changes (45), root development (10) and ability of fruit to ripen (4, 6). A completely different ranking order was found between two different criteria of adaptation to low temperature: germination at $8.5°C$ and fruit set at $4.5°C$ night. Some bean lines with high tolerance to chilling after seedling emergence appeared to lack tolerance during germination (36). Results with tomato also illustrate that different rankings of chilling tolerance can be obtained when different criteria (35) are used (Table 2). This indicates that chilling sensitivity in the broad sense may not have a single cause.

TABLE II.   *Effects of Low Temperature on Seed Germination, Seedling Growth and Fruit Set of Tomato Varieties. The Numbers in Parentheses are the Order of Ranking. Note Different Ranking Order Obtained with Different Criteria of Chilling. Adapted from Kemp (35).*

	Germination at $8.5°C$ (%)	Seedling growth $21°C/10°C$ (Scale 1 to 10)	Fruit set at $4.5°C$ night (%)
Earlinorth	64 (4)	3 (4)	58 (2)
Rocket	69 (3)	3 (4)	28 (3)
Borita	52 (5)	8 (2)	69 (1)
Fireball	0 (6)	3 (4)	6 (7)
Hilda 153	96 (1)	6 (3)	8 (6)
Early Rutgers	0 (6)	2 (5)	0 (8)
PI-280597	79 (2)	9 (1)	10 (5)
Cold Set	52 (5)	6 (3)	20 (4)

Much work has been reported on the determination of storage conditions of fresh fruit commodities. These results are generally presented as visual criteria such as the percentage of damaged fruit. Often cultivars of a species differ in resistance. For example, Yolo Wonder *Capsicum* was particularly sensitive compared to some other cultivars (25). Similar variations have been found for cucumbers (2) and bananas (57). Tests on the harvested product involve considerable time and expense in carrying the crop to marketing stage. It would be preferable to be able to make the selection at an early growth stage.

B. Laboratory Measures

1. Tissue Culture. This technique offers the advantage of being able to subject large numbers of cells to selection after mutagenesis or introduction of genes using protoplast fusion, which offers the possibility of overcoming sterility barriers between species and genera. Dix and Street (20, 21) exposed cell cultures of *Nicotiana sylvestris* and *Capsicum annuum* to chill selection with and without mutagen (see also Dix, this volume). The development of this technique and its subsequent applications will depend on the ability to regenerate plants from the tissue cultures.

2. Plant Tissue Tests. Chilling eventually disrupts the cellular integrity of plants. The subsequent leakage of cell contents, loss of active transport and changes in the levels of tissue metabolites are possible to measure and could be used to separate lines of different chilling resistance.

Leakage of cell contents has been used to assay chilling injury. Katz and Reinhold (33) found by direct observation that chilled epidermal cells of *Coleus* leaves plasmolyze more slowly than unchilled cells. An effusate conductivity assay has been applied to corn (15), cucumber (74, 82), cotton (12), and to a collection of *Passiflora* species (58). The results with the *Passiflora* collection showed that it was possible to rank *Passiflora* species for chilling tolerance on the basis of time to 50% electrolyte loss at $0^{\circ}C$. All of the above assays were done following leakage periods of longer than one day. Tatsumi and Murata (74) showed that pitting in cucumber fruit occurred before an increase in leakage was apparent. It therefore appears that gross leakage may not always be sensitive enough to detect the changes which are visually apparent. We studied leakage of leaf cells of the tomato species *L. hirsutum* and *L. esculentum.* Using the ratio of the rate of leakage at $1^{\circ}C$ to that at $20^{\circ}C$, the ecotype of *L. hirsutum* from 3200 m altitude showed a relatively lower rate of leakage at $1^{\circ}C$ than did that from 30 m altitude. For *L. esculentum,* the low temperature germinating selection PI-341985 showed a lower rate at $1^{\circ}C$ than did the variety Rutgers (Table 3).

TABLE III. Ratio of Leakage Rate at $1^{o}C$ and $20^{o}C$ for the First Six
Hours after Exposure to Temperature of Rubidium-86
Preloaded Leaf Cells of Lycopersicon spp. and Passiflora
spp.

Species		Altitude of collection (M) or chilling sensitivity	Ratio $\dfrac{1^{o}}{20^{o}C}$ rates
L. hirsutum race	MG	30	2.2
	BG	400	2.9
	PC	1000	1.68
	AC	1500	0.93
	AF	3100	1.32
	Ric	3300	0.6
L. esculentum	Rutgers	-	1.29
	PI-341985	-	0.82

Organic solute loss from chilled tissue has been considered as a
selection procedure. Christiansen (11), for example compared root
exudation from several genetic lines of cotton at chilling temperatures
and showed differences which may be a basis for genetic selection.

An interesting variant of the conductivity leakage measure is that of
Katz and Reinhold (33) who found an increase in the electrical
conductivity of *Coleus* leaves before damage was visible. This has not
been followed up as a possible breeder's selection method for chilling
tolerance.

Active transport properties of plant cells are intrinsic membrane
properties. Paull *et al* (61) showed that leucine uptake by leaf
fragments of the tomato Rutgers was more sensitive than the low
temperature germinating PI-341985. Uptake into the low altitude form
of *L. hirsutum* was more sensitive to chilling than into the high altitude
form. The assay was slow and hence impractical as a mass selection
procedure. It could be simplified by using just the ratio of uptake rate at
$20^{o}C$ and $1^{o}C$. Other active transport properties such as a direct
measurement of electrical potential may be good criteria but have not
yet been applied to chilling stress.

The use of vital stains such as neutral red or dyes excluded by intact
membrane (e.g. Evan's Blue, fluorescein diacetate) to detect uninjured,
reversibly injured and dead cells has not been developed into an assay for
chilling injury. Widholm (80) tested a number of stains and found
fluorescein diacetate for viable cells and phenosafranine for dead cells
the most reliable after freezing suspension cultures of a number of
chilling sensitive plants. Tomato culture cells held for 6 days below $10^{o}C$
to $12^{o}C$ showed a large increase in the number of cells which did not stain
with fluorescein diacetate (5). These tests are rapid, simple and
applicable to plants which can be regenerated from cultures.

Chilling temperatures cause a disproportionate decrease in respiration in isolated mitochondria from chilling sensitive species (44). Sweet potato mitochondria isolated after five weeks of storage at 7.5°C showed a decline in oxidation and phosphorylation activity when assayed at 25°C (40). Similar changes in respiration have been found in whole tissues; cucumber leaves have a Q_{10} of 5.7 below 12°C, 1.7 above 12°C (51). At non-chilling temperatures the rate of CO_2 production for whole fruit decreases with storage, whereas at chilling temperatures the rate increases with time to a plateau followed by a decline. In cucumber fruit the increased rate occurs at the same time as the onset and development of chilling injury (23). These results point out the need to specify the stage at which the assay was performed, otherwise running a selection method based upon respiration is impractical. An assay could possibly be developed using the reduction of triphenyl tetrazolium chloride (TTC) (72). Although the method requires only small amounts of material variable results with the TTC test have been reported (71) and it may not be practical for large numbers of samples. This test has been applied to tomato fruit (14). Breidenbach and Waring (5) used it on tomato cell culture and found that reductive capacity decreased below 10°C to 12°C. The method may be a rapid indicator of ability to respire and could be developed as a supplementary assay.

Photosynthesis is severely reduced by short exposures to chilling temperatures (17). Though there is no clear relationship between chilling sensitivity of a species and the presence or absence of a change in the temperature dependence of the Hill activity at low temperatures (16), an assay can be based on the rate of loss of chloroplast activity of detached leaves of susceptible species stored at 0°C (69). After various periods of chilling the isolated chloroplasts can be assayed at 23°C for photoreduction of ferricyanide in the presence of the uncoupler gramicidin D. The rate of decline in activity was taken as a measure of sensitivity.

Patterson *et al* (63) described a simple assay based upon the minimum temperature of chloroplast development. This could be applied to parental and bulk beeding lines. The seeds were germinated in the dark and the seedlings then exposed to dim red light to promote yellow etioplast development. The yellow plants were then transferred to a thermogradient bar. After temperature equilibration, the leaves were illuminated with white light for up to 6 days. The chlorophyll was extracted and determined or the seedlings subjectively ranked as to whether or not greening had occurred. This assay showed that a high altitude wild tomato was better able to green at low temperature than low altitude forms.

Chlorophyll fluorescence is an intrinsic probe of membranes. Murata and Fork (53) and Melcarek and Brown (47) showed that fluorescence monitoring of intact leaves is a simple method to specify differential responses to chilling stress. The delayed fluorescence yield of chilling sensitive plants is reduced to a greater extent than that of chilling resistant species (48). Chlorophyll fluorescence has been used as a rapid screening technique for photosynthetic mutants of higher plants

TABLE IV. *Summary of Reports on the Inheritance of Chilling Tolerance in Sensitive Crops*

Crop	Character study	Maternal effect	Genetic effect Number of genes	Dominance tolerance	References
Corn	Low temperature seed germination	yes	several	some	Pinnell (63)
		yes	multigenic	--	Helgason (29)
		yes	--	--	Ventura (78)
		--	multigenic	some	Grogan (26)
		yes	multigenic	some	Pesev (62)
	Imbibitional chilling injury	yes	--	--	Cal and Obendorf (7)
Cotton	Low temperature seed germination	yes	multigenic	partial to complete	Marani and Dag (46)
	Imbibitional chilling injury	yes	--	--	Christiansen & Lewis (13)
Tomato	Low temperature seed germination	--	single gene	recessive	Cannon et al. (8)
		yes	multigenic minimum 3 to 5	recessive	Ng & Tigchelaar (55)
		--	multigenic 24 pairs	additive	El Sayed & John (24)
	Fruit set	--	single gene	recessive	Kemp (34)
Beans	Low temperature seed germination	--	multigenic	low heritability	Dickson (18)
Muskmelon	Low temperature seed germination	--	multigenic	dominant	Kubicki (39)
Rice	Flower sterility after low temperature exposure	--	multigenic	dominant	Toriyama & Futsuhara (76)

(49). The fluorescence can be recorded either on photographic film, or by using a portable solid state fluorometer (67) or with a light meter. This technique has been applied to leaf discs of altitudinal varieties of the wild tomato *L. hirsutum* which were exposed to 1°C for 2 days. It was possible to rank the altitudinal forms visually and by the ratio of the peak to steady-state fluorescence in a similar order to that obtained for the seedling survival shown in Figure 1. The technique is rapid, having relatively low labor requirement and could be adapted to handle large numbers of samples with a low equipment requirement. The technique gave positive results with tomato, beans, egg plant, papaya, and capsicum.

Chilled tropical plants show large changes in the levels of certain metabolites. Disproportionate change in the velocity of the glycolytic and respiratory pathways of metabolism (43) may be at least in part responsible for such metabolic imbalances. These changes indicate another possible basis for a selection procedure. Starch accumulates in chloroplasts of the grass *Digitaria decumbens* moved from 30°C/30°C to 30°C/10°C (30). The difference in starch levels after two consecutive nights at 10°C was ten times that at 30°C. As yet, this has not been shown with other species. Other metabolites shown to change at chilling temperatures in various plants are decreases in total sugars and ascorbic acid (50, 77), and increases in glutamate, aspartate and chlorogenic acid (22, 41); and accumulation of α -ketoglutarate, pyruvate, ethanol, aldehydes, and α -farnesene (54, 68, 81). These changes are not universal, for while ascorbic acid decreased in chilled pineapple fruit, banana and sweet potato (28, 40, 50) it did not in tomato (14).

Tyrosine supplied to chilled pineapple fruit (73) intensified the symptoms of endogenous brown spot, which is a chilling injury. The formation of a similar disorder in lettuce (russet spotting) was preceded by an increase in phenylalanine ammonia lyase (PAL) whereas in the resistant varieties much less PAL was developed (32). Another enzyme of the chlorogenic acid phenylpropanoid pathway, hydroxycinnanyl-transferase, increases in chilled tomatoes (64). Tomato tissue cultures visibly darkened below 9°C presumably by oxidation of phenolics.

This group of assays holds considerable promise as a rapid assay for chilling tolerance.

III.   GENETICS OF CHILLING RESISTANCE

Few reports deal with inheritance of chilling resistance. The results of some of the genetic studies are summarized in Table 4. For the germination of cotton and tomato, maternal effects are important and for maize, this has been associated with the double genetic contribution from the female side to the endosperm (63). This would not be the case for cotton and tomato, as the endosperm is almost completely absorbed during seed development. The maternal effect could be attributable to the maternally derived non-nuclear genes. With the exception of low temperature fruit set in tomatoes, which was reported (34) to be recessive and simply inherited as was tomato seed germination (8), chilling tolerance appears to be an additive, multigeneric character. However, it should be borne in mind that different genetic mechanisms apparently control resistance or tolerance at different development stages (27).

IV. CONCLUSIONS

Many of the above assays correlate with the final performance of the plant as a whole to chilling stress. A major problem with any assay is that it may not indicate the chilling response at a different physiological or developmental stage.

A number of assays show promise for use in a plant breeding program for chilling tolerance. The final test will be in the field, but laboratory measurement will be useful during the initial selection. Loss of fluorescence and changes in concentrations of certain metabolites and enzymes show promise as rapid, reproducible and quantitative assays for chilling tolerance.

V. REFERENCES

1. Adair, C. R. *Crop Sci. 8,* 264-265 (1968).
2. Apeland, J. *In* "Studies on Storage of Fruits and Vegetables." International Instit. of Refrigeration, p. 325-333)1966).
3. Austin, R. B., and Maclean, M. S. M. *J. Hort. Sci. 47,* 279-290 (1972)
4. Biale, J. B. *Amer. Soc. Hort. Sci. Proc. 39,* 137-142 (1941).
5. Breidenbach, R. W., and Waring, A. J. *Plant Physiol. 60,* 190-192 (1977).
6. Burgis, D. S. *Proc. Florida State Hort. Soc. 83,* 135-137 (1970).
7. Cal, J. P., and Obendorf, R. L. *Crop Sci. 12,* 369-373 (1972).
8. Cannon, O. S., Gatherum, D. W., and Miles, W. E. *HortScience 8,* 404-405 (1973).
9. Chang, H-H, Ku, Y-L, and Lai, K-L. *J. Agric. Assoc. China N.S. 86,* 19-27 (1974).
10. Christiansen, M. N. Beltwide Cotton Prod. Res.Conf., 1971 (1971).
11. Christiansen, M. N. Beltwide Cotton Prod. Res. Conf., 1972 (1972).
12. Christiansen, M. N., Carns, H. R., and Slyter, D. L. *Plant Physiol. 46,* 53-56 (1970).
13. Christiansen, M. N., and Lewis, C. F. *Crop Sci. 13,* 210-212 (1973).
14. Crafts, C. C., and Heinze, P. H. *Proc. Amer. Soc. Hort. Sci. 64,* 343-350 (1954).
15. Creencia, R. P., and Bramlage, W. J. *Plant Physiol. 47,* 389-392 (1971).
16. Critchley, C., Smillie, R. M., and Patterson, B. D. *Aust. J. Plant*

Physiol. 5, 443-448 (1978).
17. Crookston, R. K., O'Toole, J., Lee, R., Ozbun, J. L., and Wallace, D. H. *Crop Science 14,* 457-464 (1974).
18. Dickson, M. H. *Crop Sci. 11,* 848-850 (1971).
19. Dickson, M. H. *HortScience 8,* 410 (1973).
20. Dix, P. J. *Z. Pflanzenphysiol. 84,* 223-226 (1977).
21. Dix, P. J., and Street, H. E. *Ann. Bot. 40,* 903-910 (1976).
22. Duke, S. H., Schrader, L. E., Miller, M. G., and Niece, R. L. *Plant Physiol. 62,* 642-647 (1978).
23. Eaks, I. L., and Morris, L. L. *Plant Physiol. 31,* 308-314 (1956).
24. El Sayed, M. N., and John, C. A. *J. Amer. Soc. Hort. Sci. 98,* 440-443 (1973).
25. Gorini, F. L., Zerbini, P. E., and Uncini, L. *Acta Hort. 62,* 131-149 (1977).
26. Grogan, C. O. *Proc. Ann. Corn-Sorghum Res. Conf. 25,* 90-98 (1970).
27. Gupta, G., and Kovacs, I. *Z. Acker-und Pflanzenbau 143,* 196-203 (1976).
28. Harris, P. L., and Poland, G. L. *Food Res. 4,*17-327 (1939).
29. Helgason, S. B. *Diss. Abst. 13,* 461 (1953).
30. Hilliard, J. H., and West, S. H. *Science 168,* 494-496 (1970).
31. Holbert, J. R., Burlison, W. L., and Johnson, A. G. *U.S.D.A. Circular* No. 285 (1933).
32. Hyodo, H., Kuroda, H., and Yang, S. F. *Plant Physiol. 62,* 31-35 (1978).
33. Katz, S., and Reinhold, L. *Isr. J. Botany 13,* 105-114 (1964).
34. Kemp, G. A. *Amer. Soc. Hort. Sci. Proc. 86,* 565-568 (1965).
35. Kemp, G. A. *Can. J. Plant Sci. 48,* 281-286 (1968).
36. Kemp, G. A. *Can. J. Plant Sci. 58,* 169-174 (1978).
37. Knight, R. J. *HortScience 6,* 15-18 (1971).
38. Kooistra, E. *Euphytica 20,* 208-213 (1971).
39. Kubicki, B. *Genet. Polan. 3,* 265-274 (1962).
40. Lieberman, M., Craft, C. C., Audia, W. V., and Wilcox, M. S. *Plant Physiol. 33,* 307-311 (1958).
41. Lieberman, M., Craft, C. C., and Wilcox, M. S. *Proc. Amer. Soc. Hort. Sci. 74,* 642-648 (1959).
42. Ludwig, C. A. *J. Agric. Res. 44,* 367-380 (1932).
43. Lyons, J. M. *Ann. Rev. Plant Physiol. 24,* 445-466 (1973).
44. Lyons, J. M., and Raison, J. K. *Plant Physiol. 45,* 386-389 (1970).
45. McWilliams, E. L., and Smith, C. W. *HortScience 13,* 179-180 (1978).
46. Marani, A., and Dag, J. *Crop Sci. 2,* 243-245 (1962).
47. Melcarek, P. K., and Brown, G. A. *Plant and Cell Physiol. 18,* 1099-1107 (1977a).
48. Melcarek, P. K., and Brown, G. A. *Plant Physiol. 60,* 822-825 (1977b).
49. Miles, C. C., and Daniel, D. J. *Plant Sci. Letters 1,* 237-240 (1973).
50. Miller, E. V. *Plant Physiol. 26,* 66-75 (1951).
51. Minchin, A., and Simon, E. W. *J. Exp. Bot. 24,* 1231-1235 (1973).

52. Mock, J. J., and Bakri, A. A. *Crop Sci. 16,* 230-233 (1976).
53. Murata, N., and Fork, D. C. *Plant Physiol. 56,* 791-796 (1975).
54. Murata, T. *Physiol. Plant. 22,* 401-411 (1969).
55. Ng, T. J., and Tigchelaar, E. C. *Jour. Amer. Soc. Hort. Sci. 98,* 314-316 (1973).
56. Ormrod, D. P., and Bunter, W. A. *Agron. J. 53,* 133-134 (1961).
57. Pantastico, E. B., Grierson, W., and Soule, J. *Amer. Soc. Hort. Sci. Trop. Reg. Proc. 11,* 82-91 (1967).
58. Patterson, B. D., Murata, T., and Graham, D. *Aust.   J.   Plant Physiol. 3,* 435-442 (1976).
59. Patterson, B. D., and Graham, G. *J. Exp. Bot. 28,* 736-743 (1977).
60. Patterson, B. D., Paull, R. E., and Smillie, R. M. *Aust.   J.   Plant Physiol. 5,* 609-617 (1978).
61. Paull, R. E., Patterson, B. D., and Graham, D. *Aust.   J.   Plant Physiol.* (In press) (1979).

62. Pesev, N. V. *Theor. Appl. Genetics 40,* 351-356 (1970).
63. Pinnell, E. L. *Agron. J. 41,* 562-568 (1949).
64. Rhodes, M. J. C., and Wooltorton, L. S. C. *Phytochem. 16,* 655-659 (1977).
65. Rick, C. M. *In* "Genes, Enzymes and Population" (A. M. Srb. ed.), pp. 255-269, Plenum Press, New York (1973).
66. Robinson, W., and Kowalewski, E. *Proc. XIX Inter. Hort. Congr. 18,* 716 (1974).
67. Schreiber, U., Groberman, L., and Vidaver, W. *Rev. Sci. Instrum. 46,* 538-542 (1975).
68. Smagula, J. M., and Bramlage, W. J. *HortScience 12,* 200-203 (1977).
69. Smillie, R. M., and Nott, R. *Plant Physiol. 63,* 796-801 (1979). 70.
70. Smith, O. F. *J. Amer. Soc. Agron. 27,* 467-479 (1935).
71. Stergios, B. G. and Howell, G. S. *J. Amer. Soc. Hort. Sci. 98,* 325-330 (1973).
72. Steponkus, P. L., and Lamphear, F. O. *Plant Physiol. 42,* 1423-1426 (1967).
73. Sun, S-K. *Plant Prot. Bull. (China) 13,* 39-48 (1971).
74. Tatsumi, Y., and Murata, T. *J. Japan Soc. Hort. Sci. 47,* 105-110 (1978).
75. Toole, V. K., Wester, R. E., and Toole, E. H. *Proc.   Amer.   Soc. Hort. Sci. 58,* 153-159 (1951).
76. Toriyama, K., and Futsuhara, Y. *Jap. J. Breeding 10,* 143-152 (1960).
77. Van Lelyveld, L. J., and DeBruyn, J. A. *Agrochemophysica 8,* 65-68 (1976).
78. Ventura, Y. *Diss. Abst. 20,* 1985-1986 (1959).
79. Wheaton, T. A., and Morris, L. L. *Proc. Amer. Soc. Hort. Sci. 91,* 529-533 (1967).

80. Widholm, J. M. *Stain Technology 47*, 189–194 (1972).
81. Wills, R. B. H., McBailey, W., and Scott, K. *Plant Physiol. 56*, 550–551 (1975).
82. Wright, M., and Simon, E. W. *J. Exp. Botany 24*, 400–411 (1973).

Much of the data developed for study of temperature related phenomena in plants, particularly that related to the "membrane hypothesis" has been presented as Arrhenius plots. As discussed more fully in PART VI. EPILOGUE, it may not be appropriate to apply Arrhenius law to some of the events occurring in response to temperature, but more importantly the drawing of lines, and hence conclusions, must be approached more cautiously than generally exists in the current literature.

The Seminar prompted several to approach this question of statistical tests to determine the "best-fit" of straight lines, straight line segments and curves to Arrhenius plots and other data. Several such analyses are presented in this section and they provide useful approaches for consideration. There is some redundancy and overlap in the presentations but rather than attempting to synthesize them into one consensus, they are each presented intact to allow the reader to choose the method best suited to his or her needs. It was obvious from discussions in the Seminar that some statistical rigor should be applied to Arrhenius-type data and the tests described here provide that opportunity.

BREAKS OR CURVES? A VISUAL AID
TO THE INTERPRETATION OF DATA

Mary E. Willcox

CSIRO Division of Mathematics and Statistics
North Ryde, N.S.W. 2113, Australia

Brian D. Patterson

CSIRO Division of Food Research
North Ryde, N.S.W. 2113, Australia

The visual contrast between smooth curves and intersecting straight lines in Arrhenius plots can be emphasized by changing the angle between the x and y axes. This has the effect of compressing the data in a specified direction.

Figure 1 shows generated points which lie along three intersecting straight lines (A) or on a smooth curve (B), although the difference is not obvious to the eye. Figure 2 shows the same data plotted with the angle between the axes increased to give a compression of ten times along the diagonal, and then enlarged to have the same dimensions as the original data. This treatment makes it obvious which set of points lie on a smooth curve, and which set lie along three intersecting straight lines. Table I shows a worked example of the method.

In the preceding paper (1), Figure 1 shows an Arrhenius plot of the temperature dependence of spin label motion in tomato lipids. Figure 2 of the same paper shows the data after compression. The data are well represented as a smooth curve tending to linearity at either end.

Note that the linearity of straight lines is maintained during compression, while the scatter of data points increases proportionately as the curvature is increased.

M. E. Willcox and B. D. Patterson

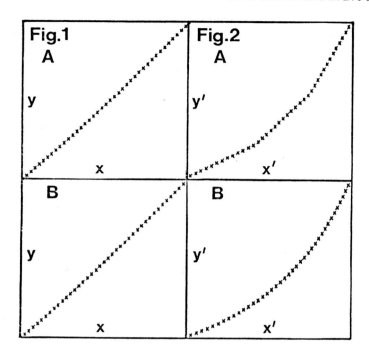

FIGURE 1. Points generated to lie on (A), three straight lines; (B), a smooth curve.

FIGURE 2. The data of A and B in Figure 1 compressed and enlarged ten times. It is now obvious that A consists of three straight lines and that B is a curve. When the data are transformed to lie between zero and 1, the formulae used for calculating the new coordinates are:--

$$x' = n[x - \frac{n-1}{2n} (x + y)]$$

$$y' = n[y - \frac{n-1}{2n} (x + y)]$$

where n is the degree of compression.

TABLE I.   *Worked Example of a Compression of Five Times (n = 5)*

Original coordinates		Coordinates transformed to lie between 0 and 1		New coordinates for n = 5	
x-axis	y-axis	x	y	x'	y'
10.0000	20.0000	0.0000	0.0000	0.0000	0.0000
10.6000	20.5000	0.1000	0.0833	0.1333	0.0500
11,1000	21.0000	0.1833	0.1667	0.2167	0.1333
11.6000	21.5000	0.2667	0.2500	0.3000	0.2167
12.2000	22.0000	0.3667	0.3333	0.4333	0.2667
12.7000	22.4000	0.4500	0.4000	0.5500	0.3000
13.2000	22.8000	0.5333	0.4667	0.6667	0.3333
13.6000	23.3000	0.6000	0.5500	0.7000	0.4500
14.0000	23.7000	0.6667	0.6167	0.7667	0.5167
14.3000	24.1000	0.7167	0.6833	0.7833	0.6167
14.6000	24,5000	0.7667	0.7500	0.8000	0.7167
14.9000	24.9000	0.8167	0.8167	0.8167	0.8167
15.4000	25.4000	0.9000	0.9000	0.9000	0.9000
16.0000	26.0000	1.0000	1.0000	1.0000	1.0000

The original coordinates were transformed to lie between 0 and 1, giving the values of x and y to substitute in the equations of Figure 2. The new coordiantes x' and y' were then calculated for n = 5.

REFERENCES

1.   Patterson, B. D., Paull, R., and Graham, D. This volume. (1979).

STATISTICAL TESTS TO DECIDE BETWEEN STRAIGHT LINE SEGMENTS AND CURVES AS SUITABLE FITS TO ARRHENIUS PLOTS OR OTHER DATA

Joe Wolfe

Department of Applied Mathematics
Research School of Physical Sciences
Australian National University
Canberra, ACT 2600, Australia

David Bagnall

Division of Plant Industry, CSIRO,
Canberra, ACT 2601, Australia

I.   INTRODUCTION

A popular and powerful method of extracting information from a set of pairs of measurements is to plot one function of the variables against another, choosing two functions which, according to the theory applied, are linearly related. The intercept and slope of a straight line fitted to this plot are related to parameters of the system investigated. Thus, for example, a plot of the length of a pendulum against the square of its period should yield a straight line with slope $g/4\pi^2$, where g is gravitational acceleration. In chemistry, Arrhenius (1, 2) showed that, in certain cases, a plot of ln (reaction rate) against the reciprocal of the absolute temperature should yield a straight line with slope $-E_a/k_B$ where E_a is the activation energy (per molecule) of the activated complex and k_B is Boltzmann's constant.

In some cases the law which predicts the linear relation may hold only over a certain range of values, and thus only part of the plot resembles a straight line, or a law may hold, but with different values of the parameters, over different ranges. In the latter case, two or more straight lines may be fitted (by minimizing the sum of the squares of displacements from the relevant line) over various regions on the plot. The coordinates of a point of intersection acquire considerable significance since it is at this point presumably that the system changes from one operating regime to another.

When intersecting straight lines are fitted, and when the angles between those lines are small, it may be difficult to be sure that the points really are well represented by straight lines rather than a curve, particularly if the number of points is small or the possible error in measurement is large. Contentious examples are the fitting of straight lines [usually three but sometimes as many as five,] to Arrhenius plots of enzyme catalyzed reaction rates or other physiological rates. Crozier (7) proposed a theory to explain the existence of such lines, but this theory has been criticized as simplistic (4, 8, 9). Several authors (3, 9, 16, 20, 21) have proposed that the log of the rate of biological reactions should have a curvilinear dependence on inverse temperature, but intersecting straight lines appear on Arrhenius plots of biological data in many recent papers. Often it is not clear that straight lines are an appropriate fit, but we know of only one case where the appropriateness was tested statistically, and in that case curves were found to fit the data equally well (11).

Attempts to fit a series of straight lines to Arrhenius plots of, for example, the rate of biological reactions are made only over a limited temperature range. This rate is zero, and hence its log is infinitely negative, outside a finite range of temperature (usually about $-5^{o}C$ to $45^{o}C$) and Arrhenius' law does not allow a zero reaction rate at a finite temperature. Further, most authors [some exceptions being (6, 12)] decline to fit straight lines to that region of the plot with positive slope, and thus avoid the awkward concept of negative activation energies. The result is that only that region of the plot in which the slope changes slowly is considered and this, combined with the spread usually present in biological data, explains why there is some ambiguity about whether or not straight lines or curves better represent the data.

To limit the range of applicability of the fit to a certain range implicitly employs two parameters – the high and low temperature limits of that range. Fitting N straight lines involves a further 2N independently adjustable parameters (the intercepts and slopes of those lines). Often the number of points is not much larger than 2N (e.g. Ref 5) and so the fit is nearly perfect. Even when the number of data points is substantially greater than 2N, the residual sum of squares is often no larger than one would expect from just the inherent spread of the data. However, the same is true of many arbitrary curves involving as many or fewer adjustable parameters (e.g. $A + B\sqrt{x} + Ce^{x} + D \cosh(E + Fx)$, or equally arbitrarily, $A + Bx + Cx^{2} + Dx^{3} + Ex^{4} + Fx^{5}$). Bagnall and Wolfe (3) for instance, found that 5th order and 3rd order polynomials gave smaller residues than 3 straight lines when fitted to their Arrhenius plots, but in all cases these residues were too small to reject the curve on that basis alone.

There are, however, other criteria which one can apply to distinguish between N straight lines and some smooth curve (not necessarily a polynomial, of course).

TABLE I.   Table of values of b for sets of p points such that, when q is determined as described in the text, $|q - p/2| > -\;|$ allows the rejection of the hypothesis that the points are well fitted by straight lines at the indicated confidence level

No. of points (p)	10*	20*	30	40	50	60
Confidence level = 95% b =	2*	3*	4	5	6	7
Confidence level = 98% b =	3	4	5	6	7	8

*This test is not likely to be very useful for sets of 20 or fewer data points.   However, any attempt to determine six or eight parameters from a set of 20 or fewer points is likewise ambitious.

Table I gives values of b which are solutions to this equation for various values of p and at confidence levels of 95% and 98%. Thus one can test the straight line hypothesis at these confidence levels by evaluation of p and q, then comparing $|q-p/2|$ with the value of b given in the appropriate column of Table I. If $|q-p/2| > b$ one can reject the hypothesis.

This test can be made more sensitive by weighting each point according to its vertical displacement from the lines and its horizontal distance from a boundary between regions I and II.   It can also be improved by choosing the boundaries of regions I and II on the basis of the data, rather than the arbitrary division used here. However, the test in the simple form works quite well and the possible improvements would make its implementation much more tedious.

Applying this test to Arrhenius plots in the literature one can in most cases reject the hypothesis that straight lines are a suitable fit. However, failure to reject the straight line fit does not imply that it can be confidently accepted. One could employ a similar test to attempt to reject a specific curve fit also, but rejecting only one curve does not necessarily imply that a straight line fit must be accepted.

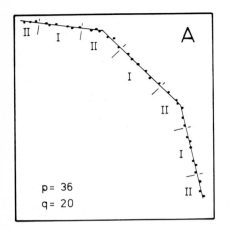 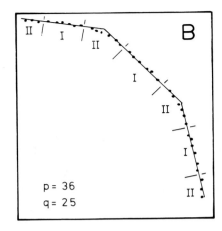

FIGURE 1. Two sets of data are shown fitted with straight line segments, and divided into regions for the application of test (A) (see text). At the 95% and 98% confidence levels, the straight line fit can be rejected for B but not for A.

II.   TESTS

A.   Is There a Pattern in the Displacement of Points from the Fitted Straight Lines?

Figure 1 shows 3 straight lines appropriately and inappropriately fitted to two sets of points. In the case where the points are better represented by a curve, the points tend to be below the lines near the ends of the segments, and above the lines in the middle. This discrimination procedure can be roughly quantified as follows: Divide each segment into 3 sections, the middle twice the length of the ends; the segment middle sections are called regions (I) and the other regions (II). Suppose that q is the number of points which lie below the lines in regions II plus the number of points which lie above the lines in regions I, and p is the total number of points. If straight lines are an appropriate fit, then the most likely value for q is p/2, and the probability of any particular q is $^{p}C_{q}(\frac{1}{2})^{p}$. Now there is a 95% change that q will lie in a range$(\frac{p}{2} - b) < q < (\frac{p}{2} + b)$ where

$$p/2 + b$$

$$\sum\ ^{p}C_{q}\ (\frac{1}{2})^{p} = 95\%$$

$$q = p/2 - b$$

B.   Are the "Breaks" or Intersection Points Dependent on the Range over which the Fit is Considered Applicable?

If straight lines are a convincing fit, then the intersections will occur at the same place if one deletes points at either end of the range and performs a least squares segment fit on the remainder (see Fig. 2). If the intersections occur at different places, then the information obtained from such lines is as much dependent on the range of temperature the experimenter sought to investigate as it is on the system studied.   Bagnall and Wolfe (3) found with their data that a segment fitting routine produced segments of approximately equal length for any subset of the data, that is, the "break points" were determined by the range chosen for the experiment.

C.   Is N, the Number of Straight Lines, Obvious and Unique?

If N straight lines is an appropriate fit, then (N + 1) straight lines fitted will include the N lines found previously, with one very short extra segment (see Fig. 3). Further, the residues from the N line fit and the (N + 1) line fit should be very nearly equal, whereas an (N − 1) line fit should give a very much larger residue. Bagnall and Wolfe (3) found for their data that the residue decreased smoothly with increasing numbers of lines.

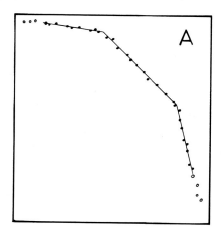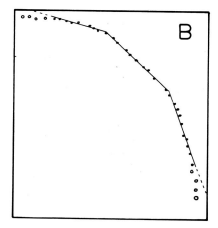

FIGURE 2.   The data are the same as in Figure 1, but four points at either end (O) are not included in the fitting procedure (see B in text).

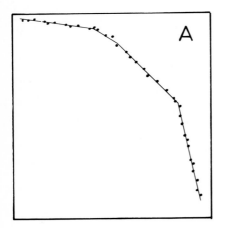 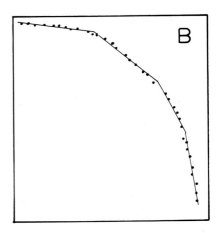

FIGURE 3. The data are the same as in Figure 1, but four straight lines have been fitted (see C in text).

D.   Is it Clear where the Straight Line Fit Ceases to Apply?

Since the plot goes infinitely negative at finite values of $1/T$, and considering the argument in (B), this is rather important.

E.   When One Shows the Points Alone to the Person Employed to Make the Tea, How Does He or She Describe it?

It is elsewhere argued that there is no theoretical reason to expect Arrhenius plots to be best fitted by straight lines. However, in the particular case of the analysis of chilling susceptibility in plants, it is sometimes argued that while there may be no theoretical reason for expecting straight lines, and while curves may fit the data at least as well as straight lines (11), the (N - 1) intersection points or breaks give useful quantitative results, i.e. one of the breaks may represent a "critical temperature" below which the plant will not survive (10, 13, 14). This is not so: First, the intersection points can be varied by the design of the experiment. The inclusion of extra points at a lower temperature will cause the lowest temperature "break" to occur at a lower temperature. Second, temperature is not the only "critical" environmental variable. One cannot ascribe to one of the "break" points the low temperature limit to survival, since the latter is dependent on irradiance, humidity and CO_2 concentration (15, 18, 19).

III. CONCLUSION

Summing up, there is often sufficient statistical evidence to decide that Arrhenius plots of biological reaction rates are highly unlikely to be represented by straight lines. Breaks calculated from such plots are mathematical or biochemical artifacts (17) and can be varied by experimental design. These "breaks" should not be used to predict a general critical survival temperature, since critical survival temperatures will be different for different conditions.

IV. REFERENCES

1. Arrhenius, S. *Z. Physik. Chem. 4*, 226-248 (1889).
2. Arrhenius, S. "Quantitative Laws in Biological Chemistry." Bell, London (1915).
3. Bagnall, D. J. and Wolfe, J. *J. Exp. Bot. 29*, 1231-1242 (1978).
4. Burton, A. C. *J. Cellular and Comp. Physiol. 9*, 1-14 (1936).
5. Champigny, M. L. and Moyse, A. *Biochim. Physiol. Pflanzen. 168*, 575-583 (1975).
6. Critchely, C., Smillie, R. M. and Patterson, B. D. *Aust. J. Plant. Physiol. 5*, 443-448 (1978).
7. Crozier, W. J. *J. Gen. Physiol. 7*, 189-216 (1924).
8. Heilbrunn, L. V. "An Outline of General Physiology." Saunders, Philadelphia (1937).
9. Johnson, F. H., Eyhring, H. and Polissar, M. J. "The Kinetic Basis of Molecular Biology." Wiley, New York (1954).
10. Lyons, J. M. *A. Rev. Pl. Physiol. 24*, 445-466 (1973).
11. Nolan, W. G. and Smillie, R. M. *Biochim. Biophys. Acta 440*, 461-475 (1976).
12. Nolan, W. G. and Smillie, R. M. *Plant Physiol. 59*, 1141-1145(1977).
13. Raison, J. K. and Chapman, E. A. *Aust. J. Pl. Physiol. 3*, 291-299 (1976).
14. Raison, J. K. and Lyons, J. M. *Pl. Physiol. 45*, 382-385 (1970).
15. Rowley, J. A. and Taylor, A. O. *New Phytol. 71*, 477-481 (1972).
16. Sharpe, P. J. H. and De Michele, D. W. *J. Theor. Biol. 64*, 649-670 (1977).
17. Silvius, J. B., Read, B. D., and McElhaney, R. N. *Science 199*, 902-904 (1978).
18. Taylor, A. O. and Rowley, J. A. *Plant Physiol. 47*, 713-718 (1971).
19. Wilson, J. M. *New Phytol. 76*, 257-270 (1976).
20. Wolfe, J. *Plant, Cell and Environment 1*, 241-247 (1978).
21. Wolfe, J. This volume (1979).

MAXIMUM LIKELIHOOD ESTIMATION OF BREAKPOINTS AND THE COMPARISON OF THE GOODNESS OF FIT WITH THAT OF CONVENTIONAL CURVES

J. F. Potter

Computer Unit, Welsh Plant Breeding Station
Plas Gogerddan, Near Aberystwyth
Dyfed, United Kingdom

G. J. S. Ross

Statistics Department
Rothamsted Experimental Station
Harpenden, Herts., United Kingdom

I. INTRODUCTION

The estimation by the method of maximum likelihood of breakpoints in multi-phase straight line functions is discussed. On one extreme it is argued that the notion of breakpoints is tenable and provides a better fit to data which might have been hitherto described by a continuous curve. On the other extreme it is shown that there is often no rigorous statistical reason for suspecting multi-phase responses. A general mathematical model for the multi-phase response is proposed, and the method of estimating its parameters is described by using the Maximum Likelihood Program (MLP).

In developing the theory of least squares, Gauss used the method of maximum likelihood. It is a very old and useful tool but lay dormant until Fisher (5) re-introduced it.

If we take minus the log of the likelihood density function for a regime having normally distributed errors, it can be shown that minimizing the residual sum of squares is sufficient to maximize the likelihood (2). However, the method may produce unsatisfactory results for small numbers of observations (12). With that reservation in mind, we find that even with small numbers of data (less than ten, say), if the errors are small and the change of gradient at a breakpoint large and well-defined, then the method works well.

A. The General Model

The mathematical model for a multiphase response having u phases and u-1 breakpoints is given by:

$$Y_i = \beta_0 + \beta_1 X_i + \delta_1 \beta_2(X_i - p_1) + \cdot\ \cdot\ \cdot\ \delta_{u-1}\ \beta_u(X_i - p_{u-1})\ \cdot\ \cdot\ \cdot\ (1)$$

where β_o is the intercept on the Y axis, β_1 is the gradient of the left-most line and further βs are gradient corrections to the left-adjacent lines.   Thus β_u is the gradient correction of phase (u - 1) to produce phase u; p_k is the kth breakpoint. δ_u is 1 if $X > p_u$, else 0.   Hence, the discontinuous response can be represented by what appears to be a continuous function. p_u is a parameter internal to the model and is thus treated as non-linear. The problem is that the function δ_u qualifies both X and p_u and is dependent on p_u.

II.   METHOD

An optimization algorithm is required to minimize the negative log likelihood function or in this case the residual sum of squares (RSS).   On finding maximum likelihood estimates (MLEs) for the breakpoints, two-dimensional 'slices' whose coordinates are p parameters taken in pairs, can be plotted through the u-dimensional parameter space.   Thus a composite picture of the likelihood in parameter space can be built up. Examination of this picture might show that an MLE corresponds to a false minimum. It may sit on a saddle point (an area of the space where a minimum in one direction is a maximum in a direction at right angles). By suitable re-adjustment of the initial estimates the iterative process is set in motion again and the MLEs for the breakpoints should eventually be found.

A.   Two Straight Lines

Possibly the commonest case is the estimation of a single assumed breakpoint between two straight lines. Equation (1) becomes simplified to:

$$Y_i = \beta_o + \beta_1 X_i + \delta\beta_2 (X_i - p)\ .\ .\ .(2)$$

The attributes that are considered important in assessing the validity of a breakpoint are as follows.

a)   The multiple correlation coefficient (R^2).
     This is

$$\frac{\text{Total sum of squares} - \text{RSS}}{\text{Total sum of squares}} \times 100\%,$$

otherwise known as the percentage variation accounted for by the fit. This merely indicates whether one model fits the data better than another.

b)   The variance ratio of

$$\frac{\text{Lack of fit mean square}}{\text{Pure error mean square}} \qquad\qquad (3).$$

This is only possible where there is replication. However, each X value need not have the same number of observations. The within-X sum of squares gives a measure of error unbiased by the fit of any model. The RSS consists of this sum of squares, added to a sum of squares representing the lack of fit of the model. Given the pure error and the RSS, the lack of fit sum of squares and the corresponding degrees of freedom can be obtained by subtraction. The F ratio,

$$\frac{\text{Lack of fit mean square}}{\text{Pure error mean square}}$$

can then be calculated.

The problem with models that are non-linear in the parameters is that the F-test is not strictly applicable. Beale (1) worked out a technique for assessing whether the non-linearity is serious or not. If Beale's non-linearity criterion is less than 0.01, linear inferences (e.g. the F-test) are not seriously wrong. Whether this is so or not, if the variance ratio is less than 1 (remember we are looking for low values), then it is safe to assume that the fit is a good one.

Methods are being considered whereby true confidence limits to the position of the breakpoint can be ascribed in a simple way. Hartley (7) has considered this problem for non-linear models in general. It may well be that discontinuous models present a further problem. Almost certainly any confidence interval calculated around the breakpoint will be asymmetric, owing to there being more scatter on one side than on the other.

c)   The number of data points supporting each line

A rule of thumb used is that three points in unreplicated and two in replicated data give sufficient support for a straight line segment. The breakpoint should really belong to both lines (or neither). In the method described here it belongs to the left-hand line; but this is just an academic point. The word 'support' used here should not be confused with the word as used by Edwards (4).

d)   The residual sum of squares

This is sufficient as a criterion for improving the fit for a given set of data, but gives no idea of confidence in the estimate.

e)   The size of the change in gradient

If it is very small then the model would be deemed overdetermined.

f)   The comparison of attributes a), b) and d) above with those for fits of empirical linear models (here restricted to linear and quadratic polynomials).

g)   Estimated variances of parameters

These are conventionally calculated from the inverse matrix of second derivatives of the log likelihood function.   For the discontinuous models under consideration their use is not recommended because the log likelihood is discontinuous.

B.   An Example of Satisfactory Breakpoint Estimation

In this example we are grateful to Dr. H. Thomas and Dr. J. L. Stoddart for supplying the data on measurements of the rate of change in concentration of certain compounds in the cell membranes of various crops in response to chilling. The familiar Arrhenius transformation, $(1/Temp\,^{\circ}K) \times 10^{4}$, was used for the X-axis. The crop used was winter oats grown at $20^{\circ}C$ prior to the chilling procedure. The compound whose concentration change was measured was chlorophyll, and the $\log_{10}$ concentration was used for the plot. From this data the following estimates were made: β_{0}, -16.5128; β_{1}, 0.5011; β_{2}, -0.2149; p, 34.8795.

TABLE I.

	Straight line	Quadratic	Two straight lines
R^2	82.28	82.99	83.78
Lack of fit MS			
Pure error MS	0.9	0.9	0.8
RSS	0.4525	0.4344	0.4142

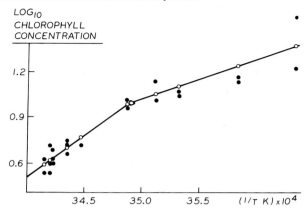

FIGURE 1. Log_{10} chlorophyll concentration on $(1/Temp.^{o}K)$ x 10^{4}. The breakpoint is shown by ∇.

The attributes for the mechanistic and the empirical models are given in Table I.

Fifteen observed data points supported each line in the two straight lines model giving good support to each line. The change of gradient at the breakpoint being -0.2149 gives good support to the notion of a breakpoint. As regards all three attributes in Table I, the two straight lines model comes out best. Now, because there is prior belief, based on the experimenters' experience, in the possibility of a breakpoint in this region, the two straight lines model can be accepted. If it made no scientific sense, the experimenter might either re-think the model, or if stubborn he might collect more data. Figure 1 shows the fitted two straight lines model through the scatter of observed data points. There are three observations per X value, but some are overwritten on the graph owing to proximity of points. The reader will note what appears to be an outlier. Surprisingly, it seems to have little effect on the position of the breakpoint (as was discovered when it was removed) but the β_2 term is affected quite seriously by it.

Responses in nature tend to follow a straight line relationship over at least some of their range. The higher order polynomials are progressive attempts to join the dots, but, if successful, the scientific meaning will be nil and extrapolation out of the question, although empirical models can be useful if handled with respect. However, a mechanistic model if it fits really well, means something and can therefore be used, e.g. for extrapolation (2).

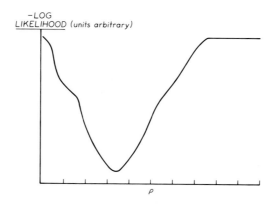

FIGURE 2. -Log likelihood plot (conditional) for values of p between 34.1 and 36.4.

C.   The Likelihood Plot

Figure 2 shows the plot of minus the log likelihood over the specified range of values for p. For each value of p, β_0, β_1 and β_2 are estimated by multiple regression and the RSS is calculated and plotted. This plot is informative. Firstly, it can be seen if the MLE for p has been achieved, and in the example cited it clearly has. The likelihood plot gives confidence in the MLE owing to its marked parabolic shape. In ill-defined examples, the likelihood plot can take on quite a bizarre shape and may have several minima indicating several possible solutions. By examining a small portion of the likelihood plot, and scaling it up, it may be possible to discover the MLE.

The work of Raison (9) and Fork *et al.* (6) are examples of a rather subjective approach to the fitting of breakpoints. The criticism of Wolfe and Bagnall (13), that if a straight line segment is not itself a least squares fit then the breakpoint is spurious, is particularly apt in the work of Fork *et al.* (6). A quick glance shows that in many cases there is more scatter on one side of a segment than on the other. In one graph, a single point is allowed to imply a dramatic break. Even if this is the mean of a set of replicated observations, it is not enough support for the line. In the work of both Raison (9) and Fork *et al.* (6), there are fairly convincing complete breakpoints where the lines 'miss' each other. This kind of discontinuity cannot be described in a single model. The two lines are distinct straight line models. But still the likelihood plot will show any tendency for this to happen. Figure 3 shows such a plot. Where the two parabolae 'crash' corresponds to where the two lines 'miss'. This sort of thing is only considered convincing if the 'crash' occurs at the same low likelihood as the two end areas of the plot. That is, it is found to be

FIGURE 3. Plot of -log likelihood for a part of p parameter space, showing 'crashing' parabolae. Data obtained from spring oats grown at $5^{o}C$ without Arrhenius transformation. Lines 'miss' at $13.56^{o}C$.

exceedingly unlikely that there is a line at this point owing to the discovery that it is exceedingly unlikely that there is a conventional breakpoint.

D.   The Maximum Likelihood Program

All the techniques used in this paper were performed using MLP, a large comprehensive non-linear model-fitting program by G. J. S. Ross *et al.* (11). As well as the standard (linear and non-linear) curves and frequencies, it enables one by means of a simple interpretative language to fit user-defined models such as the two straight lines model. The model is presented to MLP by deriving values X - p above p and zeros below. This becomes the second X variate. This and the original X variate are then fitted by multiple regression on optimizing the value of p. Optimization is a large subject and is dealt with at length in the Numerical Algorithms Group (NAG) Manual (8). Also, Ross (10) gives a full account of function minimization as a background to MLP.

III.   CONCLUSION

It is clear that breakpoints do occur in nature, but that their estimation by experimenters has been somewhat cavalier. The model given here using the MLP program is considered to be a rigorous treatment of the subject. It is hoped to produce soon a smaller dedicated program which will deal specifically with this model fitting technique. An increasing number of computer centres these days offer MLP as a package, and the manual describes fully how to estimate breakpoints.

ACKNOWLEDGMENTS

The authors are grateful to Dr. H. Thomas and Dr. J. L. Stoddart (Biochemistry Department, University College of Wales, Welsh Plant Breeding Station, Nr Aberystwyth, Dyfed, UK) and Dr. R. W. Breidenbach (Plant Growth Laboratory, University of California, Davis, USA) for posing the problem, collecting the data and creating the demand. Thanks also due to Mrs. N. Wells for her constant help.

IV. REFERENCES

1.   Beale, E. M. L. *J. Royal Statist. Soc. (B) 22, 1,* 41-88 (1960).
2.   Box, G. E. P. and Hunter, W. G. *Technometrics 7,* 23-42 (1965).
3.   Draper, N. and Smith, H. "Applied Regression Analysis", Wiley, New York (1966).
4.   Edwards, A. W. F. "Likelihood", Cambridge University Press (1972).
5.   Fisher, R. A. *Mess. Math., 41,* 155-160 (1912).
6.   Fork, D. C., Murata, N. and Sato, N. *In: Annual Report of the Director of the Department of Plant Biology,* Stanford, California (Carnegie Institution of Washington), pp. 283-289 (1978).
7.   Hartley, H. O. *Biometrika 51, 3 and 4,* 347-353 (1964).
8.   Numerical Algorithms Group. FORTRAN Library Manual (1977).
9.   Raison, J. K. *In* Rate Control of Biological Processes. *Society for Experimental Biology Symposia 27,* 485-512 (1973).
10. Ross, G. J. S. *Appl. Statist. 19, 3,* 205-221 (1970).
11. Ross, G. J. S., Jones, R. D., Kempton, R. A., Lauckner, F. B., Payne, R. W., Hawkins, Diana, White, R. P. MLP:Maximum Likelihood Program. (Revised edition), Rothamsted Experimental Station, Harpenden, Herts, UK (1979).
12. Wasan, M. T. "Parametric Estimation", McGraw-Hill, New York (1970).
13. Wolfe, J. and Bagnall, D. *In* The Proceedings of the US-Australian New Zealand Cooperative Science Program Conference on Low Temperature Stress in Crop Plants: The Role of the Membrane (1979).

EPILOGUE

The seminar "Low Temperature Stress in Crop Plants: The Role of the Membrane" was organized to examine critically the physical, biophysical and biological basis of plant responses to low temperature. Scientists with expertise in the field discussed and debated the physical and biochemical consequences of temperature effects on cellular metabolism and how these effects might be translated into dysfunction and injury. Emphasis was placed on the molecular biology of cell membranes and scrutiny of the hypothesis that the primary molecular event in the low temperature response is a temperature-induced alteration in the molecular ordering of membrane lipids. The final question posed for discussion was to consider genetic diversity in crop plants for tolerance to low temperatures and possible strategies to exploit that potential.

This short addition is not intended to represent a consensus of the seminar, but rather is provided by the editors as a means of focusing some aspects of the general discussion that occurred. The reader is left to examine each paper and the accompanying questions and answers to arrive at his or her own conclusions.

There were areas where some ambiguity and apparent contradiction in the literature was clarified and resolved; there were also areas where new questions and debate was elicited. The following brings out some of those issues and suggests areas of research needed to accelerate the general field of study.

A. Definitions Related to Chilling Injury

In the discussion, it often became evident that some confusion and differences of opinion evolved as a result of lack of common vocabulary or at least lack of uniformity in using the vocabulary. The following reiterates usage and definitions which we would hope provide some clarity:

--*chilling* is the act of exposing plant material to some low temperature (above 0°C);

--*chilling injury* is a damage to plant tissues, cells, or organs which result from exposure to some low temperature (above 0°C) for a period of time sufficient to cause permanent or irreversible damage;

--*dysfunction,* according to Webster, is a "disordered or impaired function of a system or organs";

--*primary response to temperature* is the primary event in the cell which senses the low temperature and initiates some dysfunction (which dysfunction in time will lead to injury). The primary response (or synonomously, the primary sensor) may initiate the dysfunction as an immediate response but it may involve some time period to cause the injury.

B.   Arrhenius Plots

Much attention in the seminar was focused on the use or misuse of "Arrhenius" plots in presenting temperature dependent data. The issues are important since much of the conceptual framework of the "membrane hypothesis" invoked as an explanation of the primary response to low temperature in plants revolves around the appearance of "breaks" or discontinuities in plots of some rate function plotted against the reciprocal of the absolute temperature, which according to Arrhenius theory should yield a straight line.

Debate evolved around several questions. In data presented in the literature as Arrhenius plots, are they best represented by two (or more) straight lines or are they best fit by a curve? Do the "discontinuities", the "breaks" or the "kinks" represent real events? Views were highly polarized and some of the arguments were identical to issues debated in the literature in the 1920's shortly after Arrhenius theory was applied to biological systems. The issues were perhaps no more conclusively resolved at this seminar than they were 50 years ago; but there did seem to be some axioms that emerged and are worth articulating:

--plots of the logarithm of reaction rate vs. the reciprocal of the absolute temperature can be a useful way to display temperature dependent data to search for anomalies;

--such plots may or may not be representing events which come within the concept the Arrhenius function and hence it may not be valid to calculate activation energies (E_a) from such plots. There is a limitation to the range of slopes to which Arrhenius law can be applicable and that limitation often excludes the slopes of some of the lines appearing in the literature. For example, Arrhenius theory cannot deal with slopes that go to zero as where the disassociation of some low temperature labile enzyme destroys the polymerized activation complex required for reaction. It is inappropriate to calculate an E_a for the temperature range below that where disassociation occurs. Hence, caution should be exercised and only those systems where Arrhenius theory can be applied deserve to be called "Arrhenius" plots; the others should be referred to as "plots of the logarithm of reaction rates vs. the reciprocal of the absolute temperature" and from such plots values of "apparent" E_a's may be useful;

--no longer should such plots be presented in the literature without having been subjected to appropriate statistical tests to determine the suitability of fitting various lines or curves to sets of data. Additional discussion and methods of applying such tests are presented in Part V -- Special Topics Related toThe Use of Arrhenius Plots, this volume.

C.   Characterization of Physical Phenomena in Plant Membranes

Much of the knowledge on the molecular architecture of biological membranes has been discerned by experimentation on model membranes, bilayers and the rather simple membranes from yeast and similar organisms. These results and techniques have been employed to address the question of membrane associated events in higher plants, particularly those related to temperature stress.

The main evidence presented was derived from electron spin resonance (ESR) spectroscopy using a variety of spin label probes, fluorescence polarization studies employing *trans*-paranaric acid, and differential thermal analysis (DTA) or differential scanning calorimetry (DSC). The data presented at first inspection seems to be contradictory. ESR and most of the fluorescence data indicate a change in lipid ordering at temperatures correlating with the physiology of chilling injury; some fluorescence data and the DTA (or DSC) data indicate there is no endothermic phase transition in the bulk membrane lipids in the temperature range above zero.

One issue that seemed to reflect general agreement is that a phase change from a "liquid-crystalline structure to a solid gel" in the bulk membrane lipid does not occur. This agreement would be supported by the data presented on lipid analysis (acyl chain composition, polar head group analysis, unsaturation of acyl chain, etc.) as it reflects the bulk membrane lipid.

It is clear from the flourescence polarization and ESR studies that some change occurs in the molecular ordering and fluidity of membrane lipids of chilling sensitive plants in response to low temperatures -- a change that correlates with physiological phenomena. This is one area of research that deserves increased attention to gain a better understanding of the event(s) which do occur in plant membranes at temperatures which correlate with injury and which regulate membrane and membrane associated functions. Freeze fracture techniques, expansion of the techniques using a variety of new spin label probes which discern events in the cytoplasm and the membrane continuum, and other similar powerful new techniques must be employed to elucidate these events that are real but might not be discerned by DSC or DTA.

An additional point of clarification on this subject. The terminology developed from studies of binary mixtures of phospholipids and membranes of *E. coli* has been used in designating the change in physical state where the deflection at the lower temperature is T_s and at the higher temperature, T_f. These were considered as the "start" and "finish" of a thermal phase transition in the membrane lipids. Because of discussion on the subject it would appear that the deflection at the lower temperature in the temperature response of membrane events in higher plants should simply be referred to as T_2 and at the higher temperature T_1, without any implication of denoting a "thermal phase transition".

D. Similarities between Freezing and Chilling Injury

There often occurs discussion in the literature comparing the mechanism of freezing and chilling injury, particularly as it focuses on any hypothesis that invokes possible changes in the physical state of cellular membranes in response to low temperatures as the primary event. This then leads to analysis and attempts to correlate changes in lipids and particularly levels of unsaturation in fatty acids with acclimation, conditioning and ultimate ability to withstand the freezing process. On the latter point, it is clear in these studies that (as with the chilling phenomenon) there is no consistent correlation between lipid analysis and susceptibility to freezing or chilling damage. That is, as many reports of a negative correlation of amount of unsaturation and susceptibility to low temperature injury occur as those which show a positive correlation. As the discussion at this seminar brought out, it is now apparent that lipid analysis to date can only represent a composite picture of the bulk membrane lipids: if lipids are involved in controlling the molecular order of the membrane lipids which pertain to freezing or chilling injury, this control might occur within discrete lipid domains, not in the bulk phase.

There is some difficulty with terminology in the freezing process as there is with chilling. That is, discussions often do not clearly distinguish between the *primary response* to low temperature and the *secondary* events; between the impact of low temperature on the physico-chemical events in the membrane and the translation of those events into a specific membrane lesion and ultimately cellular injury.

Agreement was not reached amongst the participants on evidence as to whether the membrane lesion and injury was the result of the freezing or the thawing process. Contributing to this problem is the fact that the specific membrane lesion is most often ill-defined and only conjecture, and the failure to consider that a spectrum of membrane lesions may result from a single freeze-thaw protocol.

E.   The "Membrane Hypothesis"

The seminar was organized to assess critically the evidence that has evolved around the proposed hypothesis on the mechanism of chilling injury. As originally stated, "a single controlling response is found in the evidence that cellular membranes in sensitive plants undergo a physical-phase transition from a normal flexible liquid-crystalline to a solid gel structure at the temperature critical for chilling injury was suggested." Furthermore, this hypothesis implied a role for the lipids as the controlling species in the membrane.

Discussion on this point was quite polarized and the probability is high that the participants were not persuaded to move too far from the position each held originally. Perhaps this reflects the fact that this particular field of science is rather juvenal -- the young bird with its first plumage of true feathers but yet lacking adult characteristics. There was insufficient evidence or hard data to resolve the opposing views. This controversy should, however, help to focus future research and hopefully develop some new insight and not simply plow old ground. Some of these points developed in the presentations and discussions as well as areas of agreement include:

— as originally stated, the membrane hypothesis assumed that a change in molecular ordering (change in "state") of the membrane would bring about a contraction, causing "cracks" or "channels" leading to increased permeability. Evidence does now exist that says this is not the case and in fact a *decrease* in membrane permeability is the immediate response to temperature below the transition in molecular ordering. The "leakage" and increase in permeability only appears after membrane damage has occurred for some time period;

— the concept of a "physical phase change from a liquid-crystalline to a solid gel" should be abandoned as far as describing events in the bulk membrane. Changes in fluidity and molecular ordering do occur but these are not discerned as "melts" of the bulk lipids. More research needs to be focused on describing events at the molecular level in membranes in response to low temperature;

— lipid analysis of the bulk membrane would suggest that there is no basis for a phase change to occur at physiological temperatures. While these analyses lead to that conclusion, it is also true that changes in fluidity and molecular ordering do occur and as we develop techniques to examine the physical state of discrete "domains" within the membrane, we must focus the chemical analysis on these same domains. (We can probably learn little more from bulk lipid analysis.);

— the "membrane hypothesis" ignores or dismisses proteins, (either soluble or membrane-associated), as a primary temperature sensor. Evidence was presented and discussed that direct effects of temperature on enzyme proteins could be discerned and that a conformational change or polymeric separation of a protein may be as plausible an explanation for the primary temperature sensor as invoking the membrane lipid as the primary sensor;

— the view was expressed by some that articulation of a single primary temperature sensor as mediating the chilling response should be abandoned. This view suggested that imposing low temperature on chilling sensitive species invokes a broad range of "primary" events which lead sequentially to the observed injury;

— the view was expressed that the primary impact of low temperature on these sensitive species was on the temperature coefficient or rate of growth and development.

Again, these points in the discussion should help focus research directions and provide a framework for areas of endeavor requiring new knowledge.

F. Germplasm and Potential for Manipulation

It is of interest that this seminar included research that has focused on development of new germplasm and exploitation of naturally occurring genetic diversity to address chilling injury in crop plants.

Traditional breeding programs have long focused on freezing tolerance and winter hardiness in cereals but little has been done with horticultural crops and chilling injury. If another seminar such as this were to be held some time in the future we would anticipate the participants would be discussing many new advances in crop improvement as the result of genetic manipulation. And much of this success will depend on accurate understanding of the mechanism(s) in low temperature stress so that presumptive and rapid screening methods can be devised.

APPENDIX

Table of Temperature Coefficients of Various Plant Responses

This table incorporates data on the temperature coefficient, of both physiological and physical parameters using whole plants, seeds,cells, organelles, membrane fragments or enzymes. Where the authors of the particular article have noted a sudden change(s) in the temperature response or temperature coefficient the temperature at which these occur have been listed as T_1 and T_2 for the higher and lower temperatures respectively. Where more than two temperatures were noted these have been included. Where the data shows an increase in temperature coefficient at low temperature but the authors have not indicated a particular temperature the change is listed as a "curve." If no change is detected, i.e., the temperature coefficient is constant over the temperature range examined it is listed as "ND" (none detected).

Abbreviations:

DCIP	2,6-dichlorophenolindophenol
DAD	2,3,5,6-tetramethyl-p-phenylenediamine
TTC	triphenyltetrazolium chloride

Note: Data on changes in the temperature coefficient of spin label motion and fluorescence intensity in lipids from a number of plants native to Australia, desert plants from Death Valley in California as well as crop plants, can be found in Chapters 13 and 21 of these proceedings.

Plant species	Tissue/organelle	Physical measurement	T_1 (°C)	T_2	Physiological	T_1 (°C)	T_2	Ref.
Apple / Malus domestica	Fruit				Ethylene synthesis		5	(1)
	Fruit mitochondria	ESR		3				(2)
Avocado / Persea americana	Fruit mitochondria				Succinate oxidation		9	(3)
Barley / Hordeum vulgare	Root	Rb^+ uptake		ND	Oxygen uptake	ND	ND	(4)
	Root	Rb^+ uptake	28	10	DCIP photoreduction	ND	ND	(5)
	Chloroplasts	Esr		10	DCIP Photoreduction with uncoupler	29 20	9	(6)
	Washed chloroplasts				DCIP photoreduction	29 ND	9	(6)
Bean / Phaseolus vulgaris	Chloroplasts	Wide angle X-ray (crystalline structure below)		-30				(16)
	microsomes			23				(16)
	Chloroplasts	Decay of delayed light emission		~15				(17)
	Whole plant				Water absorption		~14	(8)
	Chloroplasts				Reflection coefficient	11	11	(9)
Beet root / Beta vulgaris	Root-mitochondria				Succinate oxidation	ND	ND	(7)
	Leaf	Light-induced carotenoid shift		15				(10)
	Chloroplasts				H^+ efflux		5	(11)
Bell pepper / Capsicum frutescens	Fruit tissue	K^+ leakage		10				(12)
Broad bean / Vicia faba	Mitochondria (hypocotyl)				State 3:- Oxidation of malate	30	18	(18)
					Oxidation of α-ketoglutarate	30	18	(18)
					Oxidation of citrate	31	18	(18)
					Oxidation of NADH	30	10	(18)
					Oxidation of Succinate	30	10	(18)

Plant species	Tissue/organelle	Physical measurement	T_1 T_2 (°C)	Physiological	T_1 (°C)	T_2	Ref.
				State 4:-			
				Oxidation of malate	30	13	(18)
				Oxidation of α-keto-glutarate	ND	13	(18)
				Oxidation of citrate	ND	13	(18)
				Oxididation of NADH	ND	10	(18)
				Oxidation of Succinate	ND	10	(18)
				Malic enzyme	30	ND	(18)
				Malate dehydrogenase	ND	ND	(18)
Capsicum *Capsicum annuum*	Mitochondria from tissue culture			NADH oxidation of:-			
				Sensitive C$_9$ line		9	(13)
				Resistant Cv34 line		ND	(13)
Castor bean *Ricinus communis*	Seedlings			Respiration		11	(14)
	Mitochondria			Succinate oxidation		9	(14)
	Glyoxysomes			Isocitrate lyase		ND	(14)
	Glyoxysomes			Malate synthase		ND	(14)
	Glyoxysomes			Malate dehydrogenase		ND	(14)
	Glyoxysomes			β oxidation		ND	(14)
	Glyoxysomes			Glycolate oxidase		ND	(14)
	Mitochondria	ESR	11				(15)
	Glyoxysomes	ESR	11				(15)
	Proplastids	ESR	11				(15)
Cucumber *Cucumis sativus*	Seed			Germination		14	(48)
				Seed respiration		ND	(48)
	Fruit- Mitochondria			Succinate oxidation		10	(7)
	Root	Rb$^+$ uptake	15				(4)
	Fruit tissue	K$^+$ leakage	10				(12)
	Microsomes	Calorimetry	20	K$^+$-ATPase activity		10	(47)
Cauliflower *Brassic oleraceae*	Mitochondria			Succinate oxidation		ND	(7)
	Microsomes	Calorimetry	ND	K$^+$ ATPase activity		15	(47)

Plant species	Tissue/organelle	Physical measurement	T_l	T_2 (°C)	Physiological	T_1 (°C)	T_2	Ref.
Foxglove *Digitalis purpurea*	Trichome cells	Protoplasmic streaming	Curve					(19)
Jerusalem artichoke *Helianthus tuberosus*	Tuber Mitochondria	ESR (non-dormant tubers)	27	3	Succinate oxidase	27		(20)
		(dormant tubers)	9	3				(20)
Lettuce *Lactuca sativa*	Leaf (grown 15°C)	Light-induced carotenoid shift	ND					(10)
	Chloroplasts (grown 25°C)	cation-induced fluorescence shift			$H_2O \to DCIP$ Hill reaction		ND	(21)
	"	Chl *a* fluorescence in:			P700 reduction		ND	(21)
		5 mM $MgCl_2$	-31					(22)
	Chloroplasts	100 mM NaCl	-25					(22)
Maize *Zea mays*	Excised roots	R^+ uptake	10		Oxygen uptake		10	(5)
	Root mitochondria	ESR	27	12	Amino acid uptake		13	(23)
					Succinate oxidation		12	(24)
	Chloroplasts				DCIP photoreduction	27	ND	(23)
					"	39		(25)
					" uncoupled	39	11	(25)
Mango *Mangifera indica*	Fruit mitochondria (stored at low temp.)				Succinate oxidation measured at 25°		8-12	(26)
Morning glory *Pharbitis nil*	Plants				Growth rate (dry wt.)		Curve	(27)
					Leaf area increase		Curve	(27)
					Chlorophyll formation	ND	15	(27)
Mung bean *Vigna radiata*	Hypocotyl (dark)				Growth rate	28	15	(28)
	Seeds				Germination		~14	(48)

Plant species	Tissue/ organelle	Physical measurement	T₁ (°C)	T₂	Physiological	T₁ (°C)	T₂	Ref.
	Hypocotyl-mitochondria Plants	ESR	28	15	Respiration of seeds		ND	(48)
	Chloroplasts				Succination oxidation	28	15	(28)
					Growth (dry wt.)	Curve		(27)
					Leaf area increase	Curve		(27)
					Chlorophyll formation		15	(27)
					DCIP photoreduction	28	17	(25)
					DCIP photoreduction with uncoupler	28	ND	(25)
Mustard Sinapis alba	Seeds				Germination		ND	(48)
					Respiration of seeds		ND	(48)
Muskmelon Cucumis melo		Rb^+ uptake		16				(4)
Passionfruit Passiflora sp.	Chloroplasts	ESR			Photoreduction of $Fe(CN)_6^{3-}$	24	16	(29)
P. quadrangularis	Chloroplasts	ESR		9				(30)
P. flavicarpa	(Polar lipids	ESR		9	Photoreduction of $Fe(CN)_6^{3-}$		10	(30) (29)
	Chloroplasts)							
P. edulis	(Polar lipids	ESR		3	Photoreduction of $Fe(CN)_6^{3-}$	16	4	(30) (29)
	Chloroplasts)							
P. cincinnata	(Polar lipids	ESR		6	Photoreduction of $Fe(CN)_6^{3-}$	15	ND	(30)
	Chloroplasts)							
P. caerulea	Polar lipids	ESR		2	Photoreduction of $Fe(CN)_6^{3-}$	19	7	(30) (29)
	Chloroplasts							
P. flavicarpa x cincinnata	Polar lipids	ESR		7	Photoreduction of $Fe(CN)_6^{3-}$	16	1	(30) (29)
	Chloroplasts							

Plant species	Tissue/ organelle	Physical measurement	T_1 T_2 ($^\circ C$)	Physiological	T_1 ($^\circ C$)	T_2	Ref.
Pea Pisum sativum	Chloroplasts	Reflection coefficient	ND	Photoreduction of NADP			(9)
	Chloroplasts	ESR	ND			ND	(31)
				DCIP photoreduction	ND	ND	(49)
				DCIP photoreduction with uncoupler	25	14	(25)
	Chloroplasts	Decay of delayed light emission	ND				(17)
Potato	Tuber mitochondria	ESR (non-dormant tuber)	25 3				(33)
		(dormant tuber)	21 1				(33)
Solanum tuberosum	Tuber tissue	K^+ leakage	ND	Succinate oxidation			(7)
							(12)
Rice Oryza sativa	Seed (japonica and indica types)			Germination		17	(35)
	Roots	Rb^+ uptake	9				(4)
Sorghum Sorghum vulgare	Roots	Rb^+ uptake	7				(4)
Spinach	Chloroplasts			DAD reduction		5	(42)
				Methyl purple reduction		-12	(42)
Spinacia oleracea	Chloroplasts	Light-induced change at 515 nm	15				(43)
		Light-induced change at 546 nm	15				(43)
	Chloroplasts plants grown at 10° to $30^\circ C$			Photoreduction of DCIP	ND	-9	(34)
	plants grown 0° to $10^\circ C$			Photoreduction of DCIP	11	-9	(34)
	Chloroplasts	Reflection coefficient	ND				(9)
	Leaf	Light-induced carotenoid absorption shift	ND				(10)

554

Plant species	Tissue/organelle	Physical measurement	T_1 (^{o}C)	T_2	Physiological	T_1 (^{o}C)	T_2	Ref.
	Leaf	Cation-induced fluorescence shift		ND				(36)
	Chloroplasts	Chl a fluorescence in:						
		5 mM $MgCl_2$	ND	-31				(22)
		5 mM $NaCl^2$	ND	-20				(22)
Sweet potato	Root mitochondria				Succinate oxidation	10		(7)
Ipomoea batatas	Root Mitochondria	ESR	12					(32)
	Mitochondria Phospholipids	ESR	12					(32)
					Amino acid incorporation	13		(24)
Tidestromia oblongifolia	Leaf chloroplasts	Light-induced carotenoid shift	5		H^+ efflux	~7		(10) (11)

REFERENCES

1. Matoo, A. K., Baker, J. E., Chalutz, E., and Lieberman, M. *Plant and Cell Physiol. 18,* 715-719 (1977).
2. McGlasson, W. B., and Raison, J. K. *Plant Physiol. 52,* 390-392 (1972).
3. Kosiyachinda, S., and Young, R. E. *Plant Physiol. 60,* 470-474 (1977).
4. Zsoldos, F. *Z. Pflanzenernaehr Bodenk. 119,* 169-173 (1968).
5. Carey, R. W., and Berry, J. A. *Plant Physiol. 61,* 858-860 (1978).
6. Nolan, W. G., and Smillie, R. M. *Biochim. Biophys. Acta 440,* 461-475 (1976).
7. Lyons, J. M., and Raison, J. K. *Plant Physiol. 45,* 386-387 (1970).
8. Wilson, J. M. *New Phytol. 76,* 257-270 (1976).
9. Nobel, P. S. *Planta 115,* 369-372 (1974).
10. Murata, N., and Fork, D. C. *Biochim. Biophys. Acta 461,* 365-378 (1977).
11. Avron, M., and Fork, D. C. *Carnegie Inst. Year Book 76,* 231-235 (1977).
12. Tatsumi, Y., and Murata, T. *J. Japan. Soc. Hort. Sci. 47,* 105-110 (1978).
13. Dix, P. J., and Street, H. E. *Ann. Bot. 40,* 903-910 (1976).
14. Breidenbach, R. W., Wade, N. L., and Lyons, J. M. *Plant Physiol. 54,* 324-327 (1974).
15. Wade, N. L., Breidenbach, R. W., Lyons, J. M., and Keith, A. D. *Plant Physiol. 54,* 320-323 (1974).
16. McKersie, B. D. and Thompson, J. E. *Plant Physiol. 61,* 639-643.
17. Jursinic, P. and Govindjee. *Phytochem. Phytobiol. 26,* 617-628 (1977).
18. Marx, R., Brinkman, K. *Planta 144,* 359-365 (1979).
19. Patterson, B. D., and Graham, D. *J. Exp. Bot. 28,* 736-743 (1977).
20. Chapman, E. A., Wright, L., and Raison, J. K. *Plant Physiol. 63,* 363-366 (1979).
21. Murata, N., Troughton, J. H. and Fork, D. C. *Plant Physiol. 56,* 508-577 (1975).
22. Murata, N., and Fork, D. C. *Plant and Cell Physiol. 18,* 1265-1271 (1977).
23. Raison, J. K. *Chapter 13, these proceedings (1979).*
24. Towers, N. R., Kellerman, G. M., Raison, J. K., and Linnane, A. W. *Biochim. Biophys. Acta 299,* 153-161 (1973).
25. Nolan, W. G., and Smillie, R. M. *Plant Physiol. 59,* 1141-1145 (1977).
26. Kane, O., Marcellin, P., and Mazliak, P. *Plant Physiol. 61,* 634-638 (1978).
27. Bagnall, D. J., and Wolfe, J. A. *J. Exp. Bot. 29,* 1231-1242 (1978).
28. Raison, J. K., and Chapman, E. A. *Aust. J. Plant Physiol. 3,* 291-299 (1976).

29. Critchley, C., Smillie, R. M. and Patterson, B. D. *Aust. J. Plant Physiol. 5,* 443–448 (1978).
30. Patterson, B. D., Kenrick, J. R., and Raison, J. K. *Phytochemistry* 17, 1089–1092 (1978).
31. Shneyour, A., Raison, J. K., and Smillie, R. M. *Biochim. Biophys. Acta 292,* 152–161 (1973).
32. Raison, J. K., Lyons, J. M., Mehlhorn, R. J., and Keith, A. D. *J. Biol. Chem. 246,* 4036–4040 (1971).
33. Wright, L., and Raison, J. K. (Unpublished data).
34. Inoué, H. *Plant Cell Physiol. 19,* 355–363 (1978).
35. Nishiyama, I. *Plant Cell Physiol. 16,* 533–536 (1975).
36 Murata, N., and Fork, D. C. *Plant Physiol. 56,* 791 796 (1975).
37. Breidenbach, R. W., and Waring, A. *J. Plant Physiol. 60,* 190 192 (1977).
38. Melcarek, P. K., and Brown, G. N. *Plant and Cell Physiol. 18,* 1099–1107 (1977).
39. Waring, A., and Glatz, P. Chapter 26, these proceedings (1979).
40. Paull, R. E., Patterson, B. D., and Graham, D. Chapter 37, these proceedings (1979).
41. Paull, R. E. (Unpublished data).
42. Cox, R. P. *Biochim. Biophys. Acta 387,* 588–598 (1975).
43. Yamamoto, Y., and Nishimura, M. *Plant Cell Physiol. 18,* 55–66 (1977).
44. Raison, J. K., Chapman, E. A., and White, P. *Plant Physiol. 59,* 623–627 (1977).
45. Pomeroy, M. K., and Andrews, C. J. *Plant Physiol. 56,* 703–706 (1975).
46. Miller, R. W., De La Roche, I., and Pomeroy, M. K. *Plant Physiol. 53,* 426–433 (1974).
47. McMurchie, E. J. Chapter 12, these proceedings (1979).
48. Simon, E. W., Minchin, A., McMenamin, M. M., and Smith, J. M. *New Phytol.77,* 301–311 (1976).
49. Raison, J. K. *In* Rate Control of Biological Processes. Symposia for Society of Experimental Biology XXVII, p. 485–512 (1973).

INDEX